CORROSION AND CORROSION PROTECTION HANDBOOK

MECHANICAL ENGINEERING

A Series of Textbooks and Reference Books

EDITORS

L. L. FAULKNER

Department of Mechanical Engineering
The Ohio State University
Columbus, Ohio

S. B. MENKES

Department of Mechanical Engineering
The City College of the
City University of New York
New York, New York

OTHER VOLUMES IN PREPARATION

CORROSION AND CORROSION PROTECTION HANDBOOK

Edited by Philip A. Schweitzer

Chem-Pro Corporation
Fairfield, New Jersey

MARCEL DEKKER, INC. New York and Basel

Library of Congress Cataloging in Publication Data
Main entry under title:

Corrosion and corrosion protection handbook.

   (Mechanical engineering ; 19)
   Includes bibliographical references and index.
   1. Corrosion and anti-corrosives--Handbooks, manuals,
etc. I. Schweitzer, Philip A. II. Series.
TA418,74,C5928 1983   620.1'1223   83-1885
ISBN 0-8247-1705-8

MARCEL DEKKER, INC.
270 Madison Avenue, New York, New York 10016

Current printing (last digit):
10 9 8 7 6 5 4

PRINTED IN THE UNITED STATES OF AMERICA

Preface

This handbook has been designed and formulated for use by practicing engineers and other persons concerned with the problem of corrosion in their day-to-day job functions. Although the theory of corrosion is touched upon to provide a basic background for the sections on applications, the handbook is not necessarily intended for the use of metallurgical and other engineering personnel who are looking for advanced theories and techniques.

The major advantages and disadvantages of the most commonly used materials of construction are discussed and evaluated. Practical means of use are explained and their most advantageous areas of application discussed. Materials of construction considered include metals, nonmetals, plastics, coating materials, and elastomers.

Practical means of determining corrosion conditions before selection, approaches to be used for the selection of materials of construction, and means of monitoring corrosion are also included.

In general, this handbook has been designed as a guide to be used to:

1. Solve existing corrosion problems
2. Select the best material of construction for a specific application, taking into account cost and service life
3. Assist in the proper initial design so as to eliminate corrosion pitfalls from being designed into the equipment
4. Determine what corrosion problems may or may not exist
5. Establish practices to monitor corrosion of existing equipment

Based on the areas covered and the practical approach established, this handbook should also be of interest to young engineers and college students, who can profit from the vast experience of the many authors who have contributed to the text.

I wish to thank all the contributors to this work for their willingness to share the knowledge they have gained through their experience. A special thanks is extended to Zane B. Laycock of NRC Inc. for supplying data for the section on tantalum and permitting use of these data in the handbook.

Philip A. Schweitzer

Contents

Contents

Contributors

CHARLES G. ARNOLD Dow Chemical Company, Texas Division, Freeport, Texas

DEAN M. BERGER Gilbert/Commonwealth, Reading, Pennsylvania

JOHN M. CIESLEWICZ Ampco Metal, Milwaukee, Wisconsin

LOREN C. COVINGTON* Titanium Metals Corporation of America, Henderson, Nevada

GLENN W. GEORGE The Duriron Company, Inc., Dayton, Ohio

LEWIS W. GLEEKMAN† Materials and Chemical Engineering Services, Southfield, Michigan

F. GALEN HODGE High Technology Materials Division, Cabot Corporation, Kokomo, Indiana

ERNEST H. HOLLINGSWORTH‡ Alcoa Laboratories, Alcoa Center, Pennsylvania

HAROLD Y. HUNSICKER¶ Alcoa Laboratories, Alcoa Center, Pennsylvania

DONALD R. KNITTEL§ Teledyne Wah Chang, Albany, Oregon

JOHN H. MALLINSON J. H. Mallinson, P.E.& Associates Inc., Front Royal, Virginia

RONALD A. MCCAULEY Rutgers, The State University, New Brunswick, New Jersey

GENE FREDERICK RAK Exxon Company, U.S.A., Baton Rouge, Louisiana

GEORGE W. READ, JR. Sauereisen Cements Company, Pittsburgh, Pennsylvania

PHILIP A. SCHWEITZER Chem-Pro Corporation, Fairfield, New Jersey

KENNETH B. TATOR KTA-Tator, Inc., Pittsburgh, Pennsylvania

*Retired
†Deceased
‡Retired
¶Retired
§Present address: Cabot Corporation, Kokomo, Indiana

CORROSION AND CORROSION PROTECTION HANDBOOK

1
Fundamentals and Prevention of Metallic Corrosion

DEAN M. BERGER / Gilbert/Commonwealth, Reading, Pennsylvania

INTRODUCTION

Corrosion is defined as a gradual wearing away or alteration by a chemical or electrochemical oxidizing process. Since the process of corrosion returns metal to its original condition, we must consider the action *degeneration* in its true or primary form. Destruction of metals by methods other than mechanical means is therefore considered to be *corrosion failure*.

Ordinarily, iron and steel corrode in the presence of both oxygen and water, and corrosion usually does not take place in the absence of either of these. Rapid corrosion may take place in water; the rate of corrosion is accelerated by the velocity or the acidity of the water, by the motion of the metal, by an increase in temperature or aeration, by the presence of certain bacteria, and by other less prevalent factors. On the other hand, corrosion is generally retarded by protective layers (or films) consisting of corrosion products or adsorbed oxygen; high alkalinity of the water also retards the rate of corrosion on steel surfaces. Water and oxygen, however, are almost always the essential factors and the amount of corrosion is controlled by one or the other. For instance, corrosion of steel does not occur in dry air and is negligible when the relative humidity of the air is below 30% at normal or lower temperatures. Prevention of corrosion by dehumidification is based on this fact [1].

All structural metals corrode to some extent in natural environments. Bronzes, brasses, stainless steels, zinc, and aluminum corrode so slowly under the service condition in which they are placed that they are expected to survive for long periods without protection. Corrosion of structural grades of iron and steel, however, proceeds rapidly unless the metal is amply protected. This susceptibility to corrosion of iron and steel is of great concern because, owing to favorable cost and physical property considerations, vast quantities are used. Annual losses due to corrosion of steel have been variously estimated at nearly $70 billion in the United States. It is apparent that protection against corrosion of iron and steel is an indispensable phase of sound engineering.

MECHANISM OF CORROSION

Corrosion in metals, whether in the atmosphere, underwater, or underground, is caused by a flow of electricity from one metal to another metal or recipient of some kind; or from one part of the surface of one piece of metal to another part of the same metal where conditions permit the flow of electricity.

Further, a moist conductor or electrolyte must be present for this flow of energy to take place. The presence of an electrolyte is a key condition for the process of corrosion to occur. Water, therefore, especially salt water, is an excellent electrolyte.

Simply stated, energy (electricity) passes from a negative area to a positive area via the electrolyte media. So to have corrosion take place in metals, one must observe an (1) electrolyte, (2) an area or region on a metallic surface with a negative charge in relation to a second area, and (3) the second area with a positive charge in opposition to the first [2].

Sometimes the situation becomes further involved. The flow of energy (electricity) may be from one metal to another, or from one metal recipient of some kind, which might be the soil. This is due to various environments within a given soil. Soils frequently have contained dispersed metallic particles or bacteria pockets which provide a natural electrical pathway with buried metal. If an electrolyte is present and the soil is negative in relation to the metal, the electric path will occur from the metal to the soil and corrosion results.

Types of corrosion commonly identified on metal [3]:

Uniform	Corrosion fatigue
Electrochemical	Intergranular
Galvanic	Fretting
Concentration cell	Impingement

Erosion corrosion   Dezincification
Embrittlement   Graphitization
Stress corrosion   Chemical reaction
Filiform

Water readily dissolves a small amount of oxygen from the atmosphere into solution and this may become highly corrosive. When the free oxygen dissolved in water is removed, the water is practically noncorrosive unless it becomes acidic or unless anaerobic bacteria incite corrosion. If oxygen free water is maintained neutral or slightly alkaline, it will be practically noncorrosive to steel. Thus steam boilers and water supply systems are effectively protected by deaeration of the water.

Several other important facts, fundamental to the corrosion of steel, are: (1) hydrogen gas is evolved when a metal corrodes in acid and the rate of corrosion is relatively rapid; (2) surface films, usually inert and often invisible, may greatly decrease the rate of corrosion (as in the case of aluminum and stainless steel); (3) the rate of corrosion increases with temperature and velocity of motion up to a certain point; and (4) corrosion is rarely uniformly distributed over the metal surface.

Electrochemical Corrosion

The cell shown in Fig. 1 illustrates that corrosion process in its simplest form. This cell includes the following essential components:

1. A metal anode
2. A metal cathode
3. A metallic conductor between the anode and cathode
4. An electrolyte (water containing conductive salts) in contact with anode and cathode but not necessarily of the same composition at the two locations

In addition, oxygen will usually be present as a depolarizing agent. As shown in Fig. 1, these components are arranged to form a closed electrical path or circuit. In the simplest case, the anode would be one metal, perhaps iron, the cathode another, say copper, and the electrolyte might or might not have the same composition at both electrodes. Alternatively, the electrodes could be of the same metal if the electrolyte composition varied.

If the cell as shown were constructed and allowed to function, an electrical current would flow through the metallic conductor and the electrolyte. The anode would corrode (rust, if the anode were iron); chemically, this is an oxidation reaction. Simultaneously, a nondestructive chemical reaction (reduction) would proceed at the cathode, producing hydrogen gas on the cathode in most cases. When the gas

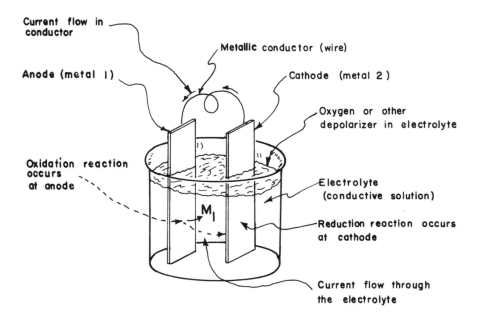

Electrochemical   Corrosion

Figure 1. Simple cell showing components necessary for corrosion.
(From Ref. 3.)

layer insulates the cathode from the electrolyte, current flow will
stop, thus polarizing the cell. However, oxygen or some other depolarizing agent is usually present to react with the hydrogen, reducing this effect, so the cell would continue to function. If the
metallic conductor were replaced with a voltmeter, a difference of
potential could be measured between the electrodes.

The difference in potential that causes these electric currents is
due mainly to contact between dissimilar metallic conductors or differences in concentration of the solution, mainly with respect to dissolved
oxygen in natural waters. Almost any lack of homogeneity on the
metal surface or in its environment may initiate attack by causing
differential in potentials which results in more or less localization of
corrosion. Many other facts are involved, but these include most of
the important facts known about ordinary corrosion and are all

satisfactorily explained by application of electrochemical principles. The metal undergoing electrochemical corrosion need not be immersed in a solution, but may be in contact with moist soil, or patches of the metal surface may be moist.

Atmospheric corrosion differs from the action that occurs in water, or underground, in that a plentiful supply of oxygen is always present. In this case, the formation of insoluble films and the presence of moisture and deposits from the atmosphere become the controlling factors; an additional factor that affects the corrosion rate is the presence of contaminants such as sulfur compounds and salt particles. Nevertheless, atmospheric corrosion is mainly electro-chemical rather than a direct result of chemical attack by the ele-ments. The anodic and cathodic areas, however, are usually quite small and close together, so that, generally, corrosion is apparently uniform, rather than in the form of severe pitting as in water or in soil.

The larger the anodic area is in relation to the cathode, the faster the rate of degeneration or corrosion It follows, therefore, that anodes and cathodes exist on all iron and steel surfaces. Sur-face imperfections, grain orientation, lack of homogeneity of the metal, variation in the environment, localized shear and torque during manu-facture, mill scale, and existing red iron oxide rust.

Technical definition of electrolyte

An electrolyte is a solution containing ions, which are particles bear-ing an electric charge. Ions are present in solutions of acids, alka-lies, and salts.

It is well known that the formation of rust by electrochemical reactions may be expressed as follows:

1. $4Fe \rightarrow 4Fe^{2+} + 8e^{-}$

    $4Fe + 3O_2 + H_2O \rightarrow 2Fe_2O_3 \cdot H_2O$ (hydrated red iron rust)

2. $4Fe + 2O_2 + 4H_2O \rightarrow 4\ Fe(OH)_2$

    $4Fe(OH)_2 + O_2 \rightarrow 2Fe_2O_3 \cdot H_2O + 2H_2O$ (hydrated red iron rust)

According to a number of investigations, the most stable form of rust is Fe_2O_3. At higher temperatures (900 to 1300°F) Fe_2O_3 reverts to Fe_3O_4. It is believed that the formation of Fe_3O_4 can occur either by heating the rusted steel to high temperatures or by oxidizing (re-ducing) operations.

In an acid environment, even without the presence of oxygen, the metal at the anode is attacked at a rapid rate while at the cathode the hydrogen gas/film is being dissolved continuously, forming a true hydrogen gas.

When corrosion by an acid results in the formation of a salt, the reaction is slowed conditionally because of the salt formation on the surface being attacked.

Galvanic Corrosion

Better known as simply *dissimilar metal corrosion*, this form of degeneration pops up in the most unusual places and often causes the most painful professional headaches.

The *Galvanic Series of Metals* guide details how the galvanic current will flow between two metals and which will corrode when they are in contact or near each other in the ground.

If rust or mill scale is present on the surface of the steel, galvanic corrosion will take place. This action is due to the dissimilarity with the metal, the base metal being the anode in this instance.

A galvanic couple may be the cause of premature failure in metal components of hydraulic structures or may be advantageously exploited. Galvanizing of iron sheets is an example of the useful application of galvanic action. In this case, iron is the cathode and is protected against corrosion at the expense of the zinc anode. Alternatively, a zinc or magnesium anode may be located in the electrolyte close to the structure and still be connected metallically to the iron or steel. This is termed *cathodic protection* of the structure. On the other hand, iron or steel becomes the anode when in contact with copper, brasses, and bronzes, and corrodes rapidly while protecting the latter metals.

The mill scale formed on steel during rolling varies with the type of operation and the rolling temperature. In general, mill scale is magnetic and contains three layers of iron oxide, but the boundaries between the oxides are not sharp. The outer layer of mill scale is essentially ferric oxide Fe_2O_3, which is relatively stable and does not easily react. The layer closest to the steel surface and sometimes intermingled with the surface crystalline structure of the steel itself is ferrous oxide FeO. This substance is unstable and the iron in ferrous oxide is easily oxidized to ferric iron, resulting in a chemical change to ferric oxide. This process, accompanied by an increase in volume, may result in the loosening of the intact mill scale. The intermediate layer of magnetic oxide is best represented by the chemical formula Fe_3O_4. The actual thickness of mill scale on structural steel, which depends on rolling conditions, varies from about 0.002 in. to about 0.020 in., and consists mainly of the magnetic oxide Fe_3O_4 and the FeO layer. Much of the mill scale formed at high initial rolling temperatures is knocked off in subsequent rolling [5].

Unfortunately, mill scale is cathodic to steel (i.e., the mill scale has a more noble potential) and an electric current can easily be

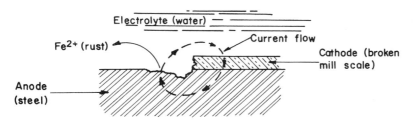

Figure 2. Mill scale is cathodic to steel, establishing a corrosion cell. (From Ref. 4.)

produced between the steel and the mill scale. This electrochemical action will corrode the steel and protect the mill scale (Fig. 2).

The driving force (difference of potential) available to promote the electrochemical corrosion reaction is reflected by the galvanic series. This is a list of a number of common metals and alloys arranged according to their tendency to corrode galvanically. Metals grouped together do not have a strong effect on each other, and the farther apart any two metals appear, the stronger the corroding effect on the one higher in the list. It is possible for certain metals to reverse their positions in some environments, but the list as given will hold generally in natural waters and the atmosphere. (The galvanic series should not be confused with the similar electromotive force series, which shows exact potentials based on highly standardized conditions that rarely exist in nature [4].)

Galvanic series

Corroded end (anodic)
    Magnesium
    Magnesium alloys
    Zinc
    Aluminum 2S
    Cadmium
    Aluminum 17ST
    Steel or iron
    Cast iron
    Chromium-iron (active)
    Ni-Resist
    18-8 Chromium-nickel-iron (active)
    18-8-3 Chromium-nickel-molybdenum-iron (active)
    Lead-tin solders
    Lead
    Tin
    Nickel (active)

Inconel (active)
Hastelloy C (active)
Brass
Copper
Bronzes
Copper-nickel alloys
Monel
Silver solder
Nickel (passive)
Inconel (passive)
Chromium-iron (passive)
18-8 Chromium-nickel-iron (passive)
18-8-3 Chromium-nickel-molybdenum-iron (passive)
Hastelloy C (passive)
Silver
Graphite
Gold
Platinum
Protected end (cathodic)

While the preceding galvanic series represents generally the driving force available to promote a corrosion reaction, the actual rate experienced may be considerably different from that predicted from the driving force alone. Electrolytes may be poor conductors, or long distances may introduce a large resistance into the corrosion cell circuit. More frequently, scale formation forms a partially insulating layer over the anode. A cathode having a layer of adsorbed gas bubbles as a consequence of the corrosion cell reaction is said to be *polarized*. The effect of such conditions is to reduce the theoretical consumption of metal by corrosion. The area relationship between the electrodes also may strongly affect the corrosion rate; a high ratio of cathode area to anode area produces rapid corrosion, but in the reverse case the cathode polarizes and the rate soon drops to a negligible level.

The passivity of stainless steels is attributed to the presence of corrosion-resistant oxide film over the surfaces. In most natural environments, they will remain in a passive state and thus tend to be cathodic to ordinary iron and steel. Change to an active state usually occurs only when chloride concentrations are high, as in seawaters or in reducing solutions. Oxygen starvation also causes the change to an active state. This occurs where there is not free access of oxygen, such as in crevices and beneath contamination on partially fouled surfaces. Accelerated corrosion of steel and iron can be produced by stray currents. Direct currents in the soil or water associated with nearby cathodic protection systems, industrial activities, or direct-current electric railways can be intercepted and carried for considerable distances by buried steel structures. Corrosion takes place where the stray currents are discharged from the steel to the environment, and damage to the structure can occur very rapidly indeed [4].

Differences in soil conditions, such as moisture content and resistivity, commonly are responsible for creating anodic and cathodic areas. Where there is a difference in the concentration of oxygen in the water or in moist soils in contact with metal at different areas, cathodes will develop at points of relatively high oxygen concentrations and anodes at points of low concentration. Strained portions of metal reportedly tend to be anodic and unstrained portions cathodic. Thus, under all ordinary circumstances where iron and steel are exposed to natural environments, the basic conditions essential to corrosion are present to a greater or lesser degree (Fig. 3).

Pitting deserves special mention because it is the type of local cell corrosion that is predominantly responsible for the functional failure of iron and steel hydraulic structures. Pitting may result in the perforation of water pipe, rendering it unserviceable, even though less than 5% of the total metal has been lost through rusting. Even where confinement of water is not a factor, pitting will cause structural failure from localized weakening effects while there is still considerable sound metal remaining.

Pitting develops when the anodic (corroding) area is small in relation to the cathodic (protected) area. For example, it can be expected where large areas of the surface are generally covered by mill scale, applied coatings, or deposits of various kinds, but breaks exist in the continuity of the protective material. Pitting may also develop on bare,

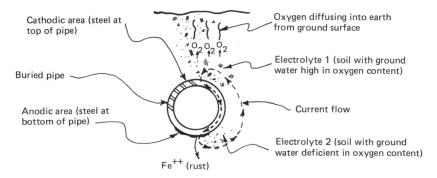

Figure 3. Metal pipe buried in moist soil may corrode on the bottom. A difference in oxygen content at different levels in the electrolyte will produce a difference of potential. Thus anodic and cathodic areas will develop, and a corrosion cell, called a concentration cell, will form. (From Ref. 4.)

clean metal surfaces because of irregularities in the physical or chemical structure of the metal. Localized, dissimilar soil conditions at the surface of steel can also create conditions that promote the pitting type of corrosion [4].

The main factors that cause and accelerate pitting are electrical contact between dissimilar materials, or between what are termed *concentration cells* (areas of the same metal where oxygen or conductive salt concentrations in water differ). Even in cleistomatic structures, these couples cause a difference of potential that results in an electric current flowing through the water, or across moist steel, from the metallic anode to a nearby cathode. The cathode may be copper or brass, mill scale, or any other portion of the metal surface that is cathodic to the more active metal areas. In practice, mill scale is cathodic to steel, and is found to be one of the more common causes of pitting. The difference of potential generated between steel and mill scale often amounts to 0.2 to 0.3 V; this couple is nearly as powerful a generator of corrosion currents as is the copper- steel couple. However, when the anodic area is relatively large compared with the cathodic area, the damage is spread out and usually negligible; when the anode is relatively small, the metal loss is concentrated and may be very serious.

On surfaces containing mill scale, the total metal loss is nearly constant as the anode area is decreased, but the degree of penetration increases. Figure 2 shows diagrammatically how a pit forms where a break occcurs in mill scale. When contact between dissimilar materials is unavoidable and the surface is painted, it is important to paint both materials—especially the cathode. If the anode only is coated, any weak points such as pinholes or holidays in the coating will probably result in intense pitting [5].

Severe corrosion, leading to pitting, is often caused by concentration cells, particularly in cases in which differences in dissolved oxygen concentration occur. When a part of the metal is in contact with water relatively low in dissolved oxygen, it is anodic to adjoining areas in contact with water higher in dissolved oxygen. This lack of oxygen may be caused by exhaustion of dissolved oxygen in a crevice (see Fig. 4). The low-oxygen area is always anodic. Figure 4 illustrates another type of concentration cell; this cell at the mouth of a crevice is due to a difference in concentration of the metal in solution. These two effects sometimes blend together as in a reentrant angle in a riveted seam.

As a pit, perhaps caused by broken mill scale, becomes deeper, an oxygen concentration cell is started by the depletion of oxygen in the pit and the rate of penetration in such cases is accelerated proportionately. Fabrication operations may crack mill scale and result in accelerated corrosion.

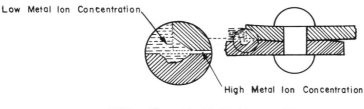

METAL ION CONCENTRATION CELL

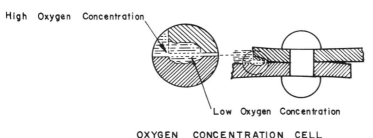

OXYGEN CONCENTRATION CELL

Figure 4. Corrosion caused at crevices by concentration cells; the attacks may occur simultaneously. (From Ref. 4.)

PREVENTING CORROSION

In the book *Designing to Prevent Corrosion in the Process Industry* [6], F. L. Whitney suggests some very basic commonsense approaches to reduce or eliminate the problem of corrosion. He suggests approaching the problem at the design stage and also through the use of good maintenance procedures after startup of the facility.

Whitney's recommendations call for the evaluation of:

1. Plant location (coastal locations are troublesome)
2. Plant layout; check prevailing winds, fallout, and related items
3. Proper fume removal and control
4. Sewer treatment of effluents
5. Design of equipment to avoid:
   a. Galvanic corrosion
   b. Crevice corrsion

For example, using Whitney's concepts in the design of tankage, one could eliminate crevice corrosion concentration cell corrosion by

Figure 5. Improper lap weld protection. (From Ref. 4.)

proper design. DO NOT LAP as shown in Fig. 5. If lap welding is
to be employed, the laps should be filled with fillet welding or a
suitable caulking compound designed to prevent crevice corrosion
(Fig. 6). The use of a proper butt weld will eliminate the need for
future maintenance (Fig. 7). This form of butt welding is especially
suitable for tank-lining work with protective coatings. Butt welding
allows for grinding down smooth and then subsequent coating with a
uniform milage of coating. Welding with dissimilar metals can establish
a corrosion cell (Fig. 8.).

The importance of organic coating system selection and application
is stressed. Adequate paint systems or lining materials exist that will
deter the corrosion rate of carbon steel surfaces. High-performance
coatings such as epoxy, polyesters, polyurethanes, vinyl, or chlorinated rubber help to satisfy the needs of the corrosion engineer.
Special primers are used to provide passivation, galvanic protection, or
barrier properties which are necessary for corrosion inhibition.

A water-soluble corrosion inhibitor will reduce galvanic action by
depassivating the metal or by laying down an insulating film on either
the anode or the cathode, or both. A very small amount of chromate,
polyphosphate, or silicate added to water acts in this way. A slightly
soluble inhibitor incorporated in the primary coat of paint may also
have a considerable protective influence. Inhibitive pigments in paint
primers act well in this manner as long as they do not dissolve sufficiently to leave holes in the paint film.

It is believed that corrosion by chemical attack is the single most
destructive force against steel surfaces. Chemicals are used everywhere and the chemical industry is rapidly expanding. Chemical attack
is more dramatic in that the steel literally dissolves and erodes away.
One observer, after inspection of a paper mill, said: "The steel beams

Figure 6. Proper lap weld protection. (From Ref. 4.)

Figure 7. Butt weld. (From Ref. 4.)

looked like they had been chewed by mice with strong teeth." The
corrosion engineer must therefore be concerned with the environment
and the chemicals surrounding a structure.

   Halogens are particularly aggressive. The galvanized roof of one
building was completely eaten away within six months after erection.
This building sat downstream of an aluminum ingot plant where fluorides
were always present in the atmosphere. Selection of materials is extremely important. In this case, galvanized steel should not have been
specified.

   Coatings help prevent corrosion by providing:

1. Sacrificial or galvanic protection
2. Passivation of the steel using inhibitive pigments
3. A barrier against the environment

Sacrificial Coatings

Zinc-rich primers are applied at 3.0 mils dry-film thickness to provide
galvanic protection. These primers are very effective, even in chemical environments, since the zinc will dissipate itself before the steel
is attacked. Adequate high-performance top coats are recommended to
prolong the life of the coating system.

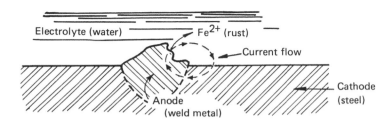

Figure 8. Weld metal may be anodic to steel, creating a corrosion cell
when immersed. (From Ref. 4.)

Passivation and Corrosion Inhibitors

Most paint primers contain an inhibitive of partially soluble pigments such
as zinc chromate which reacts with the steel substrate to form the iron
salt. These salts slow down the corrosion of steel. Chromates, phosphates, molybdates, borates, silicates, and plumbates are common.
Some pigments passivate by contributing alkalinity, thereby slowing
down the chemical attack on steel. Alkaline pigments such as meta-
borates, cement or lime, or red lead all are quite effective provided
that the environment is not too aggressive. Many new pigments have
been introduced to the paint industry. In 1976, Hammond lead offered
a zinc phosphosilicate and Silberline manufactured a zinc flake as examples.

Barrier Coatings

Protective coatings are the most widely used and recognized forms of
barrier materials in engineered and remedial construction. These protective barriers may vary in thickness from thin paint films of only a
few mils to heavy mastic coatings applied about 1/4 to 1/2 in. thick to
acid-proof brick linings several inches thick. Barrier coatings are
effective because they keep moisture and oxygen away from the steel
substrate. The lower the moisture vapor transmission rate of the
polymer, the more effective it is as a vehicle for protective coatings.
The thicker the coating, the better moisture is kept out. Protective
barrier coatings vary considerably in composition, performance, and
applied cost.

Cathodic Protection

Simply stated, cathodic protection serves to reverse electric current
flow within the corrosion cell. Cathodic protection is used to reduce or
eliminate corrosion by connecting a more active metal to the metal that
must be protected. For example, zinc and magnesium anodes are used
to protect steel in marine environments.

The use of cathodic protection to reduce or eliminate corrosion is a
long-standing successful technique employed in marine structures,
pipelines, bridge decks, sheet piling, equipment, and tankage of all
types, particularly below water or underground. Typically, zinc and
magnesium anodes are used to protect steel in marine environments.
The anodes are replaced after they are consumed.

Cathodic protection also employs the use of direct current (dc) and
a relatively inert anode. As is the case in all forms of cathodic activity,
an electrolyte is needed for current flow. Cathodic protection and the
use of protective coatings are most often employed jointly, especially in
marine applications and aboard ships, where impressed current inputs

usually do not exceed 1 v [7]. Beyond that point, many coating
systems tend to disbond. Current flow for cathodic protection in soils
is usually 1.5 to 2 V.

Choice of anodes for buried steel pipe depends on job conditions.
Magnesium is most commonly used for galvanic anodes; however, zinc
can also be used. Galvanic anodes are seldom used when the resistivity of the soil is over 3000 Ω-cm; impressed current is normally used
for these conditions. Graphite, high silicon cast iron, scrap iron,
aluminum, and platinum are used as anodes with impressed current.
The availability of low-cost power is often the deciding factor in choosing between galvanic or impressed current cathodic protection.

Protective coatings are normally used in conjunction with cathodic
protection and should not be neglected where cathodic protection is
contemplated in new construction. Since the cathodic protection current must protect only the bare or poorly insulated areas of the surface,
coatings that are highly insulating, very durable, and free of discontinuities lower the current requirements and system costs. Moreover,
a good coating enables a single impressed current installation to protect many miles of piping. Coal tar enamel, epoxy powder coatings,
and vinyl resin are examples of coatings that are most suitable for use
with cathodic protection. Certain other coatings may be imcompatible;
for instance, phenolic coatings may deteriorate rapidly in the alkaline
environment created by cathodic protection currents. Although cement
mortar initially conducts the electrical current freely, polarization (formation of an insulating film on the surface as a result of the protective
current) is believed to reduce the current requirement moderately.

Cathodic protection is being used increasingly to protect buried and
submerged metal structures in the oil, gas, and waterworks industries
and has specialized applications such as the interiors of water storage
tanks. Pipelines are routinely designed to ensure the electrical continuity necessary for effective functioning of the cathodic protection system.
Thus electrical connections or bonds are required between pipe sections
in lines utilizing mechanically coupled joints, and insulating couplings
may be employed at intervals to isolate electrically some parts of the
line from other parts. When needed, leads may be attached during construction to facilitate cathodic protection installation.

Good inspection procedures are required to maintain surveillance of
cathodic protection installations. Such service is available from
suppliers of these systems.

Good engineering practice, material selection, design consideration,
and inspection will prolong the life of metal structures.

REFERENCES

1. "Theory of Corrosion," in Volume I: *Good Painting Practice*, Steel
    Structures Painting Council, Pittsburgh, Pa., 1968, Chap. 1.

2. *Introduction to Corrosion*, Carboline Co., 1968.
3. D. M. Berger, Corrosion principles can never be forgotten in organic finishing, *Met. Finish. 72*, November 1974.
4. Jack Kiewit, *The Paint Manual*, USDI Bureau of Reclamation, Denver, Colo., 1976, Chap. III.
5. F. N. Speller, *Corrosion Causes and Prevention*, McGraw-Hill, New York, 1951.
6. F. L. Whitney, *Designing to Prevent Corrosion in the Process Industry*, 59-SA-58, American Society of Mechanical Engineers, May 1959.
7. W. A. Anderton, The aluminum vinyl for ships bottoms, *J. Oil Colour Chem. Assoc. 53*, November 1970.

2
Cathodic Protection

PHILIP A. SCHWEITZER / Chem-Pro Corporation, Fairfield, New Jersey

BACKGROUND

Cathodic protection is a major factor in corrosion control of metals. When an external electric current is applied, the corrosion rate can be reduced to practically zero. Under these conditions the metal can remain in a corrosive environment indefinitely without deterioration.

In practice, cathodic protection can be utilized with such metals as steel, copper, brass, lead, and aluminum against corrosion in all soils and almost all aqueous media. Although it cannot be used above the waterline (since the impressed electric current cannot reach areas out of the electrolyte) it can be effectively used to eliminate corrosion fatigue, intergranular corrosion, stress corrosion cracking, dezincification of brass, or pitting of stainless steels in seawater or steel in soil.

In 1982, Sir Humphry Davy reported that by coupling iron or zinc to copper, the copper could be protected against corrosion. The

British admiralty had blocks of iron attached to the hulls of copper-sheathed vessels to provide cathodic protection. Unfortunately, cathodically protected copper is subject to fouling by marine life, which reduced the speed of vessels under sail and forced the admiralty to discontinue the practice. However, the corrosion rate of the copper had been appreciably reduced. Unprotected copper supplies a sufficient number of copper ions to poison fouling organisms.

In 1829, Edmund Davy was successful in protecting the iron portions of buoys by using zinc blocks, and in 1840, Robert Mallet produced a zinc alloy that was particularly suited as a sacrificial anode. The fitting of zinc slabs to the steel hulls of vessels became standard practice as wooden hulls were replaced. This provided localized protection specifically against the galvanic action of a bronze propeller. Overall protection of seagoing vessels was not investigated again until 1950, when the Canadian Navy determined that the proper use of antifouling paints in conjunction with corrosion-resistant paints made cathodic protection of ships feasible and could reduce maintenance costs.

About 1910-1912 the first application of cathodic protection by means of an impressed electric current was undertaken in England and the United States. Since that time the general use of cathodic protection has spread widely. There are thousands of miles of buried pipelines and cables which are protected in this manner.

This form of protection is also used for water tanks, submarines, canal gates, marine piling, condensers, and chemical equipment.

THEORY

The basis of cathodic protection is shown in the polarization diagram for a Cu-Zn cell, Fig. 1. If polarization of the cathode is continued by use of an external current beyond the corrosion potential to the open-circuit potential of the anode, both electrodes reach the same potential and no corrosion of the zinc can take place. Cathodic protection is accomplished by supplying an external current to the corroding metal, on the surface of which local action cells operate as shown in Fig. 2. Current flows from the auxiliary anode and enters the anodic and cathodic areas of the corrosion cells, returning to the source of the dc current B. Local action current will cease to flow when all the metal surface is at the same potential, as a result of the cathodic areas being polarized by an external current to the open circuit potential of the anodes. As long as this external current is maintained, the metal cannot corrode.

The corrosion rate will remain at zero if the metal is polarized slightly beyond the open-circuit potential ϕ_A of the anode. However, this excess current has no value and may be injurious to amphoteric metals or coatings. For this reason, in actual practice the impressed current is maintained close to the theoretical minimum.

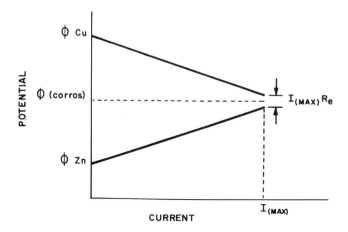

Figure 1. Polarization for copper-zinc cell.

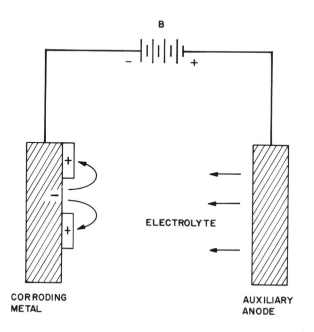

Figure 2. Cathodic protection using impressed current on local
action cell.

If the current should fall below that required for complete protection, some protection will still be afforded.

METHOD OF APPLICATION

A source of direct current and an auxiliary electrode are required to provide cathodic protection. The auxiliary electrode (anode) is usually composed of graphite or iron and is located some distance away from the structure being protected. The anode is connected to the positive terminal of the dc source and the structure being protected is connected to the negative terminal. This permits current to flow from the anode, through the electrolyte, to the structure. Applied voltages are not critical as long as they are sufficient to supply an adequate current density to all parts of the protected structure. The resistivity of the soil will determine the required applied voltage; and when the ends of a long pipeline are to be protected, the voltage will have to be increased.

Current is usually supplied by a rectifier supplying low-voltage dc of several amperes.

Sacrificial Anodes

It is possible, by selection of an anode constructed of a metal more active in the galvanic series than the metal to be protected, to eliminate the need for an external dc current. A galvanic cell will be established with the current direction exactly as described using an impressed electric current. These sacrificial anodes are usually composed of magnesium or magnesium-base alloys. Occasionally, zinc and aluminum have been used. Since these anodes are essentially sources of portable electrical energy they are particularly useful in areas where electric power is not available and/or it is uneconomical or impractical to install power lines for this purpose.

The majority of the sacrificial anodes in use in the United States are of magnesium construction. Approximately 10 million pounds of magnesium is used annually for this purpose. The open-circuit potential difference between magnesium and steel is about 1 V. This means that one anode can protect only a limited length of pipeline. However, this low voltage can have an advantage over higher impressed voltages in that the danger of overprotection to some portions of the structure is less and since the total current per anode is limited, the danger of stray-current damage to adjoining metal structures is reduced.

Magnesium anode rods have also been placed in steel hot water tanks to increase the life of these tanks. The greatest degree of protection is afforded in "hard" waters, where the conductivity of the water is greater than in "soft" waters.

Use With Coatings

Insulating coatings are advantageous to use with either impressed current or sacrificial anodes when supplying cathodic protection. These coatings need not be pore-free since the protective current flows preferentially to the exposed metal areas, which require the protection. Such coatings are useful in distributing the protective current, in reducing total current requirements, and in extending the life of the anode. For example, in a coated pipeline the current distribution is greatly improved over that of a bare pipeline, the number of anodes and total current required is less, and one anode can protect a much longer section of pipeline. Since the earth itself is a good electrical conductor and the resistivity of the soil is localized only within the region of the pipeline or electrodes, the limiting length of pipe protected per anode is imposed by the metallic resistance of the pipe itself and not the resistance of the soil.

One magnesium anode is capable of protecting approximately 100 ft (30 m) of a bare pipeline, whereas it can provide protection for approximately 5 mi (8 km) of a coated pipeline.

In a hot water tank coated with glass or an organic coating the life of the magnesium anode is extended and more uniform protection is supplied to the tank. Without the coating the tendency is for excess current to flow to the sides and insufficient current to flow to the top and bottom.

Because of these factors cathodic protection is usually provided in conjunction with coated surfaces.

CURRENT REQUIREMENTS

The specific metal and the environment will determine the current density required for complete protection. The applied current density must always exceed the current density equivalent to the measured corrosion rate under the same conditions. Therefore, as the corrosion rate increases, the impressed current density must be increased to provide protection.

For cathodically controlled corrosion rates, the corrosion potential approaches the open-circuit anode potential and the required current density is only slightly greater than the equivalent corrosion current. The required current can be considerably greater than the corrosion current for mixed control, and for anodically controlled corrosion reactions the required current is even greater.

When a protective current causes precipitation of an inorganic scale on the cathode surface, such as in hard water or seawater, the total current required is gradually reduced. This is the result of an insulating coating being formed. However, the current density at the exposed metal areas does not change; only the total current density per apparent unit area is less.

ANODE MATERIALS AND BACKFILL

The use of magnesium as a sacrificial anode has already been discussed. For use with impressed current, auxiliary anodes are usually formed of scrap iron or graphite. Scrap iron is consumed at a considerably faster rate than graphite (15 to 20 lb/A year versus 2 lb/A year); however, graphite costs more both initially and in operating expense. Graphite requires more power than scrap iron. It is also more fragile, and greater care must be taken during installation. Under certain conditions the advantage of the 8 to 10 times longer life far outweighs the added costs, particularly in areas where replacement poses problems.

Platinum clad or 2% silver-lead electrodes have been used for the protection of structures in seawater using impressed current. The latter anodes are estimated to last 10 years, whereas sacrificial magnesium anodes require replacement every 2 years. On occasion, aluminum electrodes have been used in fresh waters.

Since the effective resistivity of soil surrounding an anode is limited to the immediate area of the electrode, this local resistance is generally reduced by using backfill. For impressed current systems the anode is surrounded with a thick bed of coke mixed with three to four parts of gypsum to one part of sodium chloride. The consumption of the anode is reduced somewhat by virtue of the fact that the coke backfill is itself a conductor and carries part of the current. If the anode is immersed in a river bed, lake, or ocean, backfill is not required.

Auxiliary anodes need not be consumed in order to fulfill their purpose. Conversely, sacrificial anodes are consumed not less than is required by Faraday's law in order to supply an equivalent current.

For magnesium anodes, backfill has the advantage of reducing the resistance of insulating corrosion-product films as well as increasing the conductivity of the immediate area. A typical backfill consists of a mixture of approximately 20% bentonite (for retention of moisture), 75% gypsum, and 5% sodium sulfate.

TESTING FOR COMPLETENESS OF PROTECTION

There are several ways in which the effectiveness of protection can be checked. The first two methods are qualitative and do not provide data as to whether enough or more than enough current is being supplied. Potential measurements, the third method, is of prime importance in practice.

Coupon Tests

A metal coupon is shaped to conform to the contour of the pipe, weighed, and attached by a brazed-connected cable to the pipe. Both the cable and the surface between the coupon and pipe are coated with coal tar. The coupon is allowed to remain buried for weeks or months, recovered, cleaned, and weighed. The weight loss, if any is an indication as to whether or not the cathodic protection of the pipeline is complete.

Colormetric Tests

A piece of absorbent paper soaked in potassium ferricyanide solution is placed in contact with a cleaned section of the buried pipeline and the soil replaced. After a relatively short time the paper is retrieved. A blue color of ferrous ferricyanide indicates incomplete cathodic protection, while an absence of blue color indicates that the cathodic protection is complete.

Potential Measurements

By measuring the potential of the protected structure the degree of protection, including overprotection, can be determined quantitatively. This measurement is the generally accepted criterion and is used by corrosion engineers. The basis for this determination is the fundamental concept that cathodic protection is complete when the protected structure is polarized to the open-circuit anode potential of local action cells.

The reference electrode for making this measurement should be placed as close as possible to the protected structure to avoid and/or minimize an error caused by IR drop through the soil. IR drops through corrosion product films or insulating coatings will still be present regardless of precautions taken, tending to make the measured potential more active than the actual potential at the metal surface. For buried pipelines a compromise location is taken directly over the buried pipe at the soil surface since cathodic protection currents flow mostly to the lower surface and are minimum at the upper surface of the pipe buried a few feet below the soil surface.

The potential for steel is equal to -0.85 V versus the copper-saturated copper sulfate half-cell, or 0.53 V on the standard hydrogen scale. The theoretical open-circuit anode potential for other metals may be calculated using the Nernst equation. Several typical calculated values are shown in Table 1.

OVERPROTECTION

Overprotection of steel structures, to a moderate degree, usually does not cause any problems. The primary disadvantages are waste of electric power and increased consumption of auxiliary anodes. When

Table 1. Calculated Minimum Potential ϕ for Cathodic Protection

Metal	$E°$ (V)	Solubility product, $M(OH)_2$	ϕH_2 Scale (V)	ϕ vs. Cu-Cu SO$_4$ Reference electrode (V)
Iron	0.440	1.8×10^{-15}	-0.59	-0.91
Copper	-0.337	1.6×10^{-19}	0.16	-0.16
Zinc	0.763	4.5×10^{-17}	-0.93	-1.25
Lead	0.126	4.2×10^{-15}	-0.27	-0.59

Source: Herbert H. Uhlig, *Corrosion and Corrosion Control*, 2nd ed. Wiley, New York, 1971.

the overprotection is excessive, hydrogen can be generated at the protected structure in sufficient quantities to cause blistering of organic coatings, hydrogen embrittlement of the steel, or hydrogen cracking. Damage to steel by hydrogen absorption is more prevalent in environments where sulfides are present.

Overprotection of systems with amphoteric metals (e.g., aluminum, zinc, lead, tin) will damage the metal by causing increased attack instead of reduction of corrosion. This emphasizes the need for making potential measurements of protected structures.

ECONOMICS

The cost of cathodic protection is more than recovered by reduced maintenance costs and/or reduced installation costs. For buried pipelines the guarantee that there will be no corrosion on the soil side of the pipe has made it economically feasible to transport oil and high-pressure natural gas across the American continent. It has also permitted the use of thinner-walled pipe. Wall thicknesses need only be sufficient to withstand the internal pressures. No extra allowance has to be added for external corrosion. In some cases this savings alone has more than paid for the installation of the cathodic protection equipment.

Similarly, other cathodic protection systems have more than paid for their installation costs by reduced maintenance costs and/or longer operating periods between routine inspections and/or maintenance periods.

3
Development and Application of Corrosion — Resistant
Metals and Alloys

3.1
Carbon Steel and Low Alloy Steel

PHILIP A. SCHWEITZER / Chem-Pro Corporation, Fairfield, New Jersey

INTRODUCTION

Steel is produced from pig iron by the removal of impurities in an open-hearth furnace, a basic oxygen furnace, a Bessemer converter, or an electric furnace. In the United States over 80% of steel is produced in the basic open-hearth furnace.

As a result of the methods of production, the following elements are always present in steel: carbon, manganese, phosphorus, sulfur, silicon, and traces of oxygen, nitrogen, and aluminum. Various alloying ingredients are frequently added, such as nickel, chromium, copper, molybdenum, and vanadium. The most important of these elements in steel is carbon, and it is necessary to understand the effect of carbon on the internal structure of the steel to understand the heat treatment of carbon and low alloy steels.

CONSTITUTION AND STRUCTURE OF STEEL

When heated to 910°C (1670°F), pure iron changes its internal crystalline structure from a body-centered cubic arrangement of atoms, alpha iron, to a face-centered cubic structure, gamma iron. At 1390°C (2535°F) it changes back to the body-centered cubic structure, delta iron, and at 1538°C (2802°F) the iron melts. When carbon is added to iron, it is found that it has only slight solid solubility in alpha iron (0.001% at room temperature). However, gamma iron will hold up to 2.0% carbon in solution at 1130°C (2066°F). The alpha iron containing carbon or any other element in solid solution is called *ferrite* and the gamma iron containing elements in solid solution is called *austenite*. Usually, when not in solution in the iron, the carbon forms a compound Fe_3C (iron carbide), which is extremely hard and brittle and is known as *cementite*.

Carbon steel in equilibrium at room temperatures will have present both ferrite and cementite. The physical properties of the ferrite are approximately those of pure iron and are characteristic of the metal. The presence of cementite does not in itself cause steel to be hard; rather, it is the shape and distribution of the carbides in the iron that determine the hardness of the steel. The fact that the carbides can be dissolved in austenite is the basis of the heat treatment of steel since the steel can be heated above the critical temperature to dissolve all the carbides, and then suitable cooling through the cooling range will produce the desired size and distribution of carbides in the ferrite.

If austenite containing 0.8% carbon (eutectoid composition) is slowly cooled through the critical temperature, ferrite and cementite are rejected simultaneously, forming alternative plates or lamellae. This microstructure is called pearlite since when polished and etched it has a pearly luster. If the austenite contains less than 0.80% carbon (hypo-eutectoid composition), free ferrite will first be rejected on slow cooling through the critical temperature until the composition of the remaining austenite reaches 0.80% carbon, when the simultaneous rejection of both ferrite and carbide will again occur, producing pearlite. So a hypo-euctectoid steel at room temperature will be composed of areas of free ferrite and areas of pearlite; the higher the carbon percentage, the greater the amount of pearlite present in the steel. When austenite containing more than 0.80% carbon (hypereuctectoid composition is slowly cooled, cementite is thrown out at the austenite grain boundaries, forming a cementite network, until the austenite contains 0.80%, carbon at which time pearlite is again formed. Thus a hypereuctectoid steel when slowly cooled will have areas of pearlite surrounded by a thin carbide network.

As the cooling rate is increased the spacing between the pearlite lamellae becomes smaller; with the resulting greater dispersing of carbide preventing slip in the iron crystals, the steel becomes harder. Also, with an increase in the rate of cooling, there is less time for the separation of excess ferrite or cementite and the equilibrium amount of these constituents will not be precipitated before the austenite transforms to pearlite. Thus with a fast rate of cooling, pearlite may contain more or less carbon than that given by the eutectoid composition. When the cooling rate becomes very rapid (as obtained by quenching), the carbon does not have sufficient time to separate out in the form of carbide and the austenite transforms to a highly stressed structure supersaturated with carbon, called *martensite*. This structure is exceedingly hard but brittle and requires tempering to increase the ductility. Tempering consists of heating martensite to a temperature below critical, causing the carbide to precipitate in the form of small spheroids. The higher the tempering temperature, the larger the carbide particle size, the greater the ductility of the steel, and the lower the hardness.

In a carbon steel it is possible to have a structure consisting either of parallel plates of carbide in a ferrite matrix, the distance between the plates depending upon the rate of cooling, or of carbide spheroids in a ferrite matrix, the size of the spheroids depending on the temperature to which the hardened steel was heated.

HEAT-TREATING OPERATIONS

The following definitions of terms have been adopted by the American Society for Testing and Materials (ASTM), the Society of Automotive Engineers (SAE), and the American Society for Metals (ASM) in substantially identical form.

Heat treatment An operation, or combination of operations, involving the heating and cooling of a metal or alloy in the solid state for the purpose of obtaining certain desirable conditions or properties.

Quenching Rapid cooling by immersion in liquids or gases or by contact with metal.

Hardening Heating and quenching certain iron-base alloys from a temperature either within or above the critical range, for the purpose of producing a hardness superior to that obtained when the alloy is not quenched. Usually restricted to the formation of martensite.

Annealing Annealing is a heating and cooling operation which usually implies relatively slow cooling. The purpose of such a heat treatment may be (1) to remove stresses; (2) to induce softness; (3) to alter ductility, toughness, electrical, magnetic, or other physical properties; (4) to refine the crystalline structure; (5) to remove gases; or (6) to produce a definite microstructure. The temperature of the operation and the rate of cooling depend on the material being heat treated and the purpose of the treatment. Certain specific heat treatments coming under the comprehensive term *annealing* are as follows:

Process annealing Heating iron-base alloys to a temperature below or close to the lower limit of the critical temperatures, generally 540 to 705°C (1000 to 1300°F).

Normalizing Heating iron-based alloys to approximately 50°C (100°F) above the critical temperature range followed by cooling to below that range in still air at ordinary temperature.

Patenting Heating iron-based alloys above the critical temperature range followed by cooling below that range in air, molten lead, or a molten mixture of nitrates or nitrites maintained at a temperature usually between 425 and 555°C (800 and 1050°F), depending on the carbon content of the steel and the properties required in the finished product. This treatment is applied in the wire industry to medium or high carbon steel as a treatment to precede further wire drawing.

Spherodizing Any process of heating and cooling steel that produces a rounded or globular form of carbide. The following spheroidizing methods are used: (1) prolonged heating at a temperature just below the lower critical temperature, usually followed by relatively slow cooling; (2) in the case of small objects of high carbon steels, the spheroidizing result is achieved more rapidly by prolonged heating to temperatures alternatively within and slightly below the critical temperature range; (3) tool steel is generally spheroidized by heating to a temperature of 750 to 805°C (1380 to 1480°F) for carbon steels and higher for many alloy tool steels, holding at heat from 1 to 4 h, and cooling slowing in the furnace.

Tempering (Drawing) Reheating hardened steel to a temperature below the lower critical temperature, followed by any desired rate of cooling. Although the terms *tempering* and *drawing* are practically synonomous as used in commercial practice, the term *tempering* is preferred.

EFFECT OF ALLOYING ELEMENTS ON THE PROPERTIES OF STEEL

When relatively large amounts of alloying elements are added to steel, the characteristic behavior of carbon steel is obliterated. Most alloy steel is medium or high carbon steel to which various elements have been added to modify its properties to an appreciable extent, but it still owes its distinctive characteristics to the carbon that it contains. The percentage of alloy element required for a given purpose ranges from a few hundredths of 1% to possibly as high as 5%.

When ready for service these steels will usually contain only two constituents, ferrite and carbide. The only way that any alloying element can affect the properties of steel is to change the dispersion of carbide in the ferrite or change the properties of the carbide. The effect on the distribution of carbide is the most important factor. In large sections where carbon steels fail to harden throughout the section even under a water quench, the hardenability of the steel can be increased by the addition of any alloying element (except possibly cobalt). The elements most effective in increasing the hardenability of steel are manganese, silicon, and chromium.

Elements such as molybdenum, tungsten, and vanadium are effective in increasing the hardenability when dissolved in the austenite, but are usually present in the austenite in the form of carbides. The main advantage of these carbide-forming elements is that they prevent the agglomeration of carbides in tempered martensite. Tempering relieves the internal stresses in the hardened steel and causes spheroidization of the carbide particles, with resultant loss in hardness and strength. With these stable carbide-forming elements present, higher tempering temperatures may be employed without sacrificing strength. This permits these alloy steels to have a greater ductility for a given strength, or a greater strength for a given ductility, than plain carbon steels.

The third factor that contributes to the strength of alloy steel is the presence of the alloying element in the ferrite. Any element present in solid solution in a metal will increase the strength of the metal. The elements most effective in increasing the strength of the ferrite are phosphorus, silicon, manganese, nickel, molybdenum, tungsten, and chromium.

A final important effect of alloying elements is their influence on the austenitic grain size. Martensite, when formed from a course-grained austenite, has considerably less resistance to shock than that formed from a fine-grained austenite. Aluminum is the most effective element for fine grain-growth inhibitors.

Table 1 provides a summary of the effects of various alloying elements. This table indicates only the trends of the elements; the fact that one element has an influence on one factor does not prevent it from exerting an influence on another factor.

CASE HARDENING

The production of articles having a soft ductile interior and a very hard surface can be accomplished by carburizing a low carbon steel at an elevated temperature and then quenching. This process is known as *case hardening*.

ATMOSPHERIC CORROSION OF STEEL

Atmospheric corrosion of steel is a function of location. In country air the products of corrosion are either oxides or carbonates. In industrial atmospheres sulfuric acid is present, and near the ocean some salt is in the air. Corrosion is more rapid in industrial areas because of the

Table 1. Trends of Influence of Alloying Elements

Element	As dissolved in: Ferrite: strength	Austenite: hardenability	As dissolved carbide in austenite: fine-grain toughness	As dispersed carbide in tempering: high-temperature strength and toughness	As fine nonmetallic dispersion: fine-grain toughness
Al	Moderate	Mild	None	None	Very strong
Cr	Mild	Strong	Strong	Moderate	Slight
Co	Strong	Negative	None	None	None
Cb	Little	Strong	Strong	Strong	None
Cu	Strong	Moderate	None	None	None
Mn	Strong	Moderate	Mild	Mild	Slight
Mo	Moderate	Strong	Strong	Strong	None
Ni	Mild	Mild	None	None	None
P	Strong	Mild	None	None	None
Si	Moderate	Moderate	None	None	Moderate
Ta	Moderate	Strong	Strong	Strong	None
Ti	Strong	Strong	Very strong	Little	Moderate
W	Moderate	Strong	Strong	Strong	None
V	Mild	Very strong	Very strong	Very strong	Moderate

presence of the acid, and higher both near cities and near the ocean because of the higher electrical conductivity of the rain and the tendency to form soluble chlorides or sulfates, which cause the removal of protective scale. The rusting of steel exposed to weather is subject to many variables. The protective action of copper and other alloying elements is due to a resistant form of oxide which forms a protective coating under atmospheric conditions, but has little or no favorable effect when immersed continuously in water.

In an industrial atmosphere steel with 0.32% copper after five years will corrode only half as much as steel with 0.05% copper. A low alloy, high-strength steel having the following composition (percent) will corrode only half as much as steel having 0.32% copper:

```
C    0.12 maximum
Mn   0.20-0.50
P    0.07-0.16
S    0.05 maximum
Si   0.75 maximum
Cu   0.30-0.50
Cr   0.50-1.25
Ni   0.55 maximum
```

It will be noted that in addition to the copper, this high-strength alloy also contains notable amounts of chromium and nickel, both of which are helpful in increasing strength and adding resistance to corrosion. Phosphorus, which it also contains, is another element that aids in providing protection against atmospheric corrosion.

In general, the presence of oxygen and/or acidic conditions will promote the corrosion of carbon steel. Alkaline conditions inhibit corrosion. Steel that is embedded in concrete is protected from corrosion by the alkalinity of the products formed when cement reacts with water. If the concrete cracks or changes chemically through aging, the protection disappears.

GENERAL CORROSION RESISTANCE

The corrosion resistance of carbon steel is dependent on the formation of an oxide surface film. However, resistance to corrosion is somewhat limited. Carbon steel should not be used in contact with dilute acids; thus it is not recommended with sulfuric acid below 90%. Between 90 and 98%, steel can be used up to the boiling point, between 80 and 90% it is servicable at room temperature. Steel is not normally used with hydrochloric, phosphoric, or nitric acids.

If iron contamination is permissible, steel can be used to handle caustic soda up to approximately 75% and 100°C (212°F). Stress relieving should be employed to reduce caustic embrittlement.

Brines and seawater corrode steel at a slow rate and the metal can be used if iron contamination is not objectionable.

Steel is little affected by neutral water and most organic chemicals. Organic chlorides are an exception. Many large water tanks and storage tanks for organic solvents are fabricated from carbon steel.

CLAD STEELS

A clad steel plate is a composite plate made of mild steel with a cladding of corrosion-resistant or heat-resistant metal on one or both sides.

The cladding may consist of various grades of stainless steels, nickel, Monel, Inconel, cupronickel, titanium, or silver. The thickness of the clad material may vary from 5 to 50% of the thickness of the clad plate, but normally is held to 10 to 20%. The clad steels are available in the form of sheet, plate, and strip and may be obtained as wire.

The clad steels are used in place of solid corrosion-resistant or heat-resistant materials. They also find applicaton where corrosion is a minor problem but where freedom from contamination of the materials handled is essential. In addition to the savings in material costs, the clad steels are frequently easier to fabricate than solid plates of the cladding material. Their high heat conductivity is another reason for their selection for many applications.

Clad steels are used for processing equipment in the chemical, food, beverage, drug, paper, textile, oil, and associated industries.

METALLIC PROTECTIVE COATINGS

In order to take advantage of the physical and/or mechanical properties of carbon steel, various types of metallic coatings have been developed to provide corrosion resistance for specific types of applications.

Chrominizing

Chrominizing of low carbon steel is effective in improving corrosion resistance by developing a surface containing up to 40% chromium. Some forming operations can be carried out on chrominized material.

Galvanizing

Zinc is the least expensive of the metals used for protective coatings. It may be applied by an electroplating process but is most usually applied by dipping the metal into molten zinc.

The zinc dissolves to a small extent in the iron, and this thin film of alloy permits the outer coating of zinc to adhere firmly to the metal.

If a scratch is made on a sheet of galvanized steel, the zinc along the edges will suffer preferential corrosion and the iron will be protected even if the zinc has been removed from a strip 1/8 in. in width. Zinc-coated steel (galvanized) is widely used for wire fences, water pipes, and buckets. Pipe fittings made from cast or malleable iron are protected in a similar manner.

Tinplate, Terne Plate

Steel sheets containing less than 0.1% carbon are pickled in dilute sulfuric acid to remove mill scale and then passed through a bath of used zinc chloride, a bath of molten tin, and a bath of palm oil, all in one continuous operation. A standard grade of tinplate will have 0.7 lb of tin applied to 100 ft^2 of steel sheets.

Tin is not corroded by fruit juices or by any of the vegetables customarily used for food; consequently, tin cans are widely used in the food industry.

When tinplate is to be used for structural purposes such as roofs, an alloy of 12 to 25 parts of tin to 88 to 75 parts of lead is frequently used. This is called *terne plate*. It is less expensive and more resistant to the weather than is a pure tin coating.

3.2

Stainless Steels

PHILIP A. SCHWEITZER / Chem-Pro Corporation, Fairfield, New Jersey

INTRODUCTION

Stainless steel is probably the most widely known and most commonly used material of construction for corrosion resistance. For many years stainless steel was the only material available to provide any degree of resistance to corrosive attack.

Stainless steel is not a singular material, as its name might imply, but rather a broad group of alloys each of which exhibits its own physical and corrosion-resistance properties. There are more than 70 standard types of stainless steels and many special alloys.

These steels are produced both as cast alloys [Alloy Casting Institute (ACI) types] and wrought forms [American Iron and Steel Institute (AISI) types]. Generally, all are iron-based with 12 to 30% chromium, 0 to 22% nickel, and minor amounts of carbon, columbium, copper, molybdenum, selenium, tantalum, and titanium. They are corrosion and heat resistant, noncontaminating, and easily fabricated into complex shapes. The 70+ types can be divided into three basic groups:

1.   Austenitic
2.   Martensitic
3.   Ferritic

Table 1. Chemical Composition of Austenitic Stainless Steels

AISI type	Nominal composition, %					
	C max.	Mn max.	Si max.	Cr	Ni	Others[a]
201	0.15	7.5[b]	1.00	16.00-8.00	3.50-5.50	0.25 max. N
202	0.15	10.00[c]	1.00	17.00-19.00	4.00-6.00	0.25 max. N
205	0.25	15.50[d]	0.50	16.50-18.00	1.00-1.75	0.32/0.4 max. N
301	0.15	2.00	1.00	16.00-18.00	6.00-8.00	
302	0.15	2.00	1.00	17.00-19.00	8.00-10.00	
302B	0.15	2.00	3.00[e]	17.00-19.00	8.00-10.00	
303	0.15	2.00	1.00	17.00-19.00	8.00-10.00	0.15 min. S
303(Se)	0.15	2.00	1.00	17.00-19.00	8.00-10.00	0.15 min. Se
304	0.08	2.00	1.00	18.00-20.00	8.00-12.00	
304L	0.03	2.00	1.00	18.00-20.00	8.00-12.00	
304N	0.08	2.00	1.00	18.00-20.00	8.00-10.50	0.1/0.16 N
305	0.12	2.00	1.00	17.00-19.00	10.00-13.00	
308	0.08	2.00	1.00	19.00-21.00	10.00-12.00	
309	0.20	2.00	1.00	22.00-24.00	12.00-15.00	
309S	0.08	2.00	1.00	22.00-24.00	12.00-15.00	
310	0.25	2.00	1.50	24.00-26.00	19.00-22.00	
310S	0.08	2.00	1.50	24.00-26.00	19.00-22.00	

314	0.25	2.00	3.00[f]	23.00-26.00	19.00-22.00	
316	0.08	2.00	1.00	16.00-18.00	10.00-14.00	2.00-3.00 Mo
316F	0.08	2.00	1.00	16.00-18.00	10.00-14.00	1.75-2.50 Mo
316L	0.03	2.00	1.00	16.00-18.00	10.00-14.00	2.00/3.00 Mo
316N	0.08	2.00	1.00	16.00-18.00	10.00-14.00	2.00-3.00 Mo
317	0.08	2.00	1.00	18.00-20.00	11.00-15.00	3.00-4.00 Mo
317L	0.03	2.00	1.00	18.00-20.00	11.00-15.00	3.00-4.00 Mo
321	0.08	2.00	1.00	17.00-19.00	9.00-12.00	5XC min Cb-Ta
330	0.08	2.00	1.5[g]	17.00-20.00	34.00-37.00	0.10 TA; 0.20 Cb
347	0.08	2.00	1.00	17.00-19.00	9.00-13.00	10XC min. Cb-Ta
348	0.08	2.00	1.00	17.00-19.00	9.00-13.00	10C min. Cb-Ta

[a] Other elements in addition to those shown are as follows: Phosporus is 0.03% max. in types 201, 302, 302B, 304, 304L, 304N, 305, 308, 309, 309S, 310, 310S, 314, 316, 316N, 316L, 317, 317L, 321, 330, 347, and 348; 0.045% max. in types 301, 302; 0.06% max. in types 201 and 202; 0.20% max. in types 303, 303(Se), and 316D. Sulfur is 0.030% max, in types 201, 202, 205, 301, 302, 302B, 304, 304L, 304N, 305, 308, 309, 309S, 310, 310X, 314, 316, 316L, 316N, 317, 317L, 321, 330, 347, and 348; 0.15% min, in type 303; 0.10 min. in type 316D.

[b] Mn range 4.40 to 7.50.

[c] Mn range 7.50 to 10.00.

[d] Mn range 14.00 to 15.50.

[e] Si range 2.00 to 3.00.

[f] Si range 1.50 to 3.00.

Table 2. Physical Properties of Austenitic Stainless Steels

Stainless steel type	Form and condition	Hardness Brinell	Density, lb/in.^{3}	Specific gravity	Melting point, °F	Specific heat (32-212°F), Btu/lb °F
201	Annealed	194	0.28	7.7	255-2650	0.12
202	Annealed	184	0.28	7.7	2550-2650	0.12
205	Annealed	217				
301	Annealed	160	0.29	8.02	2550-2590	0.12
	Cold rolled	186				
302	Annealed	160	0.29	8.02	2550-2590	0.12
	Cold rolled	Up to 400				
302B	Annealed	165				
303	Annealed	165				
303(Se)	Annealed	160				
304	Annealed	160	0.29	8.02	2550-2650	0.12
	Cold rolled	Up to 400				
304L	Annealed	150	0.29	8.02	2550-2650	0.12
	Cold rolled	277				
304N	Annealed	160				
305	Annealed	156				
308	Annealed	150				
309	Annealed	165	0.29	8.02	2550-2650	0.12
	Cold rolled	275				
309S	Annealed	165	0.29	8.02	2550-2650	0.12
310	Annealed	165				
310S	Annealed	170				
314S	Annealed	170				

Thermal expansion coefficient (32-212°F), X 10^{-6} in./in. °F	Thermal conductivity (32-212°F), Btu(ft) (h) (°F/in.)	Electrical Resistivity (68°F), Ω/cir mil	Tensile modulus of elasticity, X $10^6 \psi$
	113	414	28.6
	113	414	28.6
9.4	112.8	435	28
9.6	112.8	435	28
9.6	113	435	28
9.6	113	435	28
8.3	96	470	29
8.0	96	470	29

Table 2. (Continued)

Stainless steel type	Form and condition	Hardness Brinell	Density, lb/in.^{3}	Specific gravity	Melting point, °F	Specific Heat (32-212°F), Btu/lb °F
316	Annealed	165				
	Cold rolled	275	0.29	8.02	2500-2550	0.12
316F	Annealed					
316L	Annealed	150				
	Cold drawn	275	0.29	8.02	2500-2550	0.12
316N	Annealed					
317	Annealed	160				
317L	Annealed	160				
321	Annealed	160				
	Cold rolled	300	0.286	7.92	2550-2600	0.12
330	Annealed	156				
347	Annealed	160				
	Cold rolled	300	0.286	7.92	2550-2600	0.12
348	Annealed	160				

AUSTENITIC STAINLESS STEELS

Austenitic stainless steels contain both nickel and chromium. The addition of substantial quantities of nickel to high chromium alloys stabilizes the austenite at room temperature. This group of steels contain 16 to 26% chromium and 6 to 22% nickel. The carbon content is kept low (0.08%) to minimize carbide precipitation. The most common composition is 18 Cr and 8 Ni (known as 18-8). However, many other compositions have been developed to meet special conditions and for special applications. Refer to Table 1, which gives the chemical compositions of the various austenitic stainless steels. Table 2 provides the physical properties of the austenitic stainless steels.

Type 302 is the basic alloy of this group.

Thermal expansion coefficient (32-212°F), X 10^{-6} in./in. °F	Thermal conductivity (32-212°F), Btu(ft) (h) (°F/in.)	Electrical Resistivity (68°F), Ω/cir mil	Tensile modulus of elasticity, X $10^6 \psi$
8.9	113	445	28
8.9	113	445	28
9.3	110	435	28
9.3	110	435	28

Mechanical Properties

The austenitic stainless steels cannot be hardened except by cold work. Heat treatment will not cause hardening. In the annealed condition the tensile strength is approximately 85,000 psi. Austenitic stainless steels are both tough and ductile. Table 3 lists the mechanical properties.

All plain carbon steels and low alloy steels become increasingly brittle as the temperature is reduced and should be used with caution if the operating temperature is expected to be much less than -18°C (0°F).

The austenitic stainless steels are almost exempt from this low temperature brittleness, and consequently stainless steels are widely

Table 3. Mechanical Properties of Austenitic Stainless Steels

Stainless steel type	Form and condition	Yield strength (0.2% offset), X $10^3\psi$	Tensile strength, X $10^3\psi$	Elongation in 2 in., %
201	Annealed	55	115	55
202	Annealed	50	100	60
205	Annealed	69	120	58
301	Annealed	30	100	72
	Cold rolled[a]	Up to 165	Up to 200	15[b]
302	Annealed	30	90	60
	Cold rolled[a]	Up to 165	Up to 190	8[b]
302B	Annealed	40	95	50
	Cold rolled	Up to	Up to	
303	Annealed	35	90	50
303(Se)	Annealed	35	90	50
304	Annealed	30	85	62
	Cold rolled	Up to 160	Up to 185	8[b]
304L	Annealed	30	80	60
	Cold drawn	95	125	25
304N	Annealed	48	90	50
305	Annealed	85	37	55
308	Annealed	85	35	55
309	Annealed	30	82	50
	Cold rolled	Up to 120	Up to 140	4[b]
309S	Annealed	90	40	45
310	Annealed	40	100	50
310S	Annealed	95	40	45
314	Annealed	100	50	45
316	Annealed	30	90	50
	Cold rolled	Up to 120	Up to 150	8[b]
316F	Annealed	38	85	60

Table 3. (Continued)

Stainless steel type	Form and condition	Yield strength (0.2% offset), X $10^3 \psi$	Tensile strength, X $10^3 \psi$	Elongation in 2 in., %
316L	Annealed	30	80	60
	Cold drawn	60	90	45
316N	Annealed	48	90	48
317	Annealed	90	40	50
317L	Annealed	38	86	55
321	Annealed	30	85	50
	Cold rolled	Up to 120	Up to 150	5[b]
330	Annealed	38	80	40
347	Annealed	30	85	50
	Cold rolled	Up to 120	Up to 150	5[b]
348	Annealed	92	35	50

[a]The cold-rolled properties depend on composition; types 302 and 304 are not often rolled in excess of 175,000 ψ tensile strength.

[b]The values for elongation (percent in 2 in.) are obtainable in steel cold rolled to the *maximum* stated yield strength and tensile strength. For lower values of tensile strength, elongation will be correspondingly higher.

used in equipment operating at low temperatures. The greater the percentage of nickel present, the lower the allowable operating temperature.

All standard methods of fabrication can be used to work these steels; however, the austenitic grades are difficult to machine since they work-harden and gall. Rigid machines, heavy cuts, and high speeds are necessary. Welding is readily performed, although the heat of welding may cause chromium carbide precipitation, which depletes the alloy of some chromium in the area of the weld and lowers its corrosion resistance in liquid service. High-temperature properties are not affected. This is not serious for mild service, but for severe corrosive service the carbides must be put back into solution by heat treatment, which is not always possible with field welds.

The removal of precipitated carbides from type 304

The removal of precipitated carbides from type 304 in order to restore maximum corrosion resistance can be accomplished by annealing
[at 980 to 1180°C (1800 to 2150°F)] (above the sensitizing range)
followed by rapid cooling. Stress relieving a weldment at 1500 to
1700°F will not restore corrosion resistance, and in fact may foster
carbide precipitation in stainless steels that do not have a low carbon
content or are not stabilized.

To avoid this problem, special stainless steels have been developed
by adding titanium, columbium, or tantalum to stabilize the carbon
and prevent precipitation (types 321, 347, and 346).

Alternative to the use of these alloys is to use the low carbon alloys
such as types 304L and 316L, which have maximum carbon contents of
0.03%.

Corrosion Resistance

The austenitic stainless steels perform best under oxidizing conditions
since their resistance is dependent on an oxide film that forms on the
surface of the alloy. Reducing conditions and chloride ions destroy
this film and cause rapid attack. Chloride ions, combined with high
tensile stresses, cause stress corrosion cracking.

These alloys have excellent resistance to nitric acid at practically
all concentrations and temperatures. Type 304 is used in the construction of most nitric acid plants. To handle sulfuric acid, without inhibitors, type 316 stainless steel has limited application. Below 5%
and above 85% it can be used only at temperatures below the boiling
point.

MARTENSITIC STAINLESS STEELS

The martensitic stainless steels contain 12 to 20% chromium, controlled
amounts of carbon, and other additives. Table 4 gives the chemical
composition of the various martensitic alloys, and Table 5 gives the
physical properties.

Type 410 is a typical member of this group.

Mechanical Properties

Martensitic stainless steels can be hardened by heat treatment, which
can increase the tensile strength from 80,000 to 200,000 psi. A high
hardness can be achieved by means of this heat treatment.

A hardening temperature range depends on the composition, but in
general the higher the quenching temperature, the harder the article.

Table 4. Chemical Composition of Martensitic Steels

AISI type	Nominal composition, %						
	C	Mn max.	Si max.	Cr	Ni	Other[a]	
403	0.15 max.	1.00	0.50	11.50-13.00			
410	0.15 max.	1.00	1.00	11.50-13.50			
414	0.15 max.	1.00	1.00	11.50-13.50	1.25-2.50		
416	0.15 max.	1.25	1.00	12.00-14.00		0.15 S min.	
416(Se)	0.15 max.	1.25	1.00	12.00-14.00		0.15 S min.	
420	0.15 min.	1.00	1.00	12.00-14.00			
431	0.20 max.	1.00	1.00	15.00-17.00	1.25-2.50		
440A	0.60-0.75	1.00	1.00	16.00-18.00		0.75 Mo max.	
440B	0.75-0.95	1.00	1.00	16.00-18.00		0.75 Mo max.	
440C	0.95-1.20	1.00	1.00	16.00-18.00		0.75 Mo max.	
501	0.10 min.	1.00	1.00	4.00-6.00		0.40-0.65 Mo	
502	1.10 max.	1.00	1.00	4.00-6.00		0.40-0.65 Mo	

[a]Other elements in addition to those shown below are as follows: phosphorus is 0.06% max. in types 416 and 416 (Se); sulfur is 0.03% max. in types 403, 410, 414, 420, 431, 440A, 440B, 440C, 501, and 502.

Table 5. Physical Properties of Martensitic Stainless Steels

Stainless steel type	Form and condition	Hardness Brinell	Density lb/in.^{3}	Specific gravity	Melting point, °F	Specific heat (32-212°F), Btu/lb °F
403	Annealed	155				
	Heat treated	410				
410	Annealed	150				
	Heat treated	410	0.28	7.75	2700-2790	0.11
414	Annealed	217				
	Heat treated	387	0.28	7.75	2600-2700	0.11
416	Annealed	155				
	Heat treated	410				
416(Se)	Annealed	155				
	Heat treated	410				
420	Annealed	180				
	Heat treated	480	0.28	7.75	2650-2750	0.11
431	Annealed	250				
	Heat treated	400	0.28	7.75	2600-2700	0.11
440A	Annealed	215				
	Heat treated	570				
440B	Annealed	220				
	Heat treated	590				
440C	Annealed	230				
	Heat treated	610				
501	Annealed	160				
502	Annealed	150				

Thermal expansion coefficient (32-212°F), $\times 10^{-6}$ in./in. °F	Thermal conductivity (32-212°F), Btu(ft^2) (h) (°F/in.)	Electrical resistivity (68°F), Ω/cir mil	Tensile modulus of elasticity, $\times 10^6 \psi$
5.5	173	340	29
6.1	173	420	29
5.7	173	330	29
6.5	140	430	29

Table 6. Mechanical Properties of Martensitic Stainless Steels

Stainless steel type	Form and condition	Yield strength (0.2% offset), X $10^3\psi$	Tensile strength, X $10^3\psi$	Elongation in 2 in., 5
403	Annealed	75	40	30
410	Annealed	40	75	30
	Heat treated	115	150	15
414	Annealed	80	100	22
	Heat treated	150	200	17
416	Annealed	75	40	30
416(Se)	Annealed	75	40	30
420	Annealed	60	98	28
	Heat treated	200	250	8
431	Annealed	85	120	25
	Heat treated	150	196	20
440A	Annealed	105	60	20
440B	Annealed	107	62	18
440C	Annealed	110	65	13
501	Annealed	70	30	28
502	Annealed	70	30	30

Oil quenching is preferable, but with thin and intricate shapes, hardening by cooling in air should be undertaken.

Tempering at 425°C (800°F) does not reduce the hardness of the part, and in this condition these alloys show an exceptional resistance to fruit and vegetable acids, lye, ammonia, and other corrodents to which cutlery may be subjected.

Table 6 lists the mechanical properties of the martensitic stainless steels.

Corrosion Resistance

The corrosion resistance of the martensitic stainless steels is inferior to the corrosion resistance of the austenitic stainless steels. These alloys are generally used in mildly corrosive services such as atmospheric, fresh water, and organic exposures.

Because of the ability to heat treat these alloys to a high degree of hardness and because of their resistance to oxidation, they are used extensively for cutlery, razor blades, surgical and dental instruments, springs for high temperature operations, ball valves and seats, and similar applications.

FERRITIC STAINLESS STEELS

The ferritic stainless steels contain 15 to 30% chromium with a low carbon content (0.1%). Table 7 lists the chemical composition of the members of this group, and Table 8 gives the physical properties.

Type 430 is a typical member of this group.

Mechanical Properties

The strength of ferritic stainless steels can be increased by cold working but not by heat treatment. These alloys possess considerable ductility, ability to be worked hot or cold, and excellent corrosion resistance and are relatively inexpensive.

Alloys containing 16 to 18% chromium are probably the most useful of the straight chromium steels because of their forming and medium-deep drawing properties. They are used extensively for kitchen

Table 7. Chemical Composition of Ferritic Steels

AISI type	Nominal Composition, %				
	C max.	Mn max.	Si min.	Cr	Other[a]
405	0.08	1.00	1.00	11.50-14.50	0.10-0.30 Al
403	0.12	1.00	1.00	14.00-18.00	
430F	0.12	1.25	1.00	14.00-18.00	0.15 S min.
430F(Se)	0.12	1.25	1.00	14.00-18.00	0.15 Se min.
446	0.20	1.50	1.00	23.00-17.00	0.25 max. N

[a]Other elements in addition to those shown are as follows: phosphorus is 0.06% max. in types 430F and 430F (Se); sulfur is 0.030% max. in types 405, 430, and 446; 0.15% min. in type 430F.

Table 8. Physical Properties of Ferritic Stainless Steels

Stainless steel type	Form and condition	Hardness Brinell	Density, lb/in.^{3}	Specific gravity	Melting point, °F	Specific heat (32-212°F) Btu/lb °F	Thermal expansion coefficient (32-212°F), X 10^{-6} in./in. °F	Thermal conductivity (32-212°F) Btu(ft^2)(h)(°F/in.)	Electrical resistivity (68°F), Ω/cir mil	Tensile modulus of elasticity X 10^6ψ
405	Annealed	150								
430	Annealed	165	0.28	7.75	2600-2750	0.11	6.0	180	360	29
	Cold rolled	225								
430F	Annealed	170								
430(Se)	Annealed	170								
446	Annealed	165	0.27	7.45	2600-2750	0.12	5.8	145	405	29

Table 9. Mechanical Properties of Ferritic Stainless Steels

Stainless steel type	Form and condition	Yield strength (0.2% offset), X $10^3\psi$	Tensile strength, X $10^3\psi$	Elongation in 2 in., %
405	Annealed	70	40	30
430	Annealed	75	45	30
	Cold drawn			
430F	Annealed	40	70	350
	Cold rolled	95	110	10
430(Se)	Annealed	80	55	25
446	Annealed	50	80	30

equipment, dairy machinery, interior decorative work, automobile trimmings, and chemical equipment to resist nitric acid corrosion.

The mechanical properties are given in Table 9.

Corrosion Resistance

Corrosion resistance is rated good, although ferritic alloys do not resist reducing acids such as HCl. Mildly corrosive solutions and oxidizing media are handled satisfactorily. Type 430 finds wide application in nitric acid plants. Increasing the chromium content to between 25 and 30% improves the resistance to oxidizing conditions at elevated temperatures. These alloys are useful for all types of furnace parts not subjected to high stress. Since the oxidation resistance is independent of the carbon content, soft forgeable alloys low in carbon can be rolled into plates, shapes, and sheets, and hard and wear-resistant castings can be made from higher carbon nonforgeable alloys.

3.3
Nickel and High Nickel Alloys

F. GALEN HODGE / High Technology Materials Division, Cabot Corporation,
Kokomo, Indiana

INTRODUCTION

An important group of alloys for corrosion service is based on the element nickel. With respect to usage, some of these alloys are relative newcomers but are assuming important roles in the chemical process industry. The austenitic stainless steels were developed and utilized early in the 1900s, whereas the development of the nickel-base alloys did not begin until about 1930. Initially, some of the alloys were produced only as castings, and later the wrought versions developed. Since that time there has been a steady progression of different or improved alloys emerging from the laboratories of nickel-base alloy producers. Many of these find their major usage in the high-temperature world of gas turbines, furnaces, and the like, but several are used primarily by the chemical industry for aqueous corrosion service.

Historically, the use of these alloys was typically reserved for those applications where it was adjudged that nothing else would work. At one time the primary factor in the selection of construction materials was initial cost; very little thought was given to the possible maintenance and downtime costs associated with the equipment. Today, the increasing costs of maintenance and downtime have placed greater emphasis on the reliable performance of the process equipment. Data recently published [1] demonstrate that these costs have reached significant proportions. The annual amortized cost of the equipment over the expected life is now important in the material selection.

The element nickel has some unique electrochemical properties which make it important on its own and as an alloying base. In reducing acid systems nickel quite often assumes an open-circuit potential equal to the platinum potential. However, it does not readily liberate hydrogen during corrosion. Some oxidizing ion species are therefore, required to promote the corrosion of pure nickel. However, under certain circumstances nickel has the ability to form a passive film, and thereby show good corrosion resistance to oxidizing environments. Unfortunately, the passive film formed is not very stable and nickel will pit when exposed to oxidizing chloride environments.

One of the most important attributes of nickel with respect to the formation of corrosion-resistant alloys is its metallurgical compatibility with a number of other metals, such as copper, chromium, molybdenum, and iron. A survey of the binary-phase diagrams for nickel and these other elements shows considerable solid solubility and thus one can make alloys with wide ranges in composition. Nickel alloys are, in general, all austenitic alloys; however, they can be subject to precipitation of intermetallic and carbide phases when aged. In some alloys designed for high-temperature service intermetallic and carbide precipitation reactions are encouraged to increase properties; however, for corrosion applications the precipitation of second phases usually promotes corrosion attack. The problem is rarely encountered, however, because the alloys are supplied in the annealed condition, and the service temperatures rarely approach the level required for sensitization.

In iron-chromium-nickel austenitic stainless steels, the minimization of carbide precipitation can be achieved by lowering the carbon content to a maximum of about 0.03%. As the nickel content is increased from

the nominal 8% in these alloys to that of the majority element (i.e.,
more than 50%), the nature of the carbide changes from predominantly
$M_{23}C_6$ to M_6C and the carbon solubility decreases by a factor of 10.
It was therefore very difficult in the past to produce an L-grade material because of the state of the art in melting. Many alloys attempted
to circumvent this problem by adding carbide stabilizers to tie up the
carbon, with varying degrees of success. Some of the most recent
developments in the nickel alloy field involve changes in melting technique. The transfer of alloys from air induction or vacuum induction
melting to air arc plus argon-oxygen decarburization has provided a
means for producing nickel alloys comparable to the L grades of stainless steels.

While general corrosion resistance is important, one of the major
reasons that nickel-based alloys are specified for many applications is
their excellent resistance to localized corrosion, such as pitting, crevice corrosion, and stress corrosion cracking [2]. This is a very
important design consideration since corrosion coupon data, although
helpful, may not represent the conditions encountered in actual service. In many environments, austenitic stainless steels do not exhibit
general attack but suffer from significant localized attack, often causing excessive downtime and/or expensive repair and replacement.

In general, the localized corrosion resistance of alloys is improved
by increasing their molybdenum content. However, molybdenum content alone is not the answer, as Hastelloy* alloy B-2 has the highest
molybdenum content (26.5%) and is not recommended for most localized
corrosion service. Chromium, which is present in alloy B-2 only in
residual quantities, also has an important role because the environments are normally oxidizing in nature. Because of the combination
of chromium and molybdenum, the last seven alloys in Table 1 are
employed in those environments where localized corrosion is a problem.
The resistance of these alloys increases with increasing molybdenum
content.

This chapter is an introduction to the commercially available alloys
shown in Table 1, their general corrosion resistance, mechanical
properties, typical uses, and specifications. Some typical physical
properties are shown in Table 2, mechanical properties in Table 3, and
recommended welding materials in Table 4.

COMMERCIALLY PURE NICKEL

This family is represented by nickel alloys 200 and 201. The latter is
preferred for applications over 600°F (316°C), since its low carbon
content prevents graphitization and subsequent loss of ductility.

---

*Hastelloy is a registered trademark of Cabot Corp.

Table 1. Typical Composition,[a] %

Alloy[b]	Cr	Mo	Fe	Cu	C	Si	Mn	Others
Wrought alloys								
Nickel 200	—	—	0.2	0.1	0.08	0.2	0.2	—
Nickel 201	—	—	0.2	0.1	0.01	0.2	0.2	—
Monel alloy 400	—	—	1.2	31.5	0.2	0.2	1.0	—
Monel alloy K-500	—	26.5	1.0	29.5	0.1	0.2	0.8	2.7 Al, 0.6 Ti
Hastelloy alloy B-2	0.5	26.5	1.0	—	.01[c]	.05	0.2	—
Inconel alloy 600	15.5	—	8.0	0.2	0.08	0.2	0.5	—
Incoloy alloy 800	21.0	—	46.0	0.4	0.05	0.5	0.8	0.4 Al, 0.4 Ti
Incoloy alloy 825	21.5	3.0	30.0	2.25	0.03	0.25	0.5	0.9 Ti
Hastelloy alloy G	22.5	6.5	19.5	2.0	0.03	1.50	1.50	2.0 Cb + Ta
Hastelloy alloy X	21.8	9.0	18.5	—	0.10	0.5	0.5	—
Inconel Alloy 625	21.5	9.0	2.5	—	0.05	0.25	0.25	3.65 Cb + Ta
Hastelloy alloy C-276	16.0	16.0	5.5	—	0.01[c]	0.05	0.5	3.5 W
Hastelloy alloy C-4	16.0	15.5	2.0	—	0.01[c]	0.05	0.25	0.3 Ti
Hastelloy alloy S	15.5	14.5	1.0	—	0.01	0.4	0.5	0.02 La
Cast alloys								
Alloy B	1.0[c]	26.5	5.0	—	0.12[c]	0.6	0.6	0.3 V
Chlorimet 2	1.0[c]	31.0	2.0	—	0.07[c]	0.6	0.6	—
Alloy C	16.0	16.5	6.0	—	0.12[c]	0.6	0.6	4.0 W, 0.3 V
Chlorimet 3	18.5	18.5	3.0[c]	—	0.07[c]	0.6	0.6	—
Hastelloy alloy C-4C	16.5	16.5	2.0[c]	—	0.020[c]	0.6	0.6	—

[a]Balance Ni + Co; typical analysis not to be used for specification.
[b]Some alloys are made by more than one manufacturer. The proprietary designation for such alloys has been used in this compilation. Monel, Inconel, and Incoloy are registered trademarks of International Nickel Co., Inc., Hastelloy is a registered trademark of Cabot Corp;, Chlorimet is a registered trademark of Duriron Corp.
[c]Maximum.

Table 2.   Typical Physical Properties at Room Temperature

Alloy	Density, g/cm^3	Specific heat, J/kg K	Thermal expansion, x 10^{-6} m/m K, 20-93°C	Electrical resistivity, $\mu\Omega$-m	Thermal conductivity, W/m K
Nickel 200	8.89	456	13.3	0.095	74.9
Nickel 201	8.89	456	13.3	0.076	79.3
Monel alloy 400	8.84	427	13.8	0.509	21.8
Monel alloy K-500	8.48	418	13.7	0.614	17.4
Hastelloy alloy B-2	9.22	389	10.3	1.38	12.2
Inconel alloy 600	8.42	444	13.3	1.03	14.8
Incoloy alloy 800	7.95	502	14.2	0.988	11.5
Incoloy alloy 825	8.15	—	14.0	1.13	11.1
Hastelloy alloy G	8.31	456	13.4	1.18	10.1
Hastelloy alloy X	8.23	486	13.8	1.18	9.1
Inconel alloy 625	8.45	410	12.8	1.29	9.8
Hastelloy alloy C-276	8.89	427	11.2	1.30	10.2
Hastelloy alloy C-4	8.64	406	10.8	1.25	10.1
Hastelloy alloy S	8.76	414	11.5	1.25	10.8

Table 3. Typical Mechanical Properties[a] (Annealed Sheet)

Alloy	Yield strength, ksi (MPa)	Ultimate strength, ksi (MPa)	Elongation, %	Brinell hardness
Nickel 200	22 (152)	67 (462)	47	105
Nickel 201	17 (117)	55 (379)	50	87
Monel alloy 400	37 (255)	80 (552)	48	130
Monel alloy K 500[b]	50 (345)	100 (689)	35	162
Hastelloy alloy B-2	60 (414)	130 (896)	60	210
Inconel alloy 600	40 (276)	90 (620)	45	142
Incoloy alloy 800	45 (310)	87 (600)	45	152
Incoloy alloy 825	50 (345)	95 (655)	40	150
Hastelloy alloy G	46 (317)	102 (703)	61	169
Hastelloy alloy X	52 (358)	114 (786)	43	176
Inconel alloy 625	70 (483)	135 (931)	42	192
Hastelloy alloy C-276	42 (290)	100 (689)	56	190
Hastelloy alloy C-4	53 (365)	113 (779)	55	190
Hastelloy alloy S	49 (338)	116 (800)	65	180

[a]Typical values for information—not guaranteed minimums.

[b]Respective age-hardened values are 100 (689), 147 (1013), 28, and 280.

Nickel has outstanding resistance to hot or cold alkalies, with only silver or possibly zirconium having better resistance. Nickel has been used as caustic evaporator tubes in a variety of processes because below 50% concentration the corrosion rates are typically less than 0.2 mil per year (mpy)—even in boiling solutions. As concentrations and temperatures increase, corrosion rates increase very slowly. Nickel is resistant to stress corrosion cracking in the typical chloride environments, but may be susceptible in caustic environments if severely stressed. To measure accurately the corrosion rate of nickel in aqueous or fused caustics, the corrosion test should be of a long duration, since the formation of a nickel oxide film on the surface is required to limit the corrosion.

Nickel does not readily discharge hydrogen from nonoxidizing acids; therefore, it does have some utility in dilute acids such as sulfuric,

Table 4. Recommended Welding Materials

Alloy	Bare wire (GTMA) (AWS A5.14)	Coated electrodes (SMA) (AWS A5.11)
Nickel 200	Nickel filler metal 61 (ERNi-3)	Nickel welding electrode 141 (ENi-1)
Nickel 201	Nickel filler metal 61 (ERNi-3)	Nickel welding electrode 141 (ENi-1)
Monel alloy 400	Monel filler metal 60 (ERNiCu-7)	Monel welding electrode 190 (ENiCu-2)
Monel alloy K-500	Monel filler metal 64 (ERNiCu-8)	Monel welding electrode 134 (ENiCuAl-1)
Hastelloy alloy B-2	Alloy B-2 (ERNiMo-7)	Alloy B-2 electrodes
Inconel alloy 600	Inconel filler metal 82 (ERNiCr-3)	Inconel welding electrode 182 (ENiCrFe-3)
Incoloy alloy 800	Inconel filler metal 82 (ERNiCr-3)	Inconel welding electrode 132 (ENiCrFe-1) Inco-weld A electrode (ENiCrFe-2)
Incoloy alloy 825	Incoloy filler metal 65	Incoloy welding elctrode 135
Hastelloy alloy G	Alloy G (ERNiCrMo-1)	Alloy G electrodes (ENiCrMo-1)
Hastelloy alloy X	Alloy X (ERNiCrMo-2)	Alloy X electrodes (ENiCrMo-2)
Inconel alloy 625	Inconel filler metal 625	Inconel welding electrode 112
Hastelloy alloy C-276	Alloy C-276 (ERNiCrMo-4)	Alloy C-276 electrodes (ENiCrMo-4)
Hastelloy alloy C-4	Alloy C-4 (ERNiCrMo-7)	Alloy C-4 electrodes
Hastelloy alloy S	Alloy S	—

hydrochloric, and phosphoric. The addition of oxidizers such as air or salts will significantly increase the corrosion rates. For this reason it is important to know all the components in a process stream before selecting nickel as the material of construction. Nickel is not attacked by anhydrous ammonia or ammonium hydroxide solutions of <1% concentration. Stronger concentrations can cause rapid attack through the formation of a soluble nickel-ammonium complex corrosion product.

One area where nickel receives wide usage is as a material of construction for high temperature chlorination or fluorination reactions. The pure metal appears to have better corrosion rates than any of its alloys because it is a nickel chloride or fluoride that forms and controls the corrosion reaction.

The corrosion resistance of nickel makes it particularly useful for maintaining product purity in the handling of foods, synthetic fibers, and also in structural applications where resistance to corrosion is a prime consideration. It is a general purpose material used where the special properties of the other nickel alloys are not required. In some countries it is also used for coinage. Other useful features of the alloy are its magnetic and magnetostrictive properties, and high thermal and electrical conductivity.

Applicable Specifications

Nickel 200 (UNS N 02200) and Nickel 201 (UNS N 02201)
Sheet, plate, and strip: ASTM B162-80, ASME SB 162
Rod and bar: ASTM B160-81, ASME SB 160
Fittings: ASTM B 366-77
Seamless pipe: ASTM B 161-75, ASME SB 161
Seamless tubing: ASTM B163-80, ASME SB 163

NI-CU ALLOYS

Monel Alloy 400

The alloying of 30 to 33% copper with nickel produces Monel alloy 400, which shares many of the characteristics of commercial pure nickel but improves on others. Water handling, including sea and brackish waters, is a major area of application. It gives excellent service under high velocity conditions, as in propellers, propeller shafts, pump shafts, impellers, and condenser tubes. The addition of iron to the composition improves the resistance to cavitation and erosion in condenser tube applications. The alloy can pit in stagnant seawater, as does Nickel 200; however, the rates of attack are significantly diminished. The absence of chloride stress corrosion cracking is also a factor in the selection of the alloy for this service.

The general corrosion resistance of alloy 400 in the nonoxidizing acids, such as sulfuric, hydrochloric, and phosphoric, is improved over that of pure nickel. The influence of the presence of oxidizers is, however, the same as for nickel. The alloy is not resistant to oxidizing media, such as nitric acid, ferric chloride, chromic acid, wet chlorine, sulfur dioxide, or ammonia.

Alloy 400 does have excellent resistance to hydrofluoric acid solutions at all concentrations and temperature, as shown in Fig. 1. Again aeration or the presence of oxidizing salts increases the corrosion rates. The alloy is widely used in HF alkylation, is comparatively insensitive to velocity effects, and is widely used for critical parts such as bubble caps or valves that are in contact with flowing acid. Alloy 400 is, in common with some other high nickel alloys, subject to stress corrosion cracking in moist, aerated hydrofluoric or hydrofluorosilicic acid vapor. Cracking is unlikely, however, if the metal is completely immersed in the acid.

Monel alloy K-500

Monel alloy K-500 is an age-hardenable alloy which combines the excellent corrosion resistance characteristics of Monel alloy 400 with the added advantage of increased strength and hardness. Age hardening increases its strength and hardness; however, still higher properties can be achieved when the alloy is cold worked prior to the aging treatment. Alloy K-500 has good mechanical properties over a wide range of temperatures. Strength is maintained up to about 649°C (1200°F) and the alloy is strong, tough, and ductile at temperatures as low as -235°C (-423°F). It also has low permeability and is nonmagnetic to -134°C (-210°F). Typical applications include pump shafts, impellers, doctor blades and scrapers, oil well drill collars and instruments, electronic components, springs, and valve trim.

Applicable Specifications

Monel alloy 400 (UNS N04400)
Sheet, plate, strip: ASTM B 127-80, ASME SB 127, AMS 4544C
Rod and bar: ASTM B 164-81, ASME SB 164, AMS 4675A
Fittings; ASTM B 366-77
Seamless pipe: ASTM B 165-80, ASME SB 165
Seamless tubing: ASTM B 163-80, ASME SB 163
All product forms: DIN 17743, Werkstoff No. 2.4360

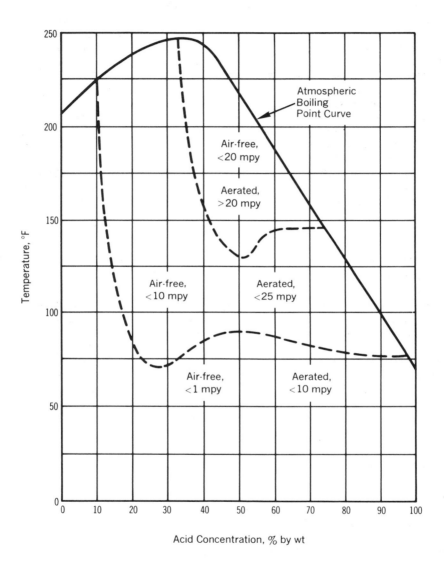

Figure 1. Isocorrosion diagram for Monel alloy 400 in hydrofluoric acid. (From Ref. 3.)

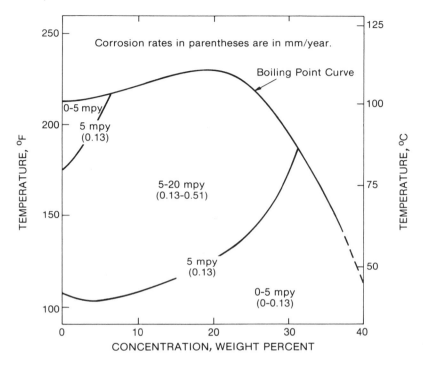

Figure 2. Isocorrosion diagram for Hastelloy alloy B-2 in hydrochloric acid. (From Ref. 4.)

NI-MO ALLOYS

Hastelloy Alloy B-2

Within the nickel-molybdenum series there is one major alloy: Hastelloy alloy B-2. Alloy B-2 is a low carbon and silicon (0.01%, 0.08% maximum) version of Hastelloy alloy B. The alloy is uniquely different from other corrosion alloys because it does not contain chromium. Molybdenum is the primary alloying element and provides significant corrosion resistance to reducing environments.

Alloy B-2 was developed to be resistant to hydrochloric acid and is used in many applications in the distillation, condensation, and handling of this acid. Alloy B-2 is recommended for service in handling all concentrations of hydrochloric acid in the temperature range 70 to 100°C (158 to 212°F) and for handling wet hydrogen chloride gas, as shown in Fig. 2.

Alloy B-2 has excellent resistance to pure sulfuric acid at all concentrations and temperatures below 60% acid and good resistance to

100°C (212°F) above 60% acid, as shown in Fig. 3. The alloy is resistant to a number of nonoxidizing environments, such as hydrofluoric and phosphoric acids, and numerous organic acids, such as acetic, formic, and cresylic. It is also resistant to many chloride-bearing salts (nonoxidizing), such as aluminum chloride, magnesium chloride, and antimony chloride.

Since alloy B-2 is nickel-rich (~70%), it is resistant to chloride-induced stress corrosion cracking (SCC). Alloy B, for example, is used as spray nozzles for a magnesium chloride dryer. By virtue of the high molybdenum content, it is highly resistant to pitting attack in most acid chloride environments.

Alloy B-2 is not recommended for elevated-temperature service except in very specific circumstances. There is no chromium in the alloy and therefore the alloy scales heavily at temperatures above 760°C (1400°F) in air. A nonprotective layer of molybdenum trioxide (MoO_3) forms and results in a heavy green oxidation scale. In a chlorine-containing environment, however, alloy B-2 has demonstrated good resistance. Alloy B-2 has excellent elevated [>900°C (>1650°F)]-temperature mechanical properties because of the high molybdenum content and has been used for mechanical components in reducing environments and vacuum furnaces. Because of the formation of the intermetallic phases Ni_3Mo and Ni_4Mo after long aging, the use of alloy B-2 in the temperature range 600 to 850°C (1110 to 1560°F) is not recommended, regardless of environment.

The major factor that limits the use of alloy B-2 is the poor corrosion resistance in oxidizing environments. Alloy B-2 has virtually no corrosion resistance to oxidizing acids such as nitric and chromic or to oxidizing salts such as ferric chloride or cupric chloride. The presence of oxidizing salts in reducing acids must also be given attention. Oxidizing salts such as ferric chloride, ferric sulfate, or cupric chloride, even when present in the ppm range, can significantly accelerate the attack in hydrochloric or sulfuric acids, as shown in Fig. 4. Even dissolved oxygen has sufficient oxidizing power to affect the corrosion rates for alloy B-2 in hydrochloric acid. Alloy B-2 also shows excellent resistance to pure phosphoric acid and has found wide acceptance in a number of key chemical processes by virtue of its resistance to corrosive catalysts. The application typically involves the halogens (Cl, F, Br) with or without the presence of reducing acids, such as hydrochloric, hydrofluoric, bromic, or sulfuric acids. The alloy has unique resistance to the aluminum chloride catalysts used in the alkylation of benzene to ethyl benzene (Friedel-Crafts processes); the ethyl benzene is used for production of styrene and cumene. The isomerization of paraffin hydrocarbons such as butane to isobutane uses antimony chloride-aluminum chloride catalysts.

The halide catalyst systems used in the newer processes for acetic acid and ethylene glycol have demonstrated significant cost reductions for producing these chemicals. Alloy B-2 is demonstrating excellent

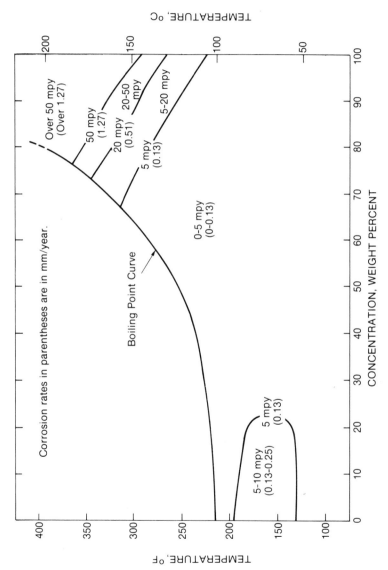

Figure 3. Isocorrosion diagram for Hastelloy alloy B-2 in sulfuric acid. (From Ref. 4.)

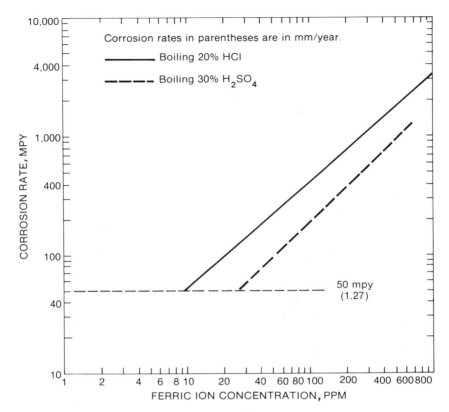

Figure 4. Effect of ferric ions on corrosion rates for Hastelloy alloy B-2. (From Ref. 4.)

corrosion resistance in this equipment and is considered to be a prime factor in making both of these viable chemical processes.

Applicable Specifications

Hastelloy alloy B-2 (UNS N10665)
Sheet, plate, and strip: ASTM B 333-77, ASME SB 333
Rod and bar: ASTM B 335-77, ASME SB 335
Fittings; ASTM B 366-77
Welded pipe: ASTM B 619-77, ASME SB 619
Welded tubing: ASTM B 626-77, ASME SB 626
Seamless pipe and tubing: ASTM B 622-77, ASME SB 622

Castings

Castings for use in conjunction with alloy B-2 equipment are available in two compositions, both of which are listed in ASTM A494. The N-12M-1 version corresponds to old alloy B, and the N-12M-2 version corresponds to Chlorimet 2. For good consistent corrosion resistance the carbon composition range for the N-12M-1 is too broad and some castings are encountered that have poor integranular corrosion resistance (but are within the specification).

The corrosion resistance of alloy B castings, in general, is not, even with the identical composition, the same as the wrought version. This is because of the significant interdendritic segregation and carbides that exist in the as-cast structure. It is less of a problem with the N-12M-2 composition becuase of the lower carbon content and the higher molybdenum content. A comparison of the isocorrosion diagram in hydrochloric acid [5] shows that there is still a minor difference.

Applicable Specifications

Alloy B: ASTM A494, Grade N-12M-1
Chlorimet 2: ASTM A494, Grade N-12M-2

NI-CR-FE Alloys

Inconel alloy 600

Inconel alloy 600 is a nickel-base alloy with about 16.0% chromium and 7.0% iron that is used primarily to resist corrosive atmospheres at elevated temperatures. The alloy is slightly less resistant than the 20% Cr-Ni alloy to oxidation, but has excellent physical properties, is readily fabricated and welded, and can be used in air up to about 1093°C (2000°F). Although resistant to oxidation, the presence of sulfur in the environment can significantly increase the rate of attack. The mode of attack is generally intergranular, and therefore the attack proceeds more rapidly and the maximum use temperature is restricted to about 315°C (600°F). Typical applications include furnace muffles, carburizing baskets, and fixtures.

Alloy 600 has excellent resistance to dry halogens at elevated temperatures, and has been used successfully for chlorination equipment at temperatures up to 538°C (1000°F). Where arrangements can be made for cooling the metal surface, the alloy can be used at even higher gas temperatures.

Resistance to stress corrosion cracking is imparted to alloy 600 by virtue of its nickel base. The alloy therefore finds considerable use in handling water environments where stainless steels fail by cracking. Becuase of its resistance to corrosion by high-purity water, it has a

number of uses in nuclear reactors, including steam generator tubing
and primary water piping. The lack of molybdenum in the alloy re-
stricts its use in applications where pitting is the primary mode of
failure.

In certain high-temperature caustic applicatons where sulfur is
present, alloy 600 is substituted for alloy 201 because of its improved
resistance. Alloy 600 is, however, subject to stress corrosion cracking in high-temperature, high-concentration alkalies. For that reason
the alloy should be stress relieved prior to use and the operating
stresses kept at a minimum. Alloy 600 is almost entirely resistant to
attack by solutions of ammonia over the complete range of temperatures
and concentrations.

The presence of chromium in the alloy provides corrosion resistance
in mildy oxidizing aqueous media, and the absence of molybdenum
limits the usefulness of the alloy in reducing acid solutions. The usefulness of the alloy is in its high-temperature applications.

Applicable Specifications

Inconel alloy 600 (UNS N06600)
Sheet, plate, and strip: ASTM B 168-75, ASME SB 168, AMS 5540H
Rod and bar: ASTM B 166-80, ASME SB 166, AMS 5665H
Fittings: ASTM B 366-77
Welded pipe: ASTM B 517-79, ASME SB 517
Welded tubing: ASTM B 516-79
Seamless pipe: ASTM B 167-80, ASME SB 167
Seamless tubing: ASTM B 163-80, ASME SB 163
Wire: AMS 5687G
All product forms: DIN 17742, Werkstoff No. 2.4816

Incoloy Alloy 800

Incoloy alloy 800 contains about 20% chromium, 32% nickel, and 46%
iron as a balance. The alloy is used primarily for its oxidation resistance and strength at elevated temperatures. It is particularly
useful for high-temperature equipment in the petrochemical industry
because the alloy does not form the embrittling sigma phase after long
exposures at 649 to 871°C (1200 to 1600°F). High creep and
rupture strengths are other factors that contribute to its performance
in many applications.

An upper use temperature for the alloy in oxidation applications is
approximately 1093°C (2000°F). Considerable experimentation has
shown that the alloy could have extensive application in the coal gasification area when the environments contain mixtures of sulfur and
oxygen. In strongly reducing mixtures of H_2S and H_2, the alloy, like
most others, has little utility. Typical applications are heat exchangers,

process piping, carburizing fixtures, and retorts. Two applications that consume significant quantities of material are electric range heating element sheathing and extruded tubing for ethylene and steam methane reforming furances. The latter is usually specified in alloy 800H, which has a slightly narrower carbon range and higher temperature anneal.

In aqueous corrosion service alloy 800 has a general resistance that falls somewhere between types 304 and 316 stainless steels. For that reason the alloy is not widely used for aqueous service. The stress corrosion cracking resistance of alloy 800 is, although not immune, better than that of the 300 series stainless steels, and substitution has been made on that basis in the past.

Applicable Specifications

Incoloy alloy 800 (UNS N08800)
Sheet, plate, and strip: ASTM B409-80, ASME SB409
Rod and bar: ASTM B408-77, ASME SB408
Fittings: ASTM B366-77
Seamless pipe: ASTM B407-77, ASME SB407
Seamless tubing: ASTM B163-80, ASME SB163
All product forms: DIN Werkstoff No. 1.4876

Incoloy alloy 825

Incoloy alloy 825 is very similar to alloy 800; however, the composition has been modified to provide for improved aqueous corrosion resistance. These modifications include 1.8% copper, 3.0% molybdenum, a higher nickel content of about 42%, and a titanium level of 0.90%.

The combination of Ni-Cr-Mo-Fe with 1.8% copper added provides good resistance for alloy 825 in pure sulfuric acid. Based on test results [3] and service experience, the alloy should have useful resistance in solutions up to 40% by weight at boiling temperatures and all concentrations at a maximum temperature of 66°C (150°F). In dilute solutions the presence of oxidizing salts such as cupric or ferric actually lowers the corrosion rates experienced. Alloy 825 has only limited use in hydrochloric acid or hydrofluoric acid. Alloy 825 has good resistance to phosphoric acid and has been used in some commercial acid applications.

The alloy is not fully resistant to stress corrosion cracking when tested in boiling magnesium chloride solution, but it has good resistance in neutral chloride environments. For this reason it is sometimes substituted for the 300 series stainless steels when localized corrosion is a problem. One present application area is that of pollution control equipment. The alloy resists stress corrosion cracking; however, the molybdenum content is the same as a type 317 stainless. The increased nickel content gives the alloy better pitting resistance, but it is not as resistant [6] as a number of other commercially available alloys.

Typical applications for alloy 825 include phosphoric acid, pickling tank heaters, pickling hooks and equipment, chemical process equipment, spent nuclear fuel element recovery, propeller shafts, and tank trucks.

Applicable Specifications

Incoloy alloy 825 (UNS N 08825)
Sheet, plate, and strip: ASTM B 424-75, ASME SB 424
Rod and bar: ASTM B 425-75, ASME SB 425
Fittings: ASTM B 366-77
Seamless pipe: ASTM B 423-75, ASME SB 423
Seamless tubing: ASTM B 163-80, ASME SB 613

NI-CR-MO-FE-CU Alloys

Hastelloy Alloy G

Hastelloy alloy G has a nickel base with additions of 22% chromium, 19% iron, 6.5% molybdenum, and 2% copper, stabilized with 2% columbium plus tantalum. Although some competitive materials use only 8 or 10 times the carbon content, a larger amount of columbium is used in alloy G to promote more satisfactory stabilization. The intention is to have alloy G used in the as-welded condition, even under the circumstance of multipass welding. The columbium addition has also been noted to provide better resistance than titanium additions in highly oxidizing environments. Because of the nickel base, the alloy is resistant to chloride-induced stress corrosion cracking.

The 2% copper addition is effective in enhancing the corrosion resistance of the alloy in reducing acids, such as sulfuric and phosphoric. The copper addition appears to alter the cathodic process by changing the overvoltage for the reaction. This in effect reduces the corrosion rate experienced by the alloy. The isocorrosion diagram for alloy G in sulfuric acid (Fig. 5) compares very favorably with that of alloy C-276 at the dilute end (<50%). Other less expensive alloys will also resist pure sulfuric acid; however, alloy G will also resist combinations of sulfuric and halides. Alloy G has excellent resistance to phosphoric acid and has been used in wet process acid evaporators, agitator shafts, pumps, and superphosphoric acid evaporators.

Applications for alloy G have increased rapidly in the past few years. In addition to its traditional application in sulfuric and phosphoric acids, alloy G has found wide acceptance in other types of chemical applications, such as HF manufacture and the processing of organic chemicals, particularly those with chlorides present. However, the largest growth of applications has been in pollution control equipment. Alloy G has demonstrated excellent performance in municipal

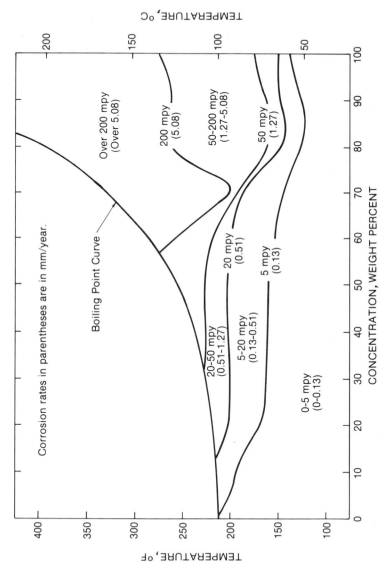

Figure 5. Isocorrosion diagram for Hastelloy alloy G in sulfuric acid. (From Ref. 4.)

garbage incinerator systems, including fans, ducts, and scrubber
equipment. The SO_2 scrubbing systems for power plants using water
or alkaline quench have been incorporating alloy G as a major material
of construction because of its resistance to sulfuric acid and conditions
where the chloride ion can concentrate. Two recent publications [7, 8]
summarize the results of corrosion tests run in a number of scrubber
environments and actual plant experience. These data show the diverse nature of attack that can occur depending on the chemistry of
the system. One scrubber environment has been encountered where,
because of low pH, high chloride levels, and the presence of oxidizing
ions, pitting attack has occurred wherever fly ash deposits collect.
In areas where the surface is kept clean, much less attack has occurred
As discussed earlier, there are some scrubber conditions where only
alloy C-276 is resistant.

Hastelloy Alloy G-3

An improved version of alloy G, Hastelloy alloy G-3, was introduced
in 1979 and offers improvements in the areas of weldability and carbide precipitationin the wrought and the cast versions. Coverage is
now available in ASTM and ASME. The significant compositional differences include lower carbon (0.015 maximum versus 0.05 maximum),
lower Cb + Ta (0.3 typical versus 2.0%), and increased molybdenum
(7.0% nominal versus 6.5% nominal). Alloy G-3 has improved localized
corrosion resistance and will replace alloy G in some applications.

Applicable Specifications

Hastelloy alloy G (UNS N06007) Hastelloy alloy G-3 (UNS N06895)
Sheet, plate, and strip: ASTM B 582-81, ASME SB 582
Rod and bar: ASTM B 518-81, ASME SB 581
Fittings: ASTM B 366-81
Welded pipe: ASTM B 619-77, ASME SB 619
Welded tubing: ASTM B 626-77, ASME SB 626
Seamless pipe and tubing: ASTM B 622-77, ASME SB 622

Castings

Cast components for use with alloy G or alloy G-3 wrought equipment
can be obtained from material of the same composition. No coverage is
presently available within the ASTM, but, it has been requested and
should be available in the future. Castings in alloy G suffer from
problems similar to those encountered in other cast nickel-base alloys,
notably lower corrosion resistance than equivalent wrought alloys.
Fewer intermetallic phases form, but carbides have been observed to
form on grain boundaries, and a solution anneal following casting is
required.

NI-CR-MO-FE ALLOYS

Hastelloy Alloy X

Hastelloy alloy X is a major material of construction in the hot section
of gas turbine engines. Applications include parts such as burner
cans and transition ducts. Alloy X owes its oxidation resistance to
the formation of a complex chromium oxide spinel which provides good
resistance to temperatures of 1177°C (2150°F). Alloy X is also used in
chemical plants for its combination of high strength and excellent
oxidation resistance.

A typical application in the process industries is for nitric acid
catalyst grid supports operating at 900°C (1650°F). The high-temperature strength and resistance to warpage and distortion provide outstanding performance. Alloy X is also used as distributor plates in
the manufacture of magnesium chloride. Other high-temperature applications include flare nozzles, thermowell protection tubes, expansion
bellows, furnace internals, retorts, muffles, and trays

The catalyst regenerator for high density polyethylene is constructed of alloy X, since high temperatures and pressures are required to revitalize the catalysts. Unfortunately, this continued temperature cycling eventually reduces the room temperature ductility in
alloy X so that repair welding becomes difficult without solution annealing. The improved stability in alloy S has been demonstrated in this
application, so that future equipment to be built in areas of the world
where annealing furnaces are not available may be constructed from
alloy S.

Applicable Specifications

Hastelloy alloy X (UNS N06002)
Sheet, plate, and strip: ASTM B435-77, ASME SB435, AMS 5536G
Rod and bar: ASTM B572-77, ASME SB572, AMS 5754G
Fittings: ASTM B366-77
Welded pipe: ASTM B619-77, ASME SB619
Welded tubing: ASTM B626-77, ASME SB626
Seamless pipe and tubing: ASTM B622-77, ASME SB622
Wire: AMS 5798A

NI-CR-MO Alloys

Inconel Alloy 625

Inconel alloy 625 has been extensively used in sheet form in high-
temperature gas turbine applications. Because of its combination of
chromium, molybdenum, carbon, and columbium + tantalum, the alloy
retains its strength and oxidation resistance to elevated temperatures.

Typical applications include ducting systems, thrust reverser assemblies, and afterburners. Use of the alloy has been considered in the high-temperature gas-cooled reactor; however, after long aging in the temperature range 590 to 760°C (1100 to 1400°F), the room temperature ductility is significantly reduced.

Resistance to aqueous solutions is good in a variety of applications, including organic acid, sulfuric acid, and hydrochloric acid at temperatures below 65°C (150°F). The alloy has satisfactory resistance to hydrofluoric acid and has been used as a liner in a hydrofluoric acid generator that reacts sulfuric acid with fluorspar. Although nickel-base alloys are not generally used in nitric acid service, alloy 625 is resistant to mixtures of nitric-hydrofluoric, where stainless loses its resistance.

Alloy 625 has excellent resistance to phosphoric acid solutions. Actual field operating information has shown this alloy to have excellent resistance to commercial grades of acid that contain fluorides, sulfates, and chlorides in the production of superphosphoric acid (72% P_2O_5).

The columbium and tantalum stabilization (3.65%) makes the alloy suitable for corrosion services in the as-welded condition. The high nickel content provides good resistance to chloride SCC. Alloy 625 has a relatively high molybdenum content (9%), and field testing data place its performance somewhere between alloys G and C-276 in pitting and crevice corrosion resistance.

Recently alloy 625 has been utilized in several new pollution control applications, such as reheaters for SO_2 scrubbing systems on coal-fired power plants and bottoms of electrostatic precipitators that are flushed with seawater. In one installation, the interior of a stack has been lined to provide resistance to the wet sulfur-containing gases.

Applicable Specifications

Inconel alloy 625 (UNS N06625)
Sheet, plate, and strip: ASTM B443-75, ASME SB 443, AMS 5599B
Rod and bar: ASTM B446-75, ASME SB446, AMS 5666A
Seamless pipe and tubing: ASTM B444-75, ASME SB4444

Hastelloy Alloy C-276

Hastelloy alloy C-276 is a low carbon (0.01% maximum) and silicon (0.08% maximum) version of Hastelloy alloy C. Alloy C-276 was developed to overcome the corrosion problems associated with the welding of alloy C. When used in the as-welded condition, alloy C was often susceptible to serious intergranular corrosion attack in many oxidizing and chloride-containing environments. The low C and Si content of alloy C-276 prevents precipitation of continuous grain boundary precipitates in the weld heat-affected zone. Thus alloy C-276 can be used in most applications in the as-welded condition without suffering severe intergranular attack.

The descriptions of the applications for alloy C-276 in the process industries are extensive and diverse. The alloy is extremely versatile as it possesses good resistance in both oxidizing and reducing media, and this includes conditions with halogen ion contamination. The pitting and crevice corrosion resistance of alloy C-276 makes it an excellent choice when dealing with acid chloride salts, whether in the process or merely handling hot seawater in heat exchangers.

Isocorrosion diagrams for alloy C-276 have been developed for a number of inorganic acids [4]: for example, for sulfuric acid in Fig. 6. Rather than having one or two acid systems where the corrosion resistance is exceptional, as was the case with alloy B-2, alloy C-276 is a good compromise material for a number of systems. For example, in sulfuric acid coolers handling 98% acid from the absorption tower, alloy C-276 is not the optimum alloy for the process side corrosion but it is excellent for the water-side corrosion and allows the use of brackish water or seawater. Concentrated sulfuric acid is used to dry chlorine gas. The dissolved chlorine will accelerate the corrosion of alloy B-2, but alloy C-276 has performed quite satisfactorily in a number of chlorine drying installations.

Another example is the spray application of alloy C-276 powder to steam plant boiler tubes to prevent liquid-phase corrosion. A complex alkali-iron-sulfate that forms a liquid slag in the range 570 to 700°C (1060 to 1290°F) was attributed with causing the rapid attack experienced on the chrome-moly-steel tubes. A spray application, without removal of the tubes, of a thin coating of alloy C-276 was measured as reducing the rate of attack from 41 mpy to 3 mpy.

Many new applications for alloy C-276 have evolved as a result of increased activity in scrubbers for the pollution control industry. Induced draft fans represent a particularly aggressive environment because they are exposed to wet/dry conditions. Alloy C-276 has functioned very satisfactorily for years in an induced-draft fan for a municipal incinerator in Grosse Point, Michigan. Alloy C-276 has been indicated as a satisfactory material for scrubber construction where problems of localized attack have occurred with the Ni-Cr-Mo-Fe-Cu alloys because of pH, temperature, or chloride content.

Hastelloy alloy C-4

A new Ni-Cr-Mo alloy known as Hastelloy alloy C-4 has been developed that achieves improved stability with regard to precipitation of both carbides and intermetallic phases. The control of these secondary phases results in excellent high-temperature stability, such that the corrosion resistance and mechanical properties in the thermally aged condition are similar to the annealed condition properties.

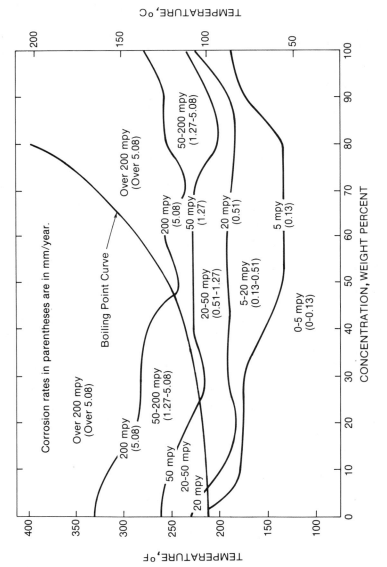

Figure 6. Isocorrosion diagram for Hastelloy alloy C-276 in sulfuric acid. (From Ref. 4.)

Within the broad scope of chemical processing, numerous examples exist where oxidizing and reducing conditions can cause serious inter-granular corrosion of a sensitized (precipitated) microstructure. The sensitized structure can result from several sources, such as (1) im-proper anneal, (2) welding, (3) thermomechanical processing such as hot forming or rolling operations, (4) stress relief or normalizing treatments required for carbon steel backing of clad materials, or (5) operation of process equipment in the sensitizing range. Alloy C-4 represents a significant improvement since the alloy can be subjected to temperatures in the normal sensitizing range 550 to 1090°C (1022 to 1994°F) for extended periods without experiencing the severe corrosion attack that plagues the common austenitic alloys.

As shown in Table 1, the composition of alloys C-4 and C-276 is, with the exception of iron and tungsten, approximately the same. Similarly, the general corrosion resistance of the two alloys is generally the same. In strongly reducing media such as hydrochloric acid, alloy C-4 has slightly higher rates than alloy C-276, but in oxidizing media the results are reversed. It would appear that these differences re-late to the effects of iron and tungsten; however, the data are not conclusive. Alloy C-4 offers good corrosion resistance to a wide variety of media, including organic acids and acid chloride solutions. Isocorrosion diagrams for alloy C-4 have also been developed for a series of inorganic acids [4].

Applicable Specifications

Hastelloy alloy C-276 (UNS N10276) and Hastelloy alloy C-4 (UNS N06455)
Sheet, plate, and strip: ASTM B575-77, ASME SB575
Rod and bar: ASTM B574-77, ASME SB574
Fittings: ASTM B366-77
Welded pipe: ASTM B619-77, ASME SB619
Welded tubing: ASTM B626-77, ASME SB626
Seamless pipe and tubing: ASTM B622-77, ASME SB622

Castings

Castings for use in conjunction with alloys C-276 or C-4 are currently available in two compositions that are listed in ASTM A494. The CW-12M-1 version corresponds to old alloy C, and the CW-12M-2 version corresponds to Chlorimet 3. For good consistent corrosion resistance the carbon composition range for the CW-12M-1 is too broad, and some castings are encountered that have poor intergranular corrosion resis-tance but are within the specification.

A further problem associated with the corrosion resistance of cast alloy C is the precipitated intermetallic phases in the interdendritic areas. These phases are not homogenized with a solution heat treat-ment and promote localized attack in some media. Figure 7 shows the

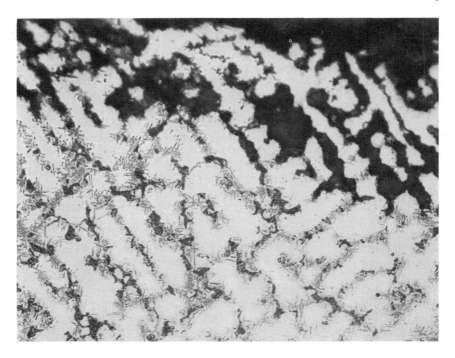

Figure 7.   Localized corrosion on a cast alloy C bolt.

interdendritic attack that occurred in cast alloy C bolts in a pulp and
paper chlorine-containing environment. Bolts made from wrought bar
stock had corrosion rates of less than 1 mpy. Similar experieinces have
been encountered with cast pipe fittings and valves.

Modification of the composition of alloy C to produce alloy C-4 has
eliminated the presence of the intermetallic phases in the wrought alloy
and minimized its presence in the cast alloy. The CW-12M-2 composition is closer to being stable by virtue of the absence of the tungsten;
however, the chromium, molybdenum, and iron levels can also produce
the metallurgical instability.

Hastelloy alloy C-4C has been developed that couples low carbon
with a balanced composition to avoid the intermetallic phases. Corrosion resistance of castings made from this alloy approaches that of

wrought alloys C-276 and C-4 in the accelerated corrosion test normally applied (ASTM G-28) to these alloys. Coverage for this alloy within ASTM A494 has been requested.

Applicable Specifications

Alloy C: ASTM A494, Grade CW-12M-1
Chlorimet 3: ASTM A494, Grade CW-12M-2
Hastelloy alloy C-4C; none

Hastelloy alloy S

Hastelloy alloy S was developed (1973) primarily for gas turbine applications requiring good alloy stability, oxidation resistance, and a low thermal expansion coefficient. Its composition is much like that of alloys C-276 and C-4 and it has similar corrosion resistance; however, the carbon content may preclude its use in some aqueous media in the as-welded condition.

Because of the combination of stability and oxidation resistance, alloy S has shown considerable promise as a welding filler metal in dissimilar metal joint applications. One example is in the welding of centrifugally cast furnace tubes to pigtails and in turn to the header. After 10,000 hours of aging in the temperature ranges typically encountered in this application, the alloy S welds exhibited 80% of their original ductility.

Applicable Specifications

Sheet, plate, and strip: AMS 5873
Rod and bar: AMS 5711
Wire: AMS 5838

REFERENCES

1. *Chemical Week*, July 7, 1976.
2. *Localized Corrosion—Cause of Metal Failure*, ASTM STP 516, American Society for Testing and Materials, Philadelphia, 1972.
3. *Resistance to Corrosion*, Bulletin S-37, Huntington Alloy Products, Division of the International Nickel Co., Inc., Huntington, W.Va.
4. T. S. Lee III and F. G. Hodge, *Mater. Perform. 15*: 29-36, September 1976.
5. *Chlorimet 2*, Bulletin A/3C, The Duriron Company Inc., Dayton, Ohio.
6. A. I. Asphahani, *Mater. Perform. 19*: 9-21, August 1980.

7. E. C. Hoxie and G. W. Tuffnell, in *Resolving Corrosion Problems in Air Pollution Control Equipment*, National Association of Corrosion Engineers, Houston, 1976, pp. 65-71.
8. W. L. Silence, P. E. Manning, A. I. Asphahani; Alloy selection for flue gas desulfurization systems, in *Solving Corrosion Problems in Air Pollution Control Equipment*, National Association of Corrosion Engineers, Houston, 1981, pp. 18/1 - 18/16.

General Information References

Publications of Stellite Division of Cabot Corporation, Kokomo, Ind.:

Bulletin 037, *Hastelloy Alloy X*
Bulletin 131, *Corrosion Resistance of Hastelloy Alloys*
Bulletin 267, *Hastelloy Alloy G*
Bulletin 356, *Hastelloy Alloy C-276*
Bulletin 526, *Hastelloy Alloy C-4*
Bulletin 532, *Hastelloy Alloy B-2*
Bulletin 489, *Hastelloy Alloy S*
Bulletin 609, *Hastelloy Alloy G-3*

Publications of the International Nickel Co., Inc., New York:

Bulletin CEB-1, *Resistance of Nickel and High-Nickel Alloys to Corrosion by Sulfuric Acid*
Bulletin CEB-2, *Corrosion Resistance of Nickel and Nickel-Containing Alloys in Caustic Soda and other Alkalies*
Bulletin CEB-3, *Resistance of Nickel and High-Nickel Alloys to Corrosion by Hydrochloric Acid, Hydrogen Chloride and Chlorine*
Bulletin CEB-4, *Corrosion Resistance of Nickel-Containing Alloys in Phosphoric Acid*
Bulletin CEB-5
    Hydrofluoric Acid, Hydrogen Fluoride and Fluorine
Bulletin CEB-6, *Corrosion Resistance of Nickel-Containing Alloys in Organic Acids and Related Compounds*

Publications of the Huntington Alloy Products Division of the International Nickel Co., Inc. Huntington, W. Va.

Bulletin T-5, *Monel Alloys*
Bulletin T-7, *Inconel Alloy 600*
Bulletin T-15, *Nickel Alloys*
Bulletin T-37, *Incoloy Alloy 825*
Bulletin T-40, *Incoloy Alloys*
Bulletin T-42, *Inconel Alloy 625*

3.4
Copper and Copper Alloys

JOHN M. CIESLEWICZ / Ampco Metal, Milwaukee, Wisconsin

INTRODUCTION

Copper is a unique material. It has excellent electrical and thermal
conductivity, and it is malleable and machinable. However, it has low
mechanical properties and must be cold worked or alloyed to obtain
strength. Copper is classed as a noble metal and is therefore corrosion
resistant. It is resistant to urban, marine, and industrial atmospheres
and finds use as sheeting, roofing, and the like, in these environments.
It is resistant to water and finds use in domestic and industrial systems
and in seawater applications.

To increase the strength of copper, and therefore its engineering
usefulness, copper is alloyed with other elements, such as zinc, aluminum, nickel, tin, and so on. There are hundreds of copper alloys.
To bring order and to identify the many alloys, the Copper Development Association [1], together with the American Society of Testing
and Materials and the Society of Automotive Engineers, developed a
five-digit system. This system is part of the Unified Numbering System for metals and alloys. In this system, the numbers C10000 through

C79999 denote the wrought alloys: The cast copper and copper alloys are numbered from C80000 through C99999.

Within the two categories noted above, the compositions are grouped into the following families: copper, 99.3% minimum copper; high copper alloys, 94% minimum copper; brasses, which contain zinc as the principal alloying element; bronzes, which contain tin, aluminum, or silicon as the principal alloying element; and copper nickels, in which nickel is the principal alloying element. Other families are nickel-silvers, leaded copper (20% or more lead), and special alloys, but these three families are of little importance in corrosion engineering. A more complete description of the families is given elsewhere [1]. We will look at these families and discuss their uses in corrosion engineering.

COPPERS

The coppers contain a minimum of 99.3% copper. The elements silver, arsenic, phosphorus, antimony, tellurium, nickel, cadmium, lead, sulfur, zirconium, magnesium, boron, and bismuth may be present singly or in combination. The primary uses of the coppers, because of their good electrical properties, are in electrical or electronic applications such as bus bars, waveguides, wire, switches, and transistor components. Since copper is a noble metal, it also finds many uses in corrosive environments. Table 1 gives the chemical composition of some of the coppers usually used in corrosion applications.

The coppers are resistant to urban, marine, and industrial atmospheres, so it is only logical that these materials are used architecturally. Coppers such as UNS numbers C11000, C12500, and C13000 are used as building fronts, downspouts, flashing, gutters, roofing, and screening. Because of their good thermal conductivity and corrosion resistance, the coppers are ideal for solar panels and related tubing and piping used in solar energy conversion.

Also, because of the excellent thermal properties, inherent resistance to various atmospheres and engine coolants, and their ease of joining by brazing and soldering, radiators are commonly made of copper. The silver-bearing coppers, such as C10400, C11300, and C12500, are used in radiator construction. Tubing for brakes and hydraulic and fuel lines are commonly made of copper.

The beverage industry makes use of the coppers in brewery and distilling operations. Brew kettles, heating vessels, heating coils, and piping are to be found in these operations.

The inherent resistance of copper to potable waters, its ease of fabrication, and its nonflammability make it an ideal material for plumbing environments, both domestic and commercial. Since 1950 [2], over 10 billion pounds of plumbing tube and pipe has gone into U.S. plumbing systems. Copper UNS numbers C10200, C10300, C10800, C12000, and C12200 are those most commonly used. These are covered by ASTM

Table 1. Chemical Composition of Coppers: Maximum Percent Composition[a]

Copper UNS No.	Cu	Ag, min.	P	As	Sb	Te	Others
C10200	99.95						
C10300	99.95		0.001-0.005				
C10400	99.95	0.027					
C10800	99.95		0.005-0.012				
C11000	99.90						
C11300	99.90	0.027					
C12000	99.90		0.004-0.012				
C12200	99.90		0.015-0.040				
C12500	99.88			0.012	0.003	0.025	0.050 Ni
							0.003 Bi
							0.004 Pb
C13000	99.88	0.085		0.012	0.003	0.025	0.050 Ni
							0.003 Bi
							0.004 Pb
C14200	99.4		0.015-0.040	0.15-0.50			

[a]Unless shown as a range.
Source: Ref. 1.

Standard Specifications B42, B88, B302, and B447. ASTM B306 covers copper drainage tube.

Occasional failures with copper plumbing tube have been encountered. Lyman and Cohen [2] reported that in a five-year period, an average of 71 incidents per year have been documented in the United States at a time when the yearly total usage of copper plumbing tube was 450 million pounds. The two most common causes of failure were pitting (24 cases) and corrosion erosion (12 cases). Most of the pitting failures occurred in cold water lines, and aggressive well waters were involved. Surface waters, unless mixed with well waters, are usually not involved in cold water pitting.

Cohen [3] reports on a computer-aided study of 120 aggressive well waters that were known to have caused pitting. The tentative conclusion of this study was that pitting waters had the following characteristics: the pH is less than 7.8, carbon dioxide is in excess of 25 ppm, the sulfate ion is over 17 ppm, and the sulfate-chloride ratio is about 3:1. There was also an apparent increase in pitting if one or more of the following ions were found: less than 42 ppm potassium, less then 25 ppm nitrate, and more than 26 ppm silicate.

Alleviation of the pitting problem is accomplished by either changing the character of the water or abandoning the aggressive well. Water characteristics can be changed by dilution of the aggressive water by a nonpitting water or by water treatment. Water treatment as simple as aeration to release the free carbon dioxide or raising the pH can alleviate the problem.

Although hot water pitting is rare, when it occurs a soft water is usually involved, and the pH is below 6.5, the CO_2 content is 10 to 50 ppm, and the water is low in chloride, sulfate, and nitrate ions.

Corrosion-erosion failure of copper water tube is the result of excessive high velocity and turbulence which develop when sudden changes in direction or cross section occur. This is easily remedied by proper design and sizing of the system so that maximum velocity is kept at 4 to 5 ft/s. Good workmanship is also essential so that a burr or an improperly prepared tube end does not cause turbulence.

HIGH COPPER ALLOYS

The high copper alloys contain a minimum of 96% copper if wrought and 94% copper in the cast form. Like the coppers, the principal use of the high coppers is in electrical and electronic applications.

The corrosion resistance of the high copper alloys is about the same as that of the coppers. When they are used in corrosion service, they are used where mechanical strength and corrosion resistance are needed. Table 2 gives the composition of some of the high copper alloys.

C19400 is essentially copper with the addition of about 2.4% iron. The iron addition enhances corrosion resistance and the alloy's first use was as seam-welded condenser tubing in desalting service.

C18000 is a new alloy developed to replace beryllium copper, C17500, in the resistance welding field. Corrosion tests in plating solutions have indicated that the alloy may be useful. Tests in a tinplating solution at room temperature gave a corrosion rate of 0.02 mil per year (mpy) for C18000, and tests at room temperature in a zinc-plating solution gave a corrosion rate of 1.4 mpy.

COPPER-ZINC ALLOYS

Brasses contain zinc as their principal alloying ingredient. Other major alloying additions are lead, tin, and aluminum. Lead is added to improve machinability and its presence does not enhance corrosion

Table 2. High Copper Alloys: Maximum Percent Composition[a]

Copper Alloy UNS No.	Cu	Fe	Ni	Co	Be	Pb	P	Zn	Sn	Cr	Si	Al
C17000	Rem	b	b	b	1.6-1.79	–	–	–	–	–	0.20	0.20
C17200	"	b	b	b	1.8-2.00	–	–	–	–	–	0.20	0.20
C17300	"	b	b	b	1.8-2.00	0.20-0.60	–	–	–	–	0.20	0.20
C18000[c]	"	0.10	2.5	–	–	–	–	–	–	0.4	0.7	–
C19200	98.7 min.	0.8-1.2	–	–	–	–	0.01-0.04	–	–	–	–	–
C19400	97.0 min.	2.1-2.6	–	–	–	0.03	0.015-0.15	0.05-0.20	–	–	–	–

[a]Unless shown as a range or minimum.

[b]Ni + Co, 0.20% min.; Ni + Fe + Co, 0.60% max.

[c]Also available in cast form as copper alloy UNS C81540.

Source: Ref. 1.

resistance. The addition of tin, nominally about 1%, increases strength and the dealloying resistance of the alloys. Aluminum is added to stabilize the protective surface film. Table 3 lists some of the brasses used in corrosion engineering. Alloys containing in excess of about 15% zinc are susceptible to dealloying in many environments. These environments include acids, both organic and inorganic, dilute and concentrated alkalies, neutral solutions of chlorides and sulfates, and mild oxidizing agents such as hydrogen peroxide. As stated above, the addition of small amounts of tin improves the dezincification resistance of these alloys. The Admiralty brasses, copper alloy UNS C44300 through C44500, and the naval brasses, C46400 through C46700, owe their dealloying resistance to the presence of tin. In addition, minor additions, usually less than 0.10%, of arsenic, antimony, or phosphorus are added to the Admiralty brasses to further enhance dealloying resistance.

The high zinc brasses, such as C27000, C28000, C44300, and C46400, resist sulfides better than do the low zinc brasses. Dry H_2S is well resisted. Because of this resistance to sulfides, alloys such as the Admiralty and naval brasses find use in petroleum refinery applications, where H_2S is commonly found.

Alloys containing 15% or less of zinc resist dealloying and are generally more corrosion resistant than the high-zinc-bearing brasses. These alloys are resistant to many acids, alkalies, and salt solutions that cause dealloying in the high zinc brasses. Dissolved air, oxidizing materials such as chlorine and ferric salts, compounds that form soluble copper complexes (e.g., ammonia), and compounds that react directly with copper (e.g., sulfur, mercury) are deleterious to the low zinc brasses. The alloys are more resistant to stress corrosion cracking than the high-zinc-containing alloys.

The brasses are widely used as plumbing hardware in valves, elbows, and piping. The inhibited brass alloy C44300 is a common condenser tube alloy, and tube sheets are C46400. Many commercial vessels use these materials, whereas most naval vessels use copper-nickels. Another excellent condenser tube alloy is aluminum brass, C68700, which contains 2% aluminum.

COPPER-TIN ALLOYS

Alloys of copper and tin are known as tin bronzes or phosphor bronzes. Tin is the principal alloying ingredient, but phosphorous is always present in small amounts, usually less than 0.5%, because of its use as a deoxidizer. Other secondary alloying elements are lead, zinc, and nickel. The microstructure of the alloys can range from solid solution alpha to complex structures of alpha plus alpha-delta eutectoid, beta, and gamma. These alloys are available in both cast and wrought form and some alloys respond to heat treatment. Tables 4A and 4B list some of the tin bronzes used in corrosion applications.

Table 3. Copper-Zinc Alloys: Maximum Percent Composition[a]

Copper Alloy UNS No.	Cu	Pb	Fe	Zn	Sn	P	Al	Others
C27000	63.0-68.5	0.10	0.07	Rem	–	–	–	–
C28000	59.0-63.0	0.30	0.07	Rem	–	–	–	–
C44300	70.0-73.0	0.07	0.06	Rem	0.8-1.2	–	–	0.02-0.10 As
C44400	70.0-73.0	0.07	0.06	Rem	0.8-1.2	–	–	0.02-0.10 Sb
C44500	70.0-73.0	0.07	0.06	Rem	0.8-1.2	0.02-0.10	–	–
C46400	59.0-62.0	0.20	0.10	Rem	0.50-1.0	–	–	–
C46500	59.0-62.0	0.20	0.10	Rem	0.50-1.0	–	–	0.02-0.10 As
C46600	59.0-62.0	0.20	0.10	Rem	0.50-1.0	–	–	0.02-0.10 Sb
C46700	59.0-62.0	0.20	0.10	Rem	0.50-1.0	0.02-0.10	–	–
C68700	76.0-79.0	0.07	0.06	Rem	–	–	1.8-2.5	0.02-0.10 As

[a]Unless shown as a range.
Source: Ref. 1.

Table 4A. Wrought Copper-Tin Alloys: Maximum Percent Composition

Copper Alloy UNS No.	Cu	Pb	Fe	Sn	Zn	P
C51000	Rem	0.05	0.10	4.2-5.8	0.30	0.03-0.35
C51100	"	0.05	0.10	3.5-4.9	0.30	0.03-0.35
C52100	"	0.05	0.10	7.0-9.0	0.20	0.03-0.35
C52400	"	0.05	0.10	9.0-11.0	0.20	0.03-0.35
C54400	"	3.5-4.5	0.10	3.5-4.5	1.5-4.5	0.01-0.50

[a]Unless shown as a range.
Source: Ref. 1.

Table 4B. Cast Copper-Tin Alloys: Maximum Percent Composition[a]

Copper Alloy UNS No.	Cu	Sn	Pb	Zn	Fe	Sb	Ni	S	P	Al	Si
C90300	86.0-89.0	7.5-9.0	0.30	3.0-5.0	0.20	0.20	1.0	0.05	0.05	0.005	0.005
C90500	86.0-89.0	9.0-11.0	0.30	1.0-3.0	0.20	0.20	1.0	0.05	0.05	0.005	0.005
C92200	86.0-90.0	5.5-6.5	1.0-2.0	3.0-5.0	0.25	0.25	1.0	0.05	0.05	0.005	0.005
C93700	78.0-82.0	9.0-11.0	8.0-11.0	0.8	0.15	0.55	1.0	0.08	0.15	0.005	0.005
C93800	75.0-79.0	6.3-7.5	13.0-16.0	0.8	0.15	0.8	1.0	0.08	0.05	0.005	0.005
C93900	76.5-79.5	5.0-7.0	14.0-18.0	1.5	0.40	0.50	0.8	0.08	1.5	0.005	0.005
C94700[b]	85.0-90.0	4.5-6.0	0.10	1.0-2.5	0.25	0.15	4.5-6.0	0.05	0.05	0.005	0.005

[a]Unless shown as a range.
[b]Mn 0.20.
Source: Ref. 1.

The tin bronzes are probably the oldest alloys known. They are the bronzes of the Bronze Age. There are many bronze artifacts in existence today, such as statues, swords, vases, and bells. These have survived hundreds of years exposure to various atmospheres, waters, and soils, testifying to the corrosion resistance of the tin bronzes to a wide variety of environments.

Modern use of tin bronzes is primarily as valves, valve components, impellers, pump casings, and so on, in water service. These alloys are used in fire protection systems because of their corrosion resistance in stagnant waters.

COPPER-ALUMINUM ALLOYS

Alloys consisting primarily of copper and aluminum are commonly referred to as aluminum bronzes. They are available in both cast and wrought form, some of them respond to heat treatment, and they can be joined by brazing and welding. The aluminum bronzes find use where combinations of strength, corrosion resistance, and wear resistance are of importance. Tables 5A and 5B list the alloys generally used in corrosion applications.

The development of the aluminum bronzes has progressed from a simple alloy of copper and aluminum to complex alloys with the addition of iron, nickel, silicon, manganese, and other elements. The microstructure of these bronzes is controlled primarily by the aluminum content. In alloys containing up to approximately 8% aluminum, the microstructure consists of the single-phase alpha structure, in which may be included some minor intermetallics. When the aluminum content is increased, other constituents appear. They may be beta, alpha-gamma eutectoid, or a combination of these constituents, depending on the aluminum content, the cooling rate from solidification, and the heat treatment. The addition of iron generally forms an iron-rich intermetallic compound, while the addition of nickel above 2% tends to introduce the additional kappa phase.

The corrosion characteristics are affected by the microstructure of the alloy. A small amount of tin added to C61300 will inhibit intergranular stress corrosion. The importance of microstructure is particularly apparent in the dealloying resistance of the alloys. Dealloying is rarely seen in all-alpha, single-phase alloys such as copper alloy UNS C60800, C61300, or C61400, and when seen, conditions of low pH and high temperature are usually present. However, duplex structures are more prone to dealloying. Proper heat treatment can control dealloying in these duplex alloys. As an example, C95400, with a structure of alpha plus eutectoid is susceptable to dealloying. A simple heat treatment above the critical temperature will transform the susceptible eutectoid to beta, which can be reatined by rapid cooling, making the alloy more resistant to dealloying. It is possible to

Table 5A. Wrought Copper-Aluminum Alloys: Maximum Percent Composition[a]

Copper Alloy UNS No.	Cu	Al	Fe	Ni	Mn	Si	Sn	Zn	Others
C60800	92.5-94.8	5.0-6.5	0.10	—	—	—	—	—	0.02-0.35 As 0.10 Pb
C61000	90.0-93.0	6.0-8.5	0.50	—	—	0.10	—	0.20	0.02 Pb
C61300	88.6-92.0	6.0-7.5	2.0-3.0	0.15	0.10	0.10	0.20-0.50	0.05	0.01 Pb
C61400	88.0-92.5	6.0-8.0	1.5-3.5	—	1.0	—	—	0.20	0.01 Pb
C61500	89.0-90.5	7.7-8.3	—	1.8-2.2	—	—	—	—	0.015 Pb
C61800	86.9-91.0	8.5-11.0	0.5-1.5	—	—	0.10	—	0.02	0.02 Pb
C62300	82.2-89.5	8.5-11.0	2.0-4.0	1.0	0.50	0.25	0.60	—	—
C63000	78.0-85.0	9.0-11.0	2.0-4.0	4.0-5.5	1.5	0.25	0.20	0.30	—
C63200	75.9-84.5	8.5-9.5	3.0-5.0	4.0-5.5	3.5	0.10	—	—	0.02 Pb

[a]Unless shown as a range.
Source: Ref. 1.

Table 5B. Cast Copper-Aluminum Alloys: Maximum Percent Composition[a]

Copper Alloy UNS No.	Cu	Al	Fe	Ni	Mn	Si	Others
C95200	86.0 min.	8.5-9.5	2.5-4.0	—	—	—	1.0 total
C95300	86.0 min.	9.0-11.0	0.8-1.5	—	—	—	1.0 total
C95400	83.0 min.	10.0-11.5	3.0-5.0	2.5	0.50	—	0.5 total
C95500	78.0 min.	10.0-11.5	3.0-5.0	3.0-5.5	3.5	—	0.5 total
C95700	71.0 min.	7.0-8.5	2.0-4.0	1.5-3.0	11.0-14.0	—	0.5 total[b]
C95800	79.0 min.	8.5-9.5	3.5-4.5[c]	4.0-5.0	0.8-1.5	0.10	0.5 total[d]

[a]Unless shown as a range or minimum.

[b]Maximum Pb 0.03.

[c]Iron content shall not exceed the nickel content.

[d]Maximum Pb 0.02.

Source: Ref. 1.

improve the alloy still further by quenching it from a high temperature 890°C (1650°F), then tempering it at a temperature, say 649°C (1200°F), above the critical. This will give a structure of martensitic beta with acicular reprecipitated alpha scattered through the beta. Table 6 gives the results of alternate immersion (aerated) laboratory tests in a 3% NaC1 solution for C95200, C95300, and C95400.

The addition of about 4% nickel coupled with a lower aluminum content is also advantageous in deterring dealloying. This addition plus heat treatment is the basis for Mil-B-24480 (Ships, January 1973), Table 7.

This alloy has been and is used in many seawater applications, such as valves, fittings, and pump casings. Pumps of temper-annealed C95800 are used in saltwater desalination units aboard both naval and commerical vessels.

The aluminum bronzes, particularly C61300, C61400, and C63000, are used as condenser tube sheets in both fossil-fueled and nuclear power plants handling fresh, brackish, and sea waters for cooling. These tube sheets are compatible with most condenser tube alloys, such as C70600, C71500, C68700, and titanium.

Recent work has indicated that aluminum bronzes may be useful in geothermal energy applications. Tests were run at various locations in the Madison Aquifer of western South Dakota [4]. Water analyses are given in Table 8 and test results in Table 9. These tests indicate that aluminum bronze, nickel-aluminum bronze, and Admiralty brass, C44300, will resist these waters. Where crevice or incipient crevice corrosion was observed, the bronzes consistently exhibited very good resistance. Copper and C71500, cupronickel, were found to be the least resistant of the copper alloys tested.

Aluminum bronze alloys are generally resistant to chloride potash solutions. Tables 10 and 11 give the results of on-site tests in several types of potash ores. As these data show, the alloys are affected by aeration and pH. This is shown in Table II. Where the aeration is 100% and the pH low (Table 11, test 1), the corrosion rate is near 50 mpy. As aeration decreases and pH increases (Table 11, tests 2 and 3), the corrosion rate drops to less than 10 mpy. A potash crystallizer was constructed of C61400 and was in service for 15 years in the New Mexico area. The original plate thickness was 12.4 mm (0.5 in.), and upon dismantling the plant, the plate thickness was 11.6 mm (0.457 in.). This is a loss of 0.8 mm (0.043 in.) in 15 years, giving a corrosion rate of 2.8 mpy. Other successful applications are in the Middle East Dead Sea area and in Spain.

Alkalies as sodium and potassium hydroxides can be handled by aluminum bronze alloys. Laboratory tests of various concentrations of sodium hydroxide (Table 12) give rates ranging from 0.1 to 15 mpy, depending on concentration, aeration, and temperature.

These alloys are resistant to nonoxidizing mineral acids such as sulfuric and phosphoric. Their resistance to these acids is controlled by

Table 6.   3% NaCl, 15 Days' Exposure, Alternate Immersion

Copper Alloy UNS no.	Thermal treatment[a]	Before cleaning
C 95200	As cast	100% of surface covered with brown-red product
C 95200	621°C (1150°F) 1 h, W. Q. 510°C (950°F) 1 h, W. Q.	90% of surface covered with brown-red product
C 95200	621°C (1150°F) 1 h, W. Q.	5% of surface covered with brown-red spots
C 95200	899°C (1650°F) 1 h, W. Q.	No brown-red spots or product
C 95300	As cast	95% of surface covered with brown-red product
C 95300	621°C (1150°F) 1 h, W. Q.	75% of surface covered with brown-red product
C 95300	621°C (1150°F) 1 h, W. Q.	No brown-red spots
C 95300	899°C (1650°F) 1 h, W. Q. 649°C (1200°F) 1 h, W. Q.	No brown-red spots
C 95400	As cast	100% of surface covered with brown-red product
C 95400	621°C (1150°F) 1 h, W. Q. 510°C (950°F) 1 h, W. Q.	75% of surface covered with with brown-red product
C 95400	621°C (1150°F) 1 h, W. Q.	10% of surface covered with brown-red product
C 95400	890°C (1650°F) 1 h, W. Q. 649°C (1200°F) 1 h, W. Q.	A few brown-red spots

[a]W. Q., water quenched.
Source: R. E. Maersch and J. M. Cieslewicz, Heat treatments reduce dealuminification of aluminum bronzes in saline solution, Mater. Prot., 3(7), 1974.

After cleaning	Fracture appearance
Red areas over 90% of surface	Irregular dealloying 1.5 mm (0.059 in.) in depth max.
Red areas over 90% of surface	Irregular dealloying 1.3 mm (0.051 in.) in depth max.
Red spots over about 5% of surface	Plug-type dealloying 0.25 mm (0.010 in.) in depth max.
No red areas	No dealloying
Red areas over 90% of surface	Irregular dealloying 1.5 mm (0.059 in.) in depth max.
Red areas over 75% of surface	Irregular dealloying 1.0 mm (0.039 in.) in depth max.
No red areas	No dealloying
No red areas	No dealloying
Red areas over 90% of surface	Irregular dealloying 1.5 mm (0.059 in.) in depth max.
Red areas over 75% of surface	Irregular dealloying 1.3 mm (0.031 in.) in depth max.
Red areas over about 10% of surface	Plug-type dealloying 0.38 mm (0.015 in.) in depth max.
A few red spots	Shallow plug-type dealloying 0.08 mm (0.003 in.) in depth max.

Table 7. Mil-B-24480 (Ships)[a], January 1973

Cu	79.0 min.
Al	8.5-9.5
Fe	3.5-4.5
Ni	4.0-5.0
Mn	0.75-1.5
Si	0.10 max.
Pb	0.03 max.

[a]The nickel content should exceed the iron content. All castings must be temper annealed at 677°C (1250°F) for at least 6 h and air cooled.
Source: Military Standard Mil-B-24480 (ships), January 1973.

Table 8. Range in Water Analysis for Madison Aquifer in South Dakota

Species	Concentration ppm except pH temperature and depth
SO_4^{2-}	120-1426
S^{2-}	0.02-0.20
Cl^-	0.6-269
HCO_3^-	103-274
TDS	189-2710
Ca^{2+}	31-572
Mg^{2+}	16-120
pH	6.5-8.13
Temperature, °C (°F)	11-71 (52-160)
Depth, m (ft)	229-1481 (751-4859)

Source: Ref. 4.

Table 9. Weight Loss Results for the Coupon Tests in Madison
Aquifer Corrosion Rate (mpy)

Copper Alloy UNS No.	Madison Aquifer test sites								
	Edgemont			Philip			Diamond Ring		
	Days			Days			Days		
	149	308	457	148	460	608	97	317	414
C61300	0.10	0.15	0.05	0.24	0.16	0.28	0.39	0.42	0.39
C61300 welded	0.15	0.16	0.09	0.43	0.17	0.24	0.59	0.47	0.48
C95800 as cast	0.13	0.16	0.04	0.51	0.35	0.48	0.33	0.18	0.57
C95800 annealed	0.36	0.15	0.14	–	–	–	0.57	0.32	0.35
C95800 welded	0.15	0.16	0.11	0.60	0.35	0.46	0.52	0.24	0.36
Copper	0.24	0.25	0.16	0.13	0.15	0.17	3.42	4.45	3.23
C44300	0.43	0.42	0.41	0.12	0.05	0.09	0.74	0.37	0.55
C71500	0.45	0.39	0.33	0.32	0.28	0.30	1.46	1.78	0.47[a]

[a] A black coating could not be removed, resulting in lower weight loss
and corrosion rate.
Source: Ref. 4.

Table 10. Aluminum-Bronze alloys in KC1: On-Site Spool Test[a]

Copper Alloy UNS No.	Corrosion rate, mpy	Remarks
C61400	2.7	Etched with machining marks discernible
C61400 welded	2.6	Etched with machining marks discernible
C95200	1.8	Etched with machining marks discernible
C95300	1.4	Three to four incipient de-alloyed plugs
C95400	1.7	Etched with machining marks discernible
C63000	0.9	Etched with machining marks discernible

[a] 48 days' exposure in submerged combustion evaporator. Vapor from
evaporation of KC1 brine. Consists of combustion products—natural
gas plus air, water-vapor-entrained KC1, NaC1, and $MgC1_2$. Average
temperature 97°C (206°F). pH slightly acid from CO_2 in combustion gas.

Table 11. Aluminum-Bronze Alloys in $MgCl_2$: On-Site Spool Tests

1. 26 days' exposure in vapor phase of submerged combustion evap-
    orator. Vapor from evaporation of $MgCl_2$ brine. Consists of
    combustion products—natural gas with 100% excess air, water
    vapor, entrained $MgCl_2$ brine with NaCl solids; HCl contaminated.
    Temperature 104°C (220°F), pH 0.5 to 3.4.

UNS no.	Corrosion rate, mpy	Remarks
C61300	47	All samples thinned with roughened surfaces
C61300 welded	46	
C95800 annealed	43	

2. 31 day's exposure in liquid in discharge position. 24% $MgCl_2$, 4%
    CaCl, 3% NaCl, 1.5% KCl at 81°C (178°F), pH 3.5 to 4.5.

UNS no.	Corrosion rate, mpy	Remarks
C61300	0.77	Etched
C61300 welded	0.77	Etched
C95200	0.70	Incipient dealloying
C95800 annealed	0.47	Etched

3. 34 day's exposure in crystallizer. Multicomponent slurry and
    brine consisting of $MgCl_2$, NaCl, KCl, $CaCl_2$, and some MgO
    and HCl, 45% solids. Temperature 67°C max. (150°F), pH 3.5
    to 5.0.

UNS no.	Corrosion rate, mpy	Remarks
C61300	9.6	Etched
C61300 welded	10.0	Etched
C95200	8.1	Etched and incipient dealloying
C95400 annealed	10.4	Etched

Table 12. Corrosion Rate (mpy) of Aluminum-Bronze Alloys C61400 and C95300 in Various Concentrations of Sodium Hydroxide

% NaOh	Immersion condition	Room temperature		52 (125)[a]		80 (175)[a]	
		C61400	C95300	C61400	C95300	C61400	C95300
1	Constant	0.6	1.4	–	–	–	–
	Alternate	0.5	0.5	–	–	–	–
3	Constant	–	0.5	–	–	–	–
	Alternate	1.8	1.2	–	–	–	–
5	Constant	0.9	1.1	–	–	–	–
	Alternate	–	–	–	–	–	–
7	Constant	1.3	0.9	–	–	–	–
	Alternate	7.2	4.2	–	–	–	–
10	Constant	1.5	0.7	4.9	3.8	8.1	8.9
	Alternate	15.7	11.4	–	–	–	–
15	Constant	0.5	0.6	3.1	2.3	7.2	3.1
	Alternate	10.3	10.2	–	–	–	–
20	Constant	0.3	0.5	1.7	0.7	7.1	1.9
	Alternate	6.7	5.8	–	–	–	–
25	Constant	0.2	0.4	0.9	0.6	4.8	0.4
	Alternate	4.6	3.8	–	–	–	–
30	Constant	0.1	0.4	0.5	0.3	7.5	0.2
	Alternate	3.7	2.4	–	–	–	–
35	Constant	0.1	0.4	0.2	0.2	3	0.1
	Alternate	0.7	0.5	–	–	–	–
45	Constant	0.1	0.3	0.1	0.3	2.1	0.3
	Alternate	0.3	0.3	–	–	–	–

[a]Temperature, °C (°F).

Table 13. C 95400 in Sulfuric Acid, Lab Tests

Percent acid	Exposure time, days	Temperature, °C (°F)	Corrosion rate, mpy
0.5	31	32.0 (90)	1.6
1.0	31	32.0 (90)	1.6
2.0	31	32.0 (90)	1.6
5.0	31	32.0 (90)	1.4
5.0-7.0	30	38.0 (100)	2.9
10.0	31	32.0 (90)	1.1
10.0	28	22.0 (72)	3.7
35.0	28	22.0 (72)	1.7
35.0	31	30.0 (86)	0.3
35.0	34	90.0 (194)	42.0
50.0	28	22.0 (72)	2.5
50.0	31	30.0 (86)	0.3

the presence of oxygen or an oxidizing agent. Alloys such as C 95300 and C 95400 are used as pickling baskets, hooks, and chain in sulfuric acid pickling operations. Table 13 gives the results of tests on C 95400 in various concentrations of sulfuric acid, and Table 14 gives results in phosphoric acid.

Aluminum bronzes are resistant to many organic acids, such as acetic, citric, formic, and lactic. In some instances, its uses in these media are limited because of the possibility of copper pickup in the finished product. This pickup, although very low, may, as an example, discolor the product. However, a polishing step such as charcoal filtration easily removes the color contaminant. Tables 15 and 16 give corrosion rates (generally low) for aluminum bronzes in various organic acids. Vessels are fabricated from alloy C 61300 for acetic acid containment. Its good corrosion resistance, strength, and heat conductivity make it a good choice for acetic acid processing.

As stated previously in the section on coppers, architectural uses of the coppers have many successful applications. Smith [5] has reported on a development of an aluminum-bronze alloy for architectural usage. This alloy is designated copper alloy UNS C 61500 and is reported to have a film resistance 20 times that of C 11000.

Table 14. Alloys in Phosphoric Acid, Lab Tests

1. C95400 in phosphoric acid, lab tests.

Percent acid	Exposure time, days	Temperature, °C (°F)	Corrosion rate, mpy
10.0	28	22.0 (72)	2.8
10.3	1[a]	81.0 (178)	31.8
25.7	1[a]	81.0 (178)	23.0
35.0	28	22.0 (72)	1.8
35.0	34	90.0 (194)	13.8
50.0	28	22.0 (72)	1.1
50.8	1[a]	81.0 (178)	20.1
Crude	1[b]	79.0 (174)	35.7

2. C95300 in phosphoric acid. Laboratory tests at 32°C (90°F). Constant immersion, 30 days. Alternate immersion, 10 days. Corrosion rate in mpy.

	Immersion condition	
Percent acid	Constant	Alternate[c]
26.0	0.8	177.5
50.0	0.6	85.5
85.0	0.3	23.4

[a]Aerated and agitated.

[b]Some agitation.

[c]Dealloying noted to varying degrees.

Source: Data from private communications from International Nickel Co., Inc., and Ampco Metal Division.

Table 15. Corrosion Rates in Acetic Acid

1. Acetic acid, 20 days, constant immersion, room temperature.

| | Alloy (corrosion rate, mpy) | | |
Percent acid	C61400	C63200	Remarks
10	2.5	2.6	Slight etch,
20	2.5	2.6	no dealloying
30	3.8	2.6	
40	3.2	2.6	

2. In experimental pressure vessel, 25% acetic acid, vapor zone, 126°C (258°F), 26 psig.

Alloy	Exposure time, days	Corrosion rate, mpy	Remarks
C61400	56	5.6	Light etch
C61400	187	2.8	
C61400 Welded	56	5.6	Light etch
C61400 Welded	187	3.1	
C95300	56	5.8	Incipient dealloying
C95300	187	2.7	
C95400	56	5.7	Incipient dealloying
C95400	187	2.7	
C95500	56	5.5	Light etch
C95500	187	2.5	

3. Acetic acid, various concentrations, alloy C95400.

Percent acid	Exposure conditions	Corrosion rate, mpy
10	31 days at 30°C (85°F)	3.1
10	28 days at 22°C (72°F)	2.2
27[a]	15 days at room temperature	0.4
35	31 days at 30°C (85°F)	4.3
35	28 days at 22°C (72°F)	3.1
35	34 days at 89°C (102°F)	19.5
50	31 days at 30°C (85°F)	7.8
50	28 days at 22°C (72°F)	3.9
Glacial[b]	15 days at 118°C (244°F)	2.3

[a] Acid solution renewed each day. No aeration or agitation.

[b] Acid solution renewed each day. Agitated by boiling.

Source: Data from private communications from International Nickel Co., Inc., and Ampco Metal Division.

Table 16. C95400 in Citric, Formic, Lactic, and Tartaric Acids

Percent acid	Exposure time, days	Temperature, °C (°F)	Corrosion rate, mpy			
			Citric	Formic	Lactic	Tartaric
10	28	22.0 (72)	2.0	2.3	1.7	0.8
35	34	89.0 (192)	1.8	18.9	16.6	1.9
35	28	22.0 (72)	1.3	2.7	1.3	0.5
50	28	22.0 (72)	0.8	2.5	1.1	0.3

Source: Data from private communications from International Nickel Co., Inc.

COPPER-NICKEL ALLOYS

The copper-nickels are single-phase solid solution alloys, with nickel being the principal alloying ingredient. Alloys containing 10% and 30% nickel are those of importance in corrosion engineering. Table 17 lists these cast and wrought alloys. Iron, manganese, silicon, and niobium may be added. Iron enhances the impingement resistance of these alloys if it is in solid solution. The presence of iron in small microprecipitates can be detrimental to corrosion resistance. Niobium is added to the cast alloys to aid weldability.

The cupronickels are useful in waters ranging from fresh to brackish to sea. However, they find their greatest use in saltwater service. Here they are used as piping, fittings, condenser tubes and plates, and pump castings. Their biofouling resistance is excellent. General corrosion rates for C70600 and C71500 in seawater are about 1 mpy. Maximum design velocities for condenser tubes are 3.6 m/s (12 ft/s) for C70600 and 4.6 m/s (15 ft/s) for C71500. Popplewell [6] gives typical pitting rates in seawater of 0.5 to 2 mpy for C70600 and 0.5 to 10 mpy for C71500. Sulfides [7] as low as 0.007 ppm in seawater can induce pitting of both C70600 and C71500, and both alloys are highly susceptible to accelerated corrosion as the sulfide concentration exceeds 0.01 ppm. Hack and Gudos [8] inhibited this attack to some degree by the use of ferrous sulfate additions to sulfide-contaminated seawater. Efrid and Lee [9] have proposed that the formation of carbon on the surface of these alloys in the presence of sulfide and oxygen in polluted waters is responsible for many cases of pitting and accelerated corrosion.

The copper-nickels are highly resistant to stress corrosion cracking. Of all the copper alloys, they are the most resistant to stress corrosion cracking in ammonia and ammoniacal environments.

Dealloying or parting corrosion has rarely been observed in the copper-nickels. When it has been observed, a condition of high temperature has existed. Parting has been reported in C71500 used as feedwater preheaters under high oxygen conditions and at "hot spots" in condenser tubing.

C71900 is a chrome-bearing cupronickel developed for naval use. The addition of chrome allows the alloy to be strengthened by spinodal decomposition. This increases the yield strength from 137.9 MPa (20.0 ksi) for C71500 to 310.3 MPa (45.0 ksi) for C71900. The alloy also has superior resistance to impingement, with possible design velocities of 7.6 to 9.1 m/s (25 to 30 ft/s). There is, however, some sacrifice in general corrosion resistance, pitting, and crevice corrosion under stagnant or low-velocity conditions.

Although these alloys are resistant to some nonoxidizing acids, alkalies, neutral salts, and organics, they are not commonly used in these environments. Possibly the high cost of the cupronickels and the availability of lower-cost materials limits their use.

Table 17. Chemical Composition of Cupronickels

Copper Alloy UNS No.	Cu	Ni	Fe	Nb	Si	Mn	Others
C70600	Bal	9.0–11.0	1.0–1.8	—	—	1.0 max.	Pb 0.05 max. Zn 1.0 max.
C71500	Bal	29.0–33.0	0.40–0.7	—	—	1.0 max.	Pb 0.05 max. Zn 1.0 max.
C71900	Bal	29.0–32.0	0.25 max.	—	—	0.5–1.0	Cr 2.6–3.2 Zr 0.08–0.2 Ti 0.02–0.08
C96200	Bal	9.0–11.0	1.0–1.8	1.0 max.	0.25 max.	1.5 max.	Pb 0.03[a] max.
C96400	Bal	28.0–32.0	0.25–1.5	1.0 max.	0.70 max.	1.5 max.	Pb 0.03[a] max.

[a]For welding grades the lead must not exceed 0.01%
Source: Ref. 1.

CONCLUSIONS

As can be seen, the available copper alloys for corrosion service are
many and there may be several alloys indicated as resistant to a particular environment. Since this may be the case, other criteria may
be needed to select the right alloy. Fortunately, the coppers and
copper alloys offer a wide range of chemical, physical, and mechanical
properties to meet most needs. Some of these materials may be available in many cast forms, such as sand, centrifugal, permanent mold,
and continuous cast. Others may be available in wrought form as
wire, rod, pipe, tube, sheet, and plate. Many can be joined by
welding, brazing, or soldering. Some are more machinable than others.
Cost will vary with composition, form, and other factors. The selection process, then, involves many criteria for obtaining optimum service from copper and its alloys.

ACKNOWLEDGMENTS

The author thanks the Management of Ampco Metal Division, Ampco-
Pittsburgh Corporation, for permission to engage in this project. He
is particularly indebted to C. W. Dralle for encouragement in this task;
to R. J. Severson, R. J. Cox, and P. A. Tully for their helpful
criticism, comments, and suggestions; and to Sharon Southart for her
patience in typing the several drafts of this contribution.

REFERENCES

1. *Application Data Sheet: Standard Designations for Copper and
    Copper Alloys*, Copper Development Association, Inc., New York.
2. W. S. Lyman and A. Cohen, Service Experience with copper
    plumbing tube, *Mater. Prot. Perform.*, 48-52, February 1972.
3. A. Cohen, Copper for hot and cold potable water systems,
    Heat./Piping/Air Cond., May 1978.
4. S. M. Howard, D. D. Carda, and J. M. Cieslewicz, Corrosion in
    geothermal waters of Western South Dakota, Paper No. 208, NACE
    Corrosion/80, Chicago, March 3-7, 1980.
5. Richard D. Smith, Staining and tarnishing of copper alloys in
    architectural applications, Paper No. 93, Houston, NACE
    Corrosion/78, March 6-10, 1978.
6. James W. Popplewell, Marine corrosion of copper alloys: an over-
    view, Paper No. 21, NACE Corrosion/78, Houston, March 6-10,
    1978.
7. John P. Gudas and Harvey P. Hack, Parametric evaluation of
    susceptibility of copper nickel alloys to sulfide induced corrosion,
    Paper No. 22, NACE Corrosion/78, Houston, March 6-10, 1978.

8. Harvey P. Hack and John P. Gudas, Inhibition of sulfide induced corrosion of copper-nickel alloys with ferrous sulfate, Paper No. 23, NACE Corrosion/78, Houston, March 6-10, 1978.

9. K. D. Efrid and T. S. Lee, The carbon mechanism for aqueous sulfide corrosion of copper base alloys, Paper No. 24, NACE Corrosion/78, Houston, March 6-10, 1978.

3.5
Aluminum Alloys

ERNEST H. HOLLINGSWORTH and HAROLD Y. HUNSICKER / Alcoa
Laboratories, Alcoa Center, Pennsylvania

INTRODUCTION

In terms of production and consumption, aluminum is second only to
iron as the most important metal of commerce in the United States today.

Its position relative to most other metals can be expected to increase as the resources of these metals are depleted further. Aluminum is the third most abundant metal in the crust of the earth, almost twice as abundant as the next metal, iron. Furthermore, resources from which aluminum can be extracted are literally inexhaustible, and even though its extraction from newer resources may be more expensive than from bauxite, the resource used presently, it will still be economically feasible. The greater cost of extraction will also be offset significantly be greater recycling of aluminum (which saves 95% of the energy required for extraction), and by new smelting processes requiring less energy [1].

In addition to low cost, aluminum and its alloys provide a high ratio of strength to weight, and they are fabricated and joined readily by most of the methods commonly used. They have a high resistance to corrosion by most atmospheres and waters and by many chemicals and other materials. Their salts are not damaging to the ecology; they are nontoxic, permitting applications with foods, beverages, and pharmaceuticals; and they are white or colorless, permitting applications with chemicals and other materials without discoloration. Other characteristics of aluminum and its alloys important for certain applications are high electrical conductivity, high thermal conductivity, high reflectivity, and noncatalytic action. Aluminum and its alloys are also nonmagnetic.

ALUMINUM WROUGHT ALLOYS AND THEIR RESISTANCE TO GENERAL CORROSION

The nominal chemical compositions of representative aluminum wrought alloys are given in Table 1. Typical tensile properties of these alloys in tempers representative of their most common usage are given in Tables 2 and 3.

Wrought alloys are of two types: non-heat-treatable of the 1XXX, 3XXX, 4XXX, and 5XXX series and heat treatable of the 2XXX, 6XXX, and 7XXX series. In the non-heat-treatable type, strengthening is produced by strain hardening, which may be augmented by solid solution and dispersion hardening. In the heat-treatable type, strengthening is produced by (1) a solution heat treatment at 460 to 565°C (860 to 1050°F) to dissolve soluble alloying elements, (2) quenching to retain them in solid solution, and (3) a precipitation or aging treatment, either naturally at ambient temperature, or more commonly, artificially at 115 to 195°C (240 to 380°F), to precipitate these elements in an optimum size and distribution.

Strengthened tempers of non-heat-treatable alloys are designated by an "H" following the alloy designation, and of heat-treatable alloys, by a "T"; suffix digits designate the specific treatment (e.g., 1100-H14 and 7075-T651). In both cases, the annealed temper representing a condition of maximum softness, is designated by an "O."

Table 1. Nominal Chemical Compositions of Representative Aluminum Wrought Alloys

Alloy	Si	Cu	Mn	Mg	Cr	Zn	Ti	V	Zr
					Percent of alloying elements				
				Non-heat-treatable alloys					
1060	99.60% minimum aluminum								
1100	99.00% minimum aluminum								
1350	99.50% minimum aluminum								
3003		0.12	1.2						
3004			1.2	1.0					
5052				2.5	0.25				
5454			0.8	2.7	0.12				
5456			0.8	5.1	0.12				
5083			0.7	4.4	0.15				
5086			0.45	4.0	0.15				
7072[a]						1.0			
				Heat-treatable alloys					
2014	0.8	4.4	0.8	0.50					
2219		6.3	0.30				0.06	0.10	0.18
2024		4.4	0.6						
6061	0.6	0.28		1.0	0.20				
6063	0.4			0.7					
7005			0.45	1.4	0.13	4.5	0.04		0.14
7050		2.3		2.2		6.2			
7075		1.6		2.5	0.23	5.6			

[a]Cladding for Alclad products.

Table 2. Typical Tensile Properties of Representative Non-Heat-Treatable Aluminum Wrought Alloys in Various Tempers[a]

Alloy and temper	Strength, MPa		Percent Elongation	
	Ultimate	Yield	In 50 mm[b]	In 5D[c]
1060 -O	70	30	43	42
-H12	85	75	16	18
-H14	100	90	12	13
-H16	115	105	8	
-H18	130	125	6	
1100 -O	90	35	35	37
-H14	125	125	9	14
-H18	165	150	5	9
3003 -O	110	40	30	
-H14	150	145	8	
-H18	200	185	4	
3004 -O	180	70	20	22
-H34	240	200	9	10
-H38	285	250	5	5
5052 -O	195	90	25	27
-H34	260	215	10	12
-H38	290	255	7	7
5454 -O	250	115	22	
-H32	275	205	10	
-H34	305	240	10	
-H111	260	180	14	
-H112	250	125	18	

Alloy	Temper				
5456	-O	310	160	22	
	-H111	325	230	16	
	-H112	310	165	20	
	-H116, H321	350	255	14	
5083	-O	290	145	20	
	-H116, H321	315	230	14	
5086	-O	260	115		22
	-H116, H32	290	205		12
	-H34	325	255		10
	-H112	270	130		14

[a] Averages for various sizes, product forms, and methods of manufacture; not to be specified as engineering requirements or used for design purposes.

[b] 1.60-mm-thick specimen.

[c] 12.5-mm-diameter specimen.

Table 3. Typical Tensile Properties of Representative Heat-Treatable Aluminum Wrought Alloys in Various Tempers[a]

Alloy and temper	Strength, MPa		Percent Elongation	
	Ultimate	Yield	In 50 mm[b]	In 5D[c]
2014 -O	185	95		16
-T4, T451	425	290		18
-T6, T651	485	415		11
2219 -O	170	75	18	
-T37	395	315	11	
-T87	475	395	10	
2024 -O	185	75	20	20
-T4, T351	470	325	20	17
-T851	480	450	6	
-T86	515	490	6	7
6061 -O	125	55	25	27
-T4, T451	240	145	22	22
-T6, T651	310	275	12	15

Alloy	Temper				
6063	-O	90	50		
	-T5	185	145	12	
	-T6	240	215	12	
	-T83	255	240	9	
7005	-O	195	85		20
	-T63, T6351	370	315		11
7050	-T76, T7651	540	485		10
	-T736, T73651	510	455		10
7075	-O	230	105	17	14
	-T6, T651	570	505	11	9
	-T76, T7651	535	470		10
	-T73, T7351	500	435		11

[a] Averages for various sizes, product forms, and methods of manufacture; not to be specified as engineering requirements or used for design purposes.

[b] 1.60-mm-thick specimen.

[c] 12.5-mm-diameter specimen.

All non-heat-treatable alloys have a high resistance to general
corrosion. Therefore, selection is usually based on other considerations. Aluminums of the 1XXX series representing unalloyed aluminum
have a relatively low strength. Alloys of the 3XXX series (Al-Mn,
Al-Mn-Mg) have the same desirable chracteristics as those of the 1XXX
series and somewhat higher strength. Practically all the manganese in
these alloys is precipitated as finely divided phases (intermetallic compounds), but corrosion resistance is not impaired because the negligible difference in electrode potential between the phases and the aluminum matrix in most environments does not create a galvanic cell.
Magnesium added to some alloys in this series provides additional
strength through solid solution hardening, but the amount is low
enough that the alloys still behave more like those with manganese
alone than like the stronger Al-Mg alloys of the 5XXX series. Alloys
of the 4XXX series (Al-Si) are low-strength alloys used for brazing
and welding products and for a cladding in architectural products
that develops a gray appearance upon anodizing; the silicon, most of
which is present in elemental form as a second-phase constituent, has
little effect on corrosion.

Alloys of the 5XXX series (Al-Mg) are the strongest non-heat-treatable aluminum alloys, and in most products, they are more economical
than alloys of the 1XXX and 3XXX series in terms of strength per unit
cost. Magnesium is one of the most soluble elements in aluminum, and
when dissolved at an elevated temperature, it is largely retained in
solution at lower temperatures, even though its equilibrium solubility
is greatly exceeded. It produces considerable solid solution hardening; additional strength is produced by strain hardening. Alloys of
this 5XXX series have not only the same high resistance to general
corrosion of other non-heat-treatable alloys in most environments, but
in slightly alkaline ones, a better resistance than that of any other
aluminum alloy. They are used widely because of their high as-welded
strength when welded with a compatible 5XXX series filler wire, reflecting the retention of magnesium in solid solution.

Among heat-treatable alloys, those of the 6XXX series, which are
moderate-strength alloys based on the quasi-binary $Al-Mg_2Si$ (magnesium silicide) system, provide a high resistance to general corrosion
equal to or approaching that of non-heat-treatable alloys. A high resistance to general corrosion is also provided by heat-treatable alloys
of the 7XXX series (Al-Zn-Mg) that do not contain copper as an alloying addition.

All other heat-treatable wrought alloys have a significantly lower
resistance to general corrosion. These include all alloys of the 2XXX
series (Al-Cu, Al-Cu-Mg, Al-Cu-Si-Mg) and those of the 7XXX series
(Al-Zn-Mg-Cu) that contain copper as a major alloying element. As
described later, the lower resistance is caused by the presence of
copper in the alloys, which are designed primarily for aeronautical
applications, where strength is a requisite and where protective
measures, wherever needed, are justified.

ALUMINUM CASTING ALLOYS AND THEIR RESISTANCE TO GENERAL CORROSION

The nominal chemical compositions of representative aluminum casting alloys are given in Table 4. Typical tensile properties of these alloys in tempers representative of their most common usage are given in Table 5.

Aluminum casting alloys are also of two types: heat treatable, corresponding to the same type of wrought alloys where strengthening is produced by dissolution of soluble alloying elements and their subsequent precipitation, and non-heat-treatable, where strengthening is produced primarily by constituent (intermetallic compounds) of insoluble or undissolved alloying elements. As with wrought alloys, tempers of heat-treatable casting alloys are designated by a "T" following the alloy designation. Those of non-heat-treatable alloys are designated by an "F". Alloys of the heat-treatable type are usually thermally treated subsequent to their casting, but in a few cases, where a significant amount of alloying elements is retained in solution during casting, they may not be given a solution heat treatment after casting; thus they may be used in both the "F" and fully strengthened "T" tempers.

Aluminum casting alloys are produced by all casting processes of which die, permanent mold, and sand casting account for the greatest production. Unlike wrought alloys, their selection involves consideration of casting characteristics as well as of properties.

As with wrought alloys, copper is the alloying element most deleterious to general corrosion. Alloys such as 356.0, A356.0, B443.0, 513.0, and 514.0 that do not contain copper as an alloying element have a high resistance to general corrosion comparable to that of non-heat-treatable wrought alloys. In other alloys, resistance becomes progressively less the greater the copper content. More so than with wrought alloys, a lower resistance is compensated by the use of thicker sections usually necessitated by requirements of the casting process.

CHEMICAL NATURE OF ALUMINUM: PASSIVITY

Aluminum is an active metal whose resistance to corrosion depends on the passivity produced by a protective oxide film. In aqueous solutions, the thermodynamic conditions under which the film develops are expressed commonly by the potential-pH diagram according to Pourbaix in Fig. 1. As this diagram shows, aluminum is passive only in the pH range of about 4 to 9. The limits of passivity depend on the temperature, the form of oxide present, and the low dissolution of aluminum that must be assumed for inertness (since this value theoretically cannot be zero for any metal). The various forms of aluminum oxide all exhibit minimum solubility at about pH 5.

Table 4. Nominal Chemical Compositions of Representative Aluminum Casting Alloys

| Alloy | Percent of alloying elements | | | | |
	Si	Cu	Mg	Ni	Zn
	Alloys not normally heat treated				
360.0	9.5		0.5		
380.0	8.5	3.5			
443.0	5.3				
514.0			4.0		
710.0		0.5	0.7		6.5
	Alloys normally heat treated				
295.0	0.8	4.5			
336.0	12.0	1.0	1.0	2.5	
355.0	5.0	1.3	0.5		
356.0	7.0		0.3		
357.0	7.0		0.5		

The protective oxide film formed in water and atmospheres at ambient temperature is only a few nanometers thick and structureless. At higher temperatures, thicker films are formed; these may consist of a thin structureless barrier layer next to the aluminum and a thicker crystalline layer next to the barrier layer. Relatively thick, highly protective films of boehmite, aluminum oxide hydroxide AlOOH, are formed in water near its boiling point, especially if it is made slightly alkaline, and thicker, more protective films are formed in water or steam at still higher temperatures.

Because the form of aluminum oxide produced depends on conditions, its identification is sometimes useful in establishing the cause of corrosion. At lower temperatures, the predominant form produced by corrosion is bayerite, aluminum trihydroxide $Al(OH)_3$, while, at higher temperatures, it is boehmite. During the complex course of the aging of aluminum hydroxide, which is first formed during corrosion in an amorphous form, still another aluminum trihydroxide, gibbsite or hydrargillite, may also be formed, especially if ions of the alkali metals are present.

Beginning at a temperature of about 230°C (445°F), a protective film no longer develops in water or steam, and reaction progresses rapidly until eventually all the aluminum exposed in these media is converted into oxide. Special alloys containing iron and nickel have

Table 5. Typical Tensile Properties of Representative Aluminum Casting Alloys in Various Tempers[a]

Alloy and temper	Type casting	Strength, MPa		Percent elongation in 50 mm[b]
		Ultimate	Yield	
295.O -T6	Sand	250	165	5
336.O -T5	Permanent mold	250	195	1
355.O -T6	Sand	240	170	3
-T6	Permanent mold	375	240	4
-T61	Sand	280	250	3
-T62	Permanent mold	400	360	1.5
356.O -T6	Sand	230	165	3.5
-T6	Permanent mold	255	185	5
-T7	Sand	235	205	2
-T7	Permanent mold	220	165	6
357.O -T6	Sand	345	295	2
-T6	Permanent mold	360	295	5
-T7	Sand	275	235	3
-T7	Permanent mold	260	205	5
360.O -F	Pressure die	325	170	3
380.O -F	Pressure die	330	165	3
443.O -F	Pressure mold	160	60	10
514.O -F	Sand	170	85	9
710-O -F	Sand	240	170	5

[a] Averages for separate cast test bars; not to be specified as engineering requirements or used for design purposes.

[b] 1.60-mm-thick specimen.

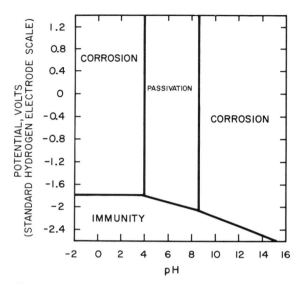

Figure 1. Potential-pH diagram according to Pourbaix for aluminum at
25°C (77°F) with an oxide film of hydrargillite. (From Ref. 2.)

been developed to retard this reaction, and these alloys may be used
up to a temperature of about 360°C (680°F) without excessive attack [3].

Reflecting its amphoteric nature, and as illustrated in Fig. 1, aluminum corrodes under both acidic and alkaline conditions, in the first
case to yield trivalent Al^{3+} ions, and in the second case to yield AlO_2^-
(aluminate) ions. There are a few exceptions, either where the oxide
film is not soluble in an acidic or alkaline solution, or where it is maintained by the oxidizing nature of the solution. Two exceptions, acetic
acid and sodium disilicate, are shown in Fig. 2. Ammonium hydroxide
above 30% concentration by weight, nitric acid above 80% concentration
by weight, and sulfuric acid of 98 to 100% concentration are also exceptions.

ELECTROCHEMICAL ASPECTS OF THE CORROSION OF ALUMINUM

As with other metals, corrosion of aluminum is controlled by the electrochemical relation of an anodic reaction (oxidation) which leads to metallic
dissolution and a cathodic reaction (reduction) of environmental species,
which does not. The relation where the anodic reaction occurs on
aluminum, and thus leads to its corrosion, is shown in Fig. 3. The
anodic polarization curve shown is typical for aluminum and its alloys

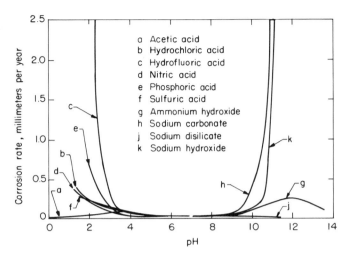

Figure 2. Relation to pH of the corrosivities toward 1100-H14 alloy sheet of various chemical solutions.

when they are polarized anodically in an electrolyte free of a readily available cathodic reactant (e.g., in a deaerated electrolyte), while the polarization curves for the cathodic reactions are schematic only. The corrosion current developed by the two reactions (which determines the rate of corrosion of the aluminum) is indicated by the intersection of the anodic polarization curve for aluminum with one of the cathodic polarization curves. Aluminum and its alloys, of course, do not corrode in the reverse case, where only a cathodic reaction occurs on them, either because of coupling to a more anodic metal or because of polarization cathodically by means of a current impressed on them from a rectifier or other external source.

Aluminum may corrode because of defects in its protective oxide film. Resistance to corrosion improves considerably as purity is increased, but the oxide film on even the purest aluminum still contains a few defects where minute corrosion can develop. In less pure aluminums of the 1XXX series and in aluminum alloys, the presence of second phases becomes the more important factor. These phases are present as an insoluble constituent of intermetallic compounds produced primarily from iron, silicon, and other impurities, and a smaller precipitate of compounds produced primarily from soluble alloying elements. Most of the phases are cathodic to aluminum, but a few are anodic. In either case, they produce galvanic cells because of the potential difference between them and the aluminum matrix.

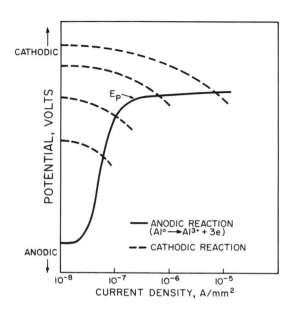

Figure 3. Typical anodic polarization curve (solid line) for an aluminum alloy in an electrolyte free of a readily available cathodic reactant
(commonly oxygen); E_p is the pitting potential of the alloy. The intersection of this curve with one of the cathodic polarization curves
(schematic) determines the corrosion current on the alloy. [From E.
H. Hollingsworth and H. Y. Hunsicker, Corrosion resistance of aluminum alloys, in *Metals Handbook; Properties and Selection: Non-
Ferrous Alloys and Pure Metals*, Vol. 2, 9th ed. (D. Benjamin, ed.),
American Society for Metals, Metals Park, Ohio, 1979, p. 205.]

Pitting Corrosion

Any corrosion of aluminum in the pH range where it is passive may be
of the pitting type, as is the case with other passive metals. Corrosion of this type is produced most commonly by halide ions, of which
chloride is the one most frequently encountered in service.

Whether pitting develops in aluminum depends on whether it is
polarized to its pitting potential [4]. As shown in Fig. 3, this potential corresponds to the plateau in the anodic polarization curve for
aluminum. Pitting develops in aluminum only when it is polarized to,
or above, its pitting potential; when it is not, pitting does not develop.
Aluminum may develop pitting in aerated solutions of halides simply
because the reactions occurring on its cathodic regions are sufficient
to polarize it to its pitting potential, as reflected by the development

of an electrode potential upon immersion in the solutions which is equal to the pitting potential. Conversely, aluminum does not develop pitting in solutions of most other salts because the reactions occurring on its cathodic regions in these solutions are not sufficient to polarize it to its pitting potential.

In all cases, the development of pitting can be prevented by removal of the reducible species required for a cathodic reaction. In neutral solutions, this species is usually oxygen. Thus its removal by deaeration will prevent the development of pitting in aluminum even in most halide solutions because, in its absence, the cathodic reactions are not sufficient to polarize aluminum to its pitting potential.

Metallurgical structure has little effect on the pitting potential of aluminum nor do second phases in the amounts present in its alloys. Severe cold work makes the potential more anodic by a few millivolts, and this change, although small, is sufficient to affect the degree to which pitting develops (e.g., in a greater degree of pitting on machined or sheared edges) [5].

Pitting of aluminum diminishes with either increasing acidity or alkalinity beyond its passive range and corrosion attack becomes more nearly uniform.

Galvanic Relations

Table 6 is a galvanic series of aluminum alloys and other metals representative of their electrochemical behavior in seawater and in most natural waters and atmospheres. Figure 4 shows the effect of alloying elements in determining the position of aluminum alloys in the series; these elements, primarily copper and zinc, affect electrode potential only when they are in solid solution.

As evident in Table 6, aluminum (and its alloys) becomes the anode in galvanic cells with most metals and corrodes sacrificially to protect them. Only magnesium and zinc are more anodic and corrode to protect aluminum. Because they have nearly the same electrode potential, neither aluminum nor cadmium corrodes sacrificially in a galvanic cell.

The degree to which aluminum corrodes when coupled to a more cathodic metal depends on the degree to which it is polarized in the galvanic cell. It is especially important to avoid contact with a more cathodic metal where aluminum is polarized to its pitting potential because, as evident in Fig. 3, a small increase in potential produces a large increase in corrosion current. In particular, contact with copper and its alloys should be avoided because of the low degree of polarization of these metals. In atmospheric and other mild environments, aluminum may be used in contact with chromium and stainless steel with only slight acceleration of corrosion; in these environments, the two metals polarize highly so that the additional corrosion current impressed onto aluminum in a galvanic cell with them is small.

Table 6. Electrode Potentials of Representative Aluminum Alloys and Other Metals[a]

Aluminum alloy or other metal[b]	Potential, V
Chromium	+0.18 to -0.40
Nickel	-0.07
Silver	-0.08
Stainless steel (300 series)	-0.09
Copper	-0.20
Tin	-0.49
Lead	-0.55
Mild carbon steel	-0.58
2219-T3, T4	-0.64[c]
2024-T3, T4	-0.69[c]
295.0-T4 (SC or PM)	-0.70
295.0-T6 (SC or PM)	-0.71
2014-T6, 355.0-T4 (SC or PM)	-0.78
355.0-T6 (SC or PM)	-0.79
2219-T6, 6061-T4	-0.80
2024-T6	-0.81
2219-T8, 2024-T8, 356.0-T6 (SC or PM), 443,O-F (PM), cadmium	-0.82
1100, 3003, 6061-T6, 6063-T6, 7075-T6,[c] 443.0-F (SC)	-0.83
1060, 1350, 3004, 7050-T73,[c] 7075-T73[c]	-0.84
5052, 5086	-0.85
5454	-0.86
5456, 5083	-0.87
7072	-0.96
Zinc	-1.10
Magnesium	-1.73

[a]Measured in an aqueous solution of 53 g of NaCl and 3 g of H_2O_2 per liter at 25°C; 0.1 N calomel reference electrode.

[b]The potential of an aluminum alloy is the same in all tempers wherever the temper is not designated.

[c]The potential varies ±0.01 to 0.02 v with quenching rate.

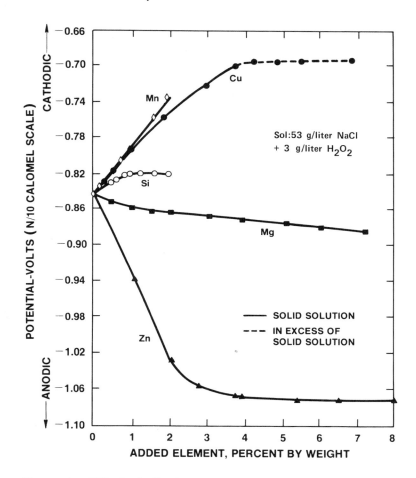

Figure 4. Effect of alloying elements on the electrode potential of aluminum.

To minimize corrosion of aluminum in contact with other metals, the ratio of the exposed area of aluminum to that of the more cathodic metal should be kept as high as possible (since such a ratio reduces the current density on the aluminum). Paints and other coatings for this purpose may be applied to both the aluminum and the cathodic metal, or to the cathodic metal alone. But they should never be applied only to the aluminum because of the difficulty in applying and maintaining them free of defects.

Galvanic corrosion of aluminum by more cathodic metals in solutions of nonhalide salts is usually less than in solutions of halide ones because the aluminum is less likely to be polarized to its pitting potential. In any solution, galvanic corrosion is reduced by removal of the cathodic reactant. Thus the corrosion rate of aluminum coupled to copper in seawater is reduced greatly when the seawater is deaerated. In closed multimetallic systems, the rate, even though it may be high initially, decreases to a low value whenever the cathodic reactant is depleted. Galvanic corrosion is also low where the electrical resistivity is low, as in high purity water.

Some semiconductors, such as graphite and magnetite, are cathodic to aluminum, and in contact with them, aluminum corrodes sacrifically.

Reduction of Ions of Other Metals by Aluminum

Aluminum reduces ions of many metals, of which copper, cobalt, lead, mercury, nickel, and tin are the ones encountered most commonly. Not only is a chemical equivalent of aluminum oxidized for each equivalent of ion reduced, but galvanic cells are also set up because the metal reduced from the ions plates onto the aluminum.

Reducible metallic ions are of most concern in acidic solutions; in alkaline solutions they are of less concern because of their greatly reduced solubilities.

Ions of copper are encountered most frequently in applications of aluminum; a common source in weathering is runoff from copper products upstream (e.g., rainwater entering aluminum gutters from roofs with copper flashing). The threshold concentration generally accepted for reduction of copper ion by aluminum is 0.02 ppm. The resistance to corrosion of aluminum alloys is impaired by the presence of more than about 0.25% copper as an alloying element because the alloys reduce the copper ions in any corrosion product from them.

Aluminum reduces ions of ferric iron, but these ions are not encountered frequently because of their reaction with oxygen and water to form insoluble oxides and hydroxides of iron. The most anodic aluminum alloys reduce ions of ferrous iron, but these ions also are not encountered frequently because they exist only in deaerated or other solutions free of oxidizing agents [6].

Mercury, whether reduced from its ions or introduced directly in the metallic form, can be severely damaging to aluminum whenever stress is present, as shown in Fig. 5. The effect is caused by amalgamation of mercury with aluminum, which, once initiated, progresses for long periods because the aluminum in the amalgam oxidizes immediately in the presence of water so that the mercury is regenerated continuously. A "safe" level of mercury is difficult to determine because of the difficulty in initiating attack of aluminum by mercury. Any concentration in a solution of more than a few parts per billion, however, should be viewed with suspicion—and in an atmosphere where attack initiates less readily, any concentration exceeding that allowed

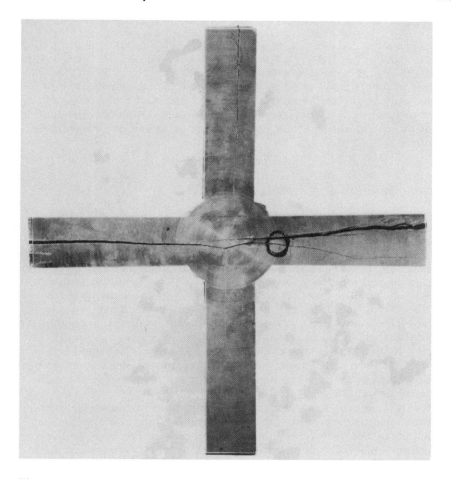

Figure 5. Cracking of a welded "cruciform" specimen of 5083-H131 alloy plate by mercury brought into contact with the plate only within the circled region. [From E. H. Hollingsworth and H. Y. Hunsicker, Corrosion resistance of aluminum and aluminum alloys, in *Metals Handbook; Properties and Selection: NonFerrous Alloys and Pure Metals*, Vol. 2, 9th ed. (D. Benjamin, ed.), American Society for Metals, Metals Park, Ohio, 1979, p. 212.]

by Environmental Protection Agency (EPA) regulations. No amount of metallic mercury should ever be allowed to come into contact with aluminum [7].

Alclad Products

Alclad products consist of a core alloy and a more anodic cladding alloy, usually representing 10% or less of the total thickness, metallurgically bonded to one or both surfaces of the core alloy. Because of the cathodic protection afforded the core by the cladding, any corrosion progresses only to the cladding/core interface and then spreads laterally. Alclad products are used extensively where perforation of a product cannot be tolerated, especially in thinner ones where it is most likely to develop.

STRESS CORROSION CRACKING

Only aluminum alloys with appreciable amounts of soluble alloying elements of copper, magnesium, silicon,and zinc are subject to stress corrosion cracking [8]. In aluminum alloys, this cracking characteristically is intergranular. According to the electrochemical theory, the requisite for the development of stress corrosion cracking is a condition along grain and other boundaries that makes them anodic to the rest of the microstructure so that corrosion propagates selectively along them [9]. This condition may be produced whenever dissolved alloying elements precipitate along grain boundaries and deplete the regions adjacent to them of these elements. Resistance to stress corrosion cracking in aluminum alloys subject to this process is achieved by metallurgical treatment either to prevent or minimize decomposition of solid solution, or more commonly, to produce it uniformly throughout the microstructure.

Research more than 40 years ago demonstrated inadequacies in the electrochemical theory of stress corrosion cracking, and the complex interaction of the factors related to the stress corrosion cracking of aluminum alloys is not yet fully understood [10]. Research in recent years has been directed toward establishing a possible effect of atomic hydrogen produced by the reaction of aluminum with water or its vapor, but the results are inconclusive [11]. Despite this and other more recent research, there is still general agreement that one factor in the stress corrosion cracking of aluminum alloys in aqueous solutions is electrochemical in nature. Strong evidence is provided by the fact that the process can be retarded greatly if not eliminated completely, by cathodic protection.

Metallurigical treatment that improves resistance to stress corrosion cracking also improves resistance to intergranular corrosion, but the optimum treatments for the two processes usually are not the same. Thus resistance to intergranular attack usually is not a satisfactory criterion for predicting resistance to stress corrosion cracking. Resistance to stress corrosion cracking also cannot be predicted reliably from an examination of microstructure even though progressive changes resulting from decomposition of solid solution may be observed as an alloy is treated to improve its resistance to stress corrosion cracking.

Factors Affecting Resistance to Stress Corrosion Cracking

As with other metals, stress corrosion cracking of an aluminum alloy in a susceptible temper is determined by the magnitude and duration of a tensile stress acting on its surface. In most wrought alloys, it is also determined by the direction of stressing, as shown in Fig. 6. The effect of stressing in different directions is caused by the highly directional grain structure shown in Fig. 7 that is typical of many wrought products [12]. As evident in these figures, resistance to stress corrosion cracking is highest for stressing parallel to the direction of elongation of grains (longitudinal) and lowest for stressing across the minimum thickness of grains (short transverse). Thus, in wrought alloys with an elongated grain structure, and in products thick enough for stressing in all directions, resistance to stress corrosion cracking in the short transverse direction may be the controlling factor in applications of the alloys.

Direction of stressing usually has little effect on casting alloys because the grains in these alloys are equiaxed.

The "threshold" values indicated by curves of the type in Fig. 6 provide valid comparisons of the relative susceptibilities of aluminum

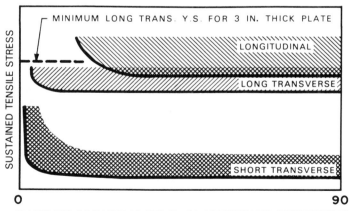

Figure 6. Effect of direction of testing on the resistance to stress corrosion cracking of 7075-T651 alloy plate. The bands indicate the range of values for the numerous specimens tested in the three principal directions. (From Ref. 12.)

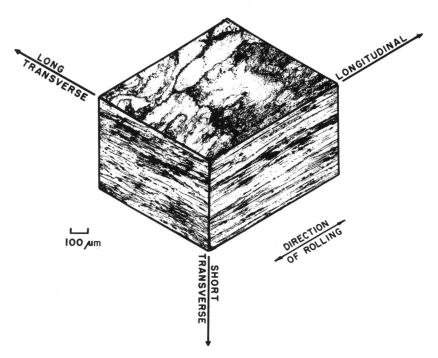

Figure 7. Composite micrograph of the three-dimensional grain structure of 38-mm-thick 7075-T6 alloy plate.

alloys to stress corrosion cracking only under the conditions of a
specific environment and test procedure. They may not be valid,
therefore, for other conditions.

Water or water vapor is a requisite for the stress corrosion cracking
of aluminum alloys, and in its absence, cracking does not occur. Among
species that accelerate cracking further, halides have the greatest
effect [10].

Alloys and Tempers

Wrought alloys of the 2XXX, 5XXX, 6XXX, and 7XXX series contain
soluble alloying elements in amounts sufficient to make them subject to
stress corrosion cracking. Stress corrosion cracking can be produced
in 6XXX series alloys by special treatment, but no case of cracking in
commercial alloys has been reported. For alloys of the other three
series, tempers providing a high to very high resistance to stress
corrosion cracking (Table 7) have been developed.

Table 7. Relative Stress Corrosion Ratings

Rating	Explanation
Very high	Stress corrosion cracking highly unlikely
High	Stress corrosion cracking unlikely at sustained tensile stresses not exceeding 50% of the minimum yield strength specified for the alloy
Intermediate	Stress corrosion cracking unlikely at sustained tensile stresses not exceeding 25% of the minimum yield strength specifed for the alloy
Low	Stress corrosion cracking likely at any sustained tensile stress in the most susceptible direction of an alloy

In naturally aged T3 and T4 tempers, thicker products of alloys of the heat-treatable 2XXX series have a low resistance in the short transverse direction, reflecting localized decomposition of solid solution during quenching. Heating all products of the alloys in these tempers for short periods may impair resistance. Heating for longer periods improves resistance with the artificial aging treatment for the T6 temper producing an intermediate to very high resistance in the short transverse direction, and that for the T8 temper, a high to very high resistance.

The same sequence holds for heat-treatable alloys of the 7XXX series that contain copper as an alloying element. Resistance improves progressively with the severity of artificial aging in the order of T6, T76, and T73 tempers, with the last temper being practically immune to stress corrosion cracking. The improvement is accompanied by a decrease in strength in the same order. Alloys of this series without copper do not respond as well to artificial aging and resistance is produced only by special heat-treating practices.

Although not heat treatable in the usual sense, alloys of the 5XXX series in most tempers do contain magnesium in a supersaturated solid solution which may decompose selectively along boundaries and result in susceptibility to stress corrosion cracking. Except under rare conditions, significant decomposition does not develop in alloys with less than about 3% magnesium. Decomposition in alloys with more magnesium depends on the magnesium content, the period and temperature of exposure, and especially on the degree of a strain hardening, as shown in Fig. 8. In strain-hardened tempers subject to such decomposition selectively along boundaries, these alloys may develop susceptibility

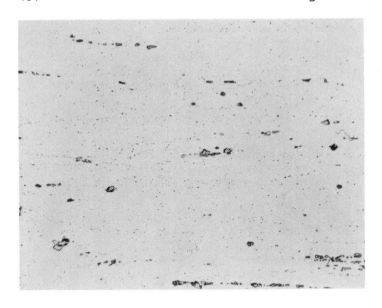

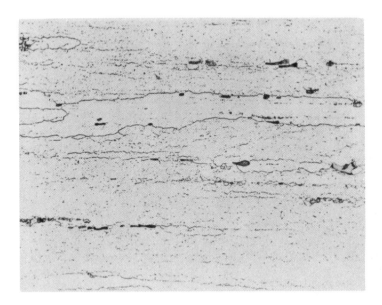

Figure 8. Microstructure of annealed 5083 alloy plate that was (a) first stretched 1% and (b) then heated 40 days at 121°C (250°F). (Original magnification 500X.)

to stress corrosion cracking after long periods at ambient temperature or shorter ones at elevated temperatures. In annealed and other strain-hardened tempers where appreciable decomposition of solid solution has been produced uniformly throughout the microstructure, the alloys do not develop susceptibility after long periods at ambient or slightly elevated temperatures.

Stress corrosion cracking of casting alloys in service has occurred only in those of the Al-Mg and Al-Zn-Mg types with large amounts of soluble alloying elements to produce the highest strength.

EXFOLIATION CORROSION

In certain tempers, wrought products of some aluminum alloys are subject to exfoliation corrosion, which causes a leafing or delamination of the products, as shown in Fig. 9. This type of corrosion developes in products that have a pronounced directional structure similar to that in Fig. 9, in which grains are elongated highly and thin relative to their length and width. Alloys of the 2XXX, 5XXX, and 7XXX series are those most prone to exfoliation corrosion. In alloys of this series, both exfoliation corrosion and stress corrosion cracking are associated with decomposition of solid solution selectively along boundaries. Thus metallurigcal treatment that improves resistance to stress corrosion cracking also improves that to exfoliation corrosion, but resistance to exfoliation corrosion usually is achieved first. As with stress corrosion cracking, resistance to intergranular attack usually is not a satisfactory criterion for predicting resistance to exfoliation corrosion.

Exfoliation corrosion is infequent and less severe in wrought alloys of the non-heat-treatable type.

OTHER TYPES OF CORROSION

The fatigue strength of aluminum alloys is lower in a corrosive medium than in a noncorrosive one, especially under low stresses for longer periods. As shown in Fig. 10, the effect is less for the more-corrosion-resistant alloys (5XXX and 6XXX) series) than for the less-corrosion-resistant ones (2XXX and 7XXX series). In contrast to stress corrosion cracking, corrosion fatigue is not affected much by a directional grain structure and the fracture is predominantly transgranular.

Reflecting their greater hardness, high-strength aluminum alloys have the highest resistance to cavitation and other types of erosion in a noncorrosive medium where the action is mechanical only. In a corrosive medium such as seawater, however, more-corrosion-resistant, lower-strength alloys may have a higher resistance because, as with fatigue, the effect of such a medium is less for those alloys.

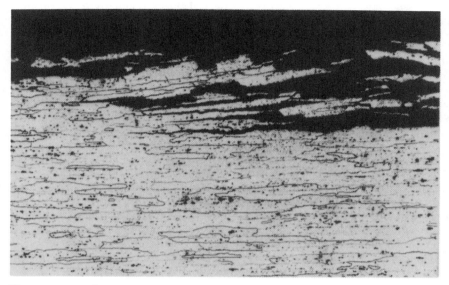

Figure 9. Exfoliation corrosion in 7178-T651 alloy plate exposed in a seacoast environment. (Original magnification 100X.) (From Ref. 13.)

As with other metals, although to a lesser degree than with some, corrosion of aluminum alloys may be promoted by conditions that develop concentration cells, most commonly those where a galvanic current is produced by a difference in the concentration of oxygen. These conditions are developed most frequently by crevices in wet mechanical joints, but they may also be developed by contact with wet porous or fibrous materials even though the materials themselves may be inert. In milder environments, the corrosion produced by such conditions usually is slight. In more corrosive environments, these conditions should be avoided insofar as possible, and where not possible, protective measures such as painting and sealing should be used.

Corrosion of aluminum alloys may also be promoted by contact with wet materials (e.g., certain types of insulation) that produce leaches more alkaline than a pH of about 9. Alkalinity sufficient to cause corrosion may also be produced by the application of an excessively high current in the cathodic protection of aluminum alloys because the usual cathodic reaction occurring on the alloys (reduction of oxygen) produces hydroxyl ions.

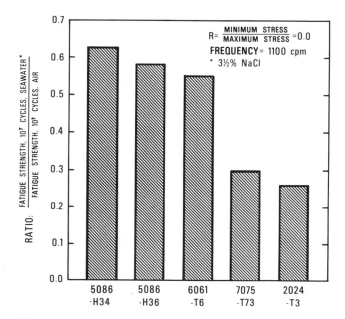

Figure 10. Axial-stress fatigue strengths of aluminum alloy sheet in air and seawater.

WELDED, BRAZED, AND SOLDERED JOINTS

The resistance to corrosion of weldments in aluminum alloys is affected by the alloy welded and by the filler alloy and welding process used [14]. Galvanic cells that cause corrosion may be created because of potential differences among the parent alloy, the filler alloy, and the heat-affected regions where microstructural changes have been produced. Incomplete removal of fluxes after welding with processes using them may also cause corrosion.

Weldments in non-heat-treatable alloys generally have a good resistance to corrosion. Microstructural changes in the heat-affected region in these alloys have little effect on potential and the filler alloys recommended have potentials close to those of the parent alloys. In some heat-treatable alloys, however, the effect on potential of microstructural changes may be large enough to cause appreciable corrosion in more aggressive environments; the corrosion is selective, either in the weld bead or in a restricted portion of the heat-affected region. To a considerable degree, the effect on corrosion of microstructural changes in the heat-affected zone can be eliminated by postweld heat treatment.

Stress corrosion cracking in weldments subject to this process usually is caused by residual stresses introduced during welding, but its occurrence is rare. Stresses sufficient to cause cracking can be avoided by the use of welding procedures that minimize restraint of the alloy being welded.

Brazed joints in aluminum alloys also have a good resistance to corrosion. Excessive corrosion usually is caused by fluxes from processes using them that are not removed completely, or that are removed by a treatment that, together with the fluxes, may cause corrosion.

Soldered joints have a resistance to corrosion satisfactory for applications in milder environments but not for those in more aggressive ones.

WEATHERING

Except for those that contain copper as a major alloying element, aluminum alloys have a high resistance to weathering in most atmospheres, as demonstrated by their extensive use for architectural applications in commercial, industrial, and residential buildings. In many cases, paints and other coatings are applied to products of the alloys to enhance appearance as well as to provide additional protection, but in many others, they are not.

As shown in Fig. 11, the depth of attack during the weathering of aluminum alloys characteristically decreases to a low rate after the initial period of exposure. The loss in strength also decreases in the same manner after the initial period, but not to as low a rate. The difference is a consequence of the fact that, while older sites of attack tend to become inactive, newer ones develop. The newer sites of attack decrease the cross-sectional area on which strength depends and the average depth of attack but not the maximum depth.

This "self-limiting" characteristic of corrosive attack during weathering is also typical of the behavior of aluminum alloys in many other environments. The characteristic, therefore, should always be appreciated whenever data from exposure tests of short duration are used to predict long-time performance. In many cases, long-time performance will be much better than predicted from short-duration data.

Numerous weathering data for aluminum alloys exposed at locations throughout the world have been published [15, 16].

WATERS

The more resistant aluminum alloys, such as wrought alloys of the 1XXX, 3XXX, and 5XXX series, have an excellent resistance to corrosion by high-purity water [16. A slight reaction takes place when the alloys are first exposed to the water, but it decreases to a low rate

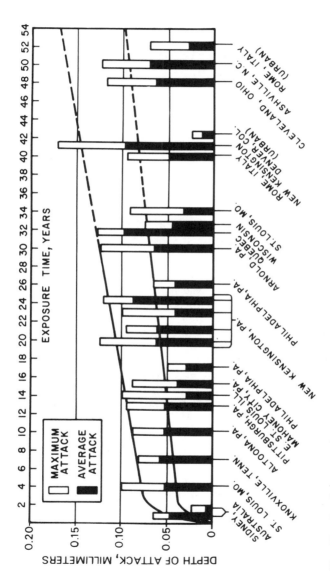

Figure 11. Weathering data for 1100, 3003, and 3004 alloys in the industrial atmosphere at New Kensington, Pennsylvania (curves), and at various other localities.

within a few days on formation of a protective oxide film of equilibrium thickness on the alloys; and pickup of aluminum by the water becomes negligible. Resistance of the alloys is not affected significantly by carbon dioxide and oxygen dissolved in the water or by the chemicals added to it to minimize corrosion of steel because of the presence of these gases.

The same types of alloys are also resistant to many natural waters [16]. In general, resistance is higher in neutral or slightly alkaline waters, and less in acidic ones, especially if they also contain ions of reducible metals such as copper; other factors affecting resistance were described earlier. Correlations of the compositions of various waters with their corrosivities to aluminum alloys have been developed, but none predict the corrosivities of other waters with dependability [17].

Numerous exposure tests throughout the world demonstrate the high resistance of the same alloys to seawater, as does service experience in pipelines, structures, ships, and other marine applications [16, 18]. Corrosion of aluminum alloys in seawater is mainly of the pitting type, as would be expected from its salinity and enough dissolved oxygen as a cathodic reactant to polarize the alloys to their pitting potentials. Rates of pitting usually range from 3 to 6 μm per/ year during the first year and from 0.8 to 1.5 μm per/year averaged over a 10-year period; the lower rate for the longer period reflects the tendency for older pits to become inactive. General corrosion is minimal, with the total loss in thickness usually amounting to only a fraction of the depth of pitting.

The corrosion behavior of aluminum alloys in deep seawater, judging from tests at 1.6 km, is generally the same as at the surface except that the effect of crevices is greater [19].

Among wrought alloys, those of the 5XXX series have the highest resistance to seawater, and considering their other desirable characteristics, they are used most widely for marine applications. Among casting alloys, those of the 356.0 and 514.0 types are used extensively for marine applications.

FOODS

The widespread usage of the more-corrosion-resistant aluminum alloys for household cooking utensils attests to their compatibility with most foods, as does their extensive usage for the handling and processing of foods commercially. In some cases, unsatisfactory performance is caused by the use of improper cleaners rather than by incompatibility with the foods themselves. Because they may cause excessive corrosion, alkaline cleaners should not be used for aluminum equipment unless they are inhibited.

Large quantities of aluminums and aluminum alloys are used for the packaging of foods and beverages, as foil, as foil laminated to plastics,

and as cans. Most cans for beverages, both those of all-aluminum construction and those with steel bodies, have easy-open ends of aluminum.
For most applications, lacquers and plastically laminated coatings are
applied to the alloys because only the minutest corrosion can be tolerated in view of the long periods of exposure, the effect of corrosion on
the food packaged, and the thinness of the container.

CHEMICALS

Aluminum alloys that do not contain copper as a major alloying element
have high resistance to corrosion by many chemicals, and they are
used extensively in the chemical process industry.

The resistance of aluminum alloys to corrosion by a large number of
chemicals, representative of practically all classifications, has been
established in laboratory tests, and in many cases, by service experience as well. Data are readily available from handbooks, from company
literature, and from publications by trade organizations. *Aluminum
and Foods*, published by the Aluminum Association, is especially useful [20].

Aluminum alloys are compatible with dry salts of most inorganic
salts, and within their passive range of pH 4 to 9, with aqueous solutions of many of them. Corrosion of the pitting type is produced by
aqueous solutions, mostly of halide salts, under conditions where the
alloys are polarized to their pitting potentials, but not in most other
solutions, where conditions that polarize the alloys to these potentials
are less likely.

Aluminum alloys are not suitable for most inorganic acids, bases,
and salts with a pH outside the passive range of the alloys, as shown
in Fig. 2 and as described earlier.

Aluminum alloys are resistant to a wide variety of organic compounds, including most aldehydes, esters, ethers, hydrocarbons,
ketones, mercaptans, and other sulfur-containing compounds, and
nitro compounds. They are also resistant to most acids, alcohols, and
phenols except when these compounds are nearly dry and near their
boiling points; carbon tetrachloride also exhibits the same behavior.
As shown by the data for phenol in Fig. 12, which are typical of the
behavior, corrosion is prevented by the presence of a trace of water
which appears to be needed to maintain a protective oxide film on the
alloys. The degree of dryness required for such corrosion is difficult
to obtain in lower molecular weight compounds.

Conditions to ensure safety should be established before aluminum
alloys are used with halogenated organic compounds. Under most conditions, especially at ambient temperature, the alloys are resistant to
these compounds, but under certain others, they may react rapidly
with some of them; under some conditions, especially when the alloys
are finely divided, a violent reaction may occur. If water is present,

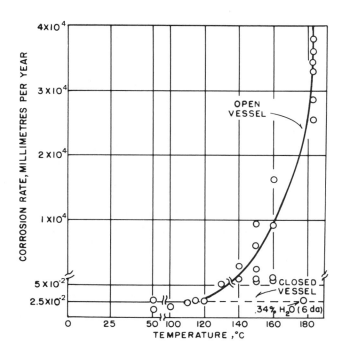

Figure 12. Resistance to corrosion of 1100-H14 alloy sheet to phenol.
Note that no corrosion occurred in a closed vessel where evaporation
of water was prevented.

the compounds may hydrolyze to produce mineral acids, which in turn
may corrode the alloys to produce aluminum halides that can act as a
catalyst for reactions of the organic compounds themselves; highly
reactive aluminum alkyls may be formed in some cases. These reactions, once initiated, may also become autocatalytic.

   Aluminum alloys are most resistant to organic compounds halogenated with fluorine, followed in order of decreasing resistance with
those of chlorine, bromine, and iodine. They are resistant to highly
polymerized compounds, reflecting the high degree of stability of
these compounds.

   It is important also to recognize that the compatibility of aluminum
alloys with mixtures of organic compounds cannot always be predicted
from their compatibility with each of the compounds. For example,
some aluminum alloys are corroded severely in mixtures of carbon
tetrachloride and methyl alcohol, even though they are resistant
to each compound alone. Caution should also be exercised in

using data for pure organic compounds to predict performances of the alloys with commercial grades that may contain contaminants. Ions of halides and reducible metals, commonly chloride and copper, frequently have been found to be the cause of excessive corrosion of aluminum alloys in commercial grades of organic chemicals that would not have been predicted from their resistance to the pure chemicals.

COATINGS

Paints and other coatings are applied to aluminum alloys for decorative purposes as well as to provide protection [14]. Almost any type of paint for metals (acrylic, alkyl, polyester, vinyl, etc.) is suitable; the performance of a particular paint, when applied properly, is better than that on steel. As with any metal, surface preparation is important. Conversion coatings, either of the chromate or phosphate type, are recommended for the preparation of aluminum alloys. For milder environments, the paint may be applied onto the conversion coating; for more aggressive environments such as those with chlorides, a chromated primer should be applied first. Aluminum alloys are especially amenable to waterborne paints, which are being used increasingly because of environmental considerations. Many products precoated in a variety of colors for agricultural, industrial, and residential applications are available commercially.

Although more expensive, and restricted to in-plant application, anodized coatings provide excellent protection to aluminum alloys [14]. They are also sometimes used as a base for paints. The many monumental buildings with outer walls of anodized aluminum alloys attest to the durability of the coatings in weathering. Anodized coatings also provide a variety of colorations, most commonly shades of gray and bronze, produced by selection of both alloy and anodizing process.

The coatings are produced by an electrolytic process in which the surface of an alloy made the anode is converted to aluminum oxide, bound as tenaciously to the alloy as the natural oxide film, but much thicker. The thickness of coatings to provide resistance to corrosion ranges from 5 to 30 μm; little or no additional protection is provided by thicker coatings. As with the alloys themselves, anodized coatings are not resistant to most environments with a pH outside the range 4 to 9. Within this range, the resistance to corrosion may be improved by an order of magnitude or more; in atmospheric weathering tests, the number of pits that developed in the base metal was found to decrease exponentially with the thickness of coating. As a general rule, however, anodized coatings do not provide protection sufficient to make aluminum alloys suitable for environments where, without the coatings, the alloys themselves are unsuitable because of their inherently poor resistance in them.

REFERENCES

1. New aluminum smelting process, *Light Met. Age 31*:8, 1973.
2. Marcel Pourbaix, *Atlas of Electrochemical Equilibria in Aqueous Solutions*, Pergamon Press, New York, 1966.
3. M. H. Brown, R. H. Brown, and W. W. Binger, Aluminum alloys for handling high temperature water, in *High Purity Water Corrosion of Metals* (N. E. Hammer, ed.), Publ. No. 60-13, National Association of Corrosion Engineers, Houston, 1960, p. 82.
4. H. Bohni and H. H. Uhlig, Environmental factors affecting the critical pitting potential of aluminum, *J. Electrochem. Soc. 116*: 906, 1969.
5. R. B. Mears and R. H. Brown, Causes of corrosion currents, *Ind. Eng. Chem. 33*: 1001, 1941.
6. E. H. Cook and F. L. McGeary, Electrodeposition of iron from aqueous solutions onto an aluminum alloy, *Corrosion 20*: 11t, 1964.
7. W. W. Binger, R. H. Brown and M. H. Brown, Mercury and its compounds—a corrosion hazard, *Corrosion 8*: 155 1952.
8. D. O. Sprowls and R. H. Brown, Stress-corrosion mechanisms for aluminum alloys, in *Fundamental Aspects of Stress-Corrosion Cracking* (R. W. Staehle, A. J. Forty, and D. Van Rooyen, eds.), National Association of Corrosion Engineers, Houston, 1969, p. 466.
9. E. H. Dix, Jr., Acceleration of the rate of corrosion by high constant stresses, *Trans. AIME 137*: 11, 1940.
10. M. O. Spiedel, Stress corrosion cracking of aluminum alloys, *Met. Trans. A 6A*: 631, 1975.
11. M. O. Spiedel, Hydrogen embrittlement of aluminum alloys, in *Hydrogen in Metals* (L. M. Bertsein and A. W. Thompson, eds.), American Society for Metals, Metals Park, Ohio, 1974, p. 249.
12. D. O. Sprowls, High strength aluminum alloys with improved resistance to corrosion and stress corrosion cracking, in *Proceedings of the 1976 Tri-Service Conference on Corrosion* (S. J. Ketcham, ed.), The Naval Advisory Council on Materials, Washington, D. C. 1976, p. 89.
13. Standard Method of Test for Exfoliation Corrosion Susceptibility in 2XXX and 7XXX Series Aluminum Alloys (EXCO Test), Standard Test Procedure G34-39, *Annual Book of ASTM Standards*, Part 10, American Society for Testing and Materials, Philadelphia, 1979.
14. *Aluminum Vol. III: Fabrication and Finishing* (K. R. Van Horn, ed.) American Society for Metals, Metals Park, Ohio, 1967.
15. W. K. Boyd and F. W. Fink, *Corrosion of Metals in the Atmosphere*, Rep. MCIC, Battelle Laboratories, Columbus, Ohio, 1974.
16. H. P. Goddard, W. B. Jepson, M. R. Bothwell, and R. L. Kane, *The Corrosion of Light Metals*, Wiley, New York, 1967.

17.   B. R. Pathak and H. P. Goddard, Equations for predicting the corrosivity of natural waters to aluminum, *Nature 218*(5144): 893, 1968.

18.   W. K. Boyd and F. W. Fink, *Corrosion of Metals in Marine Environments*, Rep. MCIC-75-245R, Battelle Laboratories, Columbus, Ohio, 1975; Rep. MCIC-78-37, 1978.

19.   F. M. Reinhart, *Corrosion of Metals and Alloys in the Deep Ocean*, Rep. R834, U.S. Naval Engineering Laboratory, Port Hueneme, Calif., 1976.

20.   *Aluminum with Food and Chemicals*, 3rd ed., The Aluminum Association, Washington, D.C., 1975.

SUGGESTED READING

E. H. Hollingsworth and H. Y. Hunsicker, Corrosion resistance of aluminum and aluminum alloys, in *Metals Handbook; Properties and Selection: NonFerrous Alloys and Pure Metals*, Vol. 2, 9th ed. (D. Benjamin, ed.), American Society for Metals, Metals Park, Ohio, 1979.

3.6
Titanium

LOREN C. COVINGTON / Titanium Metals Corporation of America,
Henderson, Nevada

INTRODUCTION

Titanium is quite plentiful in the earth's crust, being the ninth most abundant element. It is the fourth most abundant metal and is more plentiful than chromium, nickel, or copper, which are commonly employed in alloys used to resist corrosion. However, in spite of early recognition of its light weight, strength, and corrosion resistance, titanium metal is a relative newcomer to the industrial scene.

147

Commercialization of the Kroll process made titanium sponge available in about 1950. Aerospace requirements subsequently speeded development of titanium as a structural metal. The industrial uses for titanium developed more slowly at first. However, the lowering of cost, the availability of product forms and new corrosion-resisting alloys, coupled with the development of fabrication techniques, has resulted in a rapidly expanding industrial market for titanium.

The industrial utilization of titanium results mainly from the excellent corrosion resistance that this metal offers. The strength properties of titanium alloys are also utilized, but to a lesser extent.

PROPERTIES OF TITANIUM

Alloys Available

The compositions of representative titanium grades, as covered by American Society of Testing and Materials (ASTM) specifications, are given in Table 1. Unalloyed titanium, represented by ASTM Grade 2 in Table 1, is most often used for corrosion resistance. Other unalloyed grades (ASTM Grades 1, 3, and 4), containing more or less iron and oxygen than Grade 2, are available if better formability or higher strength are required.

The titanium-palladium alloy (Grade 7) offers improved corrosion resistance compared to unalloyed titanium. This alloy, as Grade 11, is also available with low oxygen and iron content for improved formability. Ti-Code 12 (ASTM Grade 12), containing 0.8% nickel and 0.3% molybdenum, is a recent Timet development, which is a low-cost alternate for Ti-Pd for some applications. The Ti-6Al-4V alloy (Grade 5) is a general purpose alloy, widely used in aerospace applications which require higher strength or fatigue resistance. The Ti-6Al-4V alloy, in general terms, has corrosion resistance somewhat inferior to that of unalloyed titanium.

All the alloys listed in Table 1 are available in various product forms covered by the ASTM specifications listed in Table 2. Heat exchanger tubing is most readily available as unalloyed titanium (Grade 2) and as the Ti-Code 12 (Grade 12) and Ti-Pd (Grade 7) alloys.

Mechanical Properties

The minimum room temperature tensile properties for the various titanium alloys, as defined by the ASTM specifications, are given in Table 3. Strength and ductility, comparable to other corrosion-resistant alloys, are available in titanium and its alloys. The Ti-Code 12 alloy (Grade 12) is seen to offer improved strength compared to Grade 2 or Grade 7 titanium.

Table 1. Chemical Composition of Titanium Alloys

Element	Composition, wt %				
	Ti-50A[a] (ASTM Grade 2)	Ti-6Al-4V[a] (ASTM Grade 5)	Ti-Pd (ASTM Grade 7)	Ti-Code 12[a] (ASTM Grade 12)	
Nitrogen, max.	0.03	0.05	0.03	0.03	
Carbon, max.	0.10	0.10	0.10	0.08	
Hydrogen, max.	0.015	0.015	0.015	0.015	
Iron, max.	0.30	0.40	0.30	0.30	
Oxygen, max.	0.25	0.20	0.25	0.25	
Aluminum	–	5.5-6.75	–	–	
Vanadium	–	3.5-4.5	–	–	
Palladium	–	–	0.12-0.25	–	
Molybdenum	–	–	–	0.2-0.4	
Nickel	–	–	–	0.6-0.9	
Titanium	Remainder	Remainder	Remainder	Remainder	

[a]Timet designation.

Table 2. ASTM Titanium Specifications

ASTM B 265-76	Titanium and Titanium Alloy Strip, Sheet and Plate
ASTM B 337-76	Seamless and Welded Titanium and Titanium Alloy Pipe
ASTM B 338-76	Seamless and Welded Titanium and Titanium Alloy Tubes for Condensers and Heat Exchangers
ASTM B 348-76	Titanium and Titanium Alloy Bars and Billets
ASTM B 363-76	Seamless and Welded Unalloyed Titanium Welding Fittings
ASTM B 367-69	(1974) Titanium and Titanium Alloy Castings
ASTM B 381-76	Titanium and Titanium Alloy Forgings

Allowable stress values from the ASME Pressure Vessel Code for titanium alloys as a function of temperature are given in Table 4. The values for the Ti-Code 12 alloy (Grade 12) are seen to fall off less rapidly relative to unalloyed titanium (Grade 2) and Ti-Pd (Grade 7) with increase in temperature. At 260°C (500°F), the design stress for Ti-Code 12 (Grade 12) is 70% greater than that of Grade 2 or Grade 7. The Ti-6Al-4V alloy is not Code-covered.

The effects of temperature on the strength, ductility, and toughness of unalloyed titanium (Grade 2) and annealed Ti-6Al-4V alloy (Grade 5) are shown in Fig. 1. Titanium and its alloys maintain excellent properties to low temperatures.

Table 3. Minimum Mechanical Properties of Annealed Titanium Alloys

Alloy	Tensile strength, ksi	0.2% Yield strength, ksi	Elongation, % (2 in.)
Ti-50A (Grade 2)	50	40	20
Ti-6A1-4V (Grade 5)	130	120	10
Ti-Pd (Grade 7)	50	40	20
Ti-Code 12 (Grade 12)	70	50	18

Table 4. Design Stresses for Titanium Plate[a]

For metal temperatures not exceeding: °F	Allowable stress values, ksi		
	Ti-50A (Grade 2)	Ti-Pd (Grade 7)	Ti-Code 12 (Grade 12)[b]
100	12.5	12.5	17.5
200	10.9	10.9	16.4
300	9.0	9.0	14.2
400	7.7	7.7	12.5
500	6.6	6.6	11.4
600	5.7	5.7	

[a] ASME Section VIII, Division 1—Pressure Vessels.
[b] Case BC78-326.

The fatigue properties of titanium are excellent. As shown in Table 5, the ratio of fatigue strength to tensile strength for titanium and its alloys at ambient temperature is high, in the range 0.5 to 0.6. In addition, unlike many metals, the fatigue properties of titanium are relatively unaffected by many corrosive media. Cotton and Downing, for instance, have shown that the fatigue limit of unalloyed titanium in distilled water or simulated seawater was actually higher than that for air [1].

Physical Properties

Physical properties for titanium are given in Table 6. Titanium is a light metal with density slightly over half that of iron or copper-base alloys. Modulus of elasticity of titanium is also about half that of steel. Specific heat and thermal conductivity are similar to those of stainless steel. Titanium has relatively high electrical resistivity and low expansion coefficient. It is important that these physical properties, particularly modulus of elasticity and coefficient of expansion, be carefully considered when designing or fabricating process equipment of titanium.

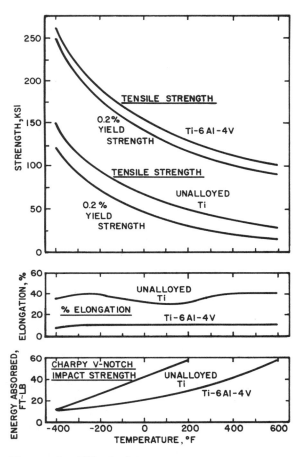

Figure 1. Effect of temperature on strength, ductility, and toughness of unalloyed titanium (Grade 2) and Ti-6Al-4V alloy (Grade 5).

Table 5. Fatigue Strength of Titanium

Alloy	Tensile strength, ksi	Fatigue strength (10^8 cycles), ksi	Ratio FS/TS
Unalloyed Ti	52.6	32.7	0.62
Unalloyed Ti	92.2	47.1	0.51
Ti-6Al-4V	140.7	85.5	0.60
Ti-6Al-4V	163.4	92.8	0.56

Table 6. Physical Properties of Titanium

Property	Ti-50A (Grade 2)	Ti-6Al-4V (Grade 5)	Ti-Pd (Grade 7)	Ti-Code 12 (Grade 12)
Modulus of elasticity, tension, 10^6 psi	14.9	16.5	14.9	15.0
Modulus of elasticity, torison, 10^6 psi	6.5	6.1	6.5	6.2
Density, lb/in.^{3}	0.163	0.160	0.163	0.164
Specific heat at 75°F, Btu/lb °F	0.125	0.135	0.125	0.130
Thermal Conductivity at 75°F, Btu/(ft^2) (h °F) (in.)	114	50	114	132
Coefficient of expansion, 32–600°F (10^{-6} in./in. °F)	5.1	5.1	5.1	5.4
Electrical resistivity at 75°F, $\mu\Omega$-cm	56	171	56.7	52

Table 7.   Relative Costs—Titanium Alloy Plate

Material	Relative costs,[a] 1 ft^2 of 1/4-in. plate
Ti-50A (ASTM Grade 2)	1.0
Ti-Code 12 (ASTM Grade 12)	1.2
Ti-Pd (ASTM Grade 7)	1.7

[a]Ratio of base price of 1 ft^2 of alloy to that of Ti-50A (ASTM Grade 2).

Material Costs

Cost is always an important consideration when selecting a material. Relative costs for titanium alloy plate are given in Table 7. Compared to other materials, titanium is more costly than stainless steels but less than nickel-base alloys. In many applications, titanium's corrosion resistance permits it to be used without a corrosion allowance. Thinner sections, coupled with decreased maintenance requirements and longer life expectancy in many applications, permit titanium equipment installations to be cost effective, despite a higher initial cost.

CORROSION RESISTANCE

Oxide Film

Titanium's corrosion resistance is due to a stable, protective, strongly adherent oxide film which covers its surface. This film forms instantly when a fresh surface is exposed to air or moisture. This film is transparent and not easily detected by visual means. A study of the corrosion resistance of titanium is basically a study of the properties of the oxide film. Additions of alloying elements to titanium affect corrosion resistance because these elements alter the composition of the oxide film.

The oxide film on titanium, although very thin, is very stable and is attacked by only a few substances, most notable of which is hydrofluoric acid. Because of its strong affinity for oxygen, titanium is capable of healing ruptures in this film almost instantly in any environment where a trace of moisture or oxygen is present. Thus we find that titanium is impervious to attack in moist chlorine gas. If the moisture content of dry chlorine gas falls below a critical level of about 0.5%, however, rapid or even catastrophic attack can occur.

Anhydrous conditions, in the absence of a source of oxygen, should be avoided with titanium, since the protective film may not be regenerated if it is damaged.

Velocity Effects

For most metals, there is a critical water velocity beyond which protective films are swept away and accelerated corrosion attack occurs. This is known as erosion corrosion. The velocity at which the protective film strips off differs greatly from one material to another. Erosion corrosion occurs on some metals at velocities as low as 2 to 3 ft/s. The critical velocity for titanium in seawater is in excess of 90 ft/s [2]. Numerous corrosion-erosion tests have been conducted and all have shown titanium to have outstanding resistance to this form of attack [3, 4]. Experiments with sand-laden, rapidly moving seawater have demonstrated titanium to be 20 times more resistant to erosion than the best copper-base alloys.

Corrosion in Waters

Titanium is immune to corrosion in all natural waters, including highly polluted seawater, at temperatures up to the boiling point. Nearly 20 years of service in such environments have now been accumulated and as far as is known, no failures due to corrosion have occurred. Titanium has replaced copper-base alloys which were corroding in the presence of sulfides, as well as stainless steels which were suffering from pitting and stress corrosion cracking due to chlorides.

Titanium also stands up to steam up to temperatures as high as 315°C (600°F) and pressures up to 2000 psi. Resistance to erosion by wet steam has been shown to be as good as that of 18Cr-8Ni stainless steels [2]. The Ti-6Al-4V (Grade 5) exhibited significantly better resistance to erosion by steam than did 18Cr-8Ni stainless steel.

Resistance to Fouling

Because of its hardness and corrosion resistance, which maintains a smoother surface, titanium has excellent resistance to fouling or deposit buildup in most environments. Observations of titanium tubing in service in natural cooling waters confirm that the resistance to fouling of titanium equals or exceeds the performance of stainless steels. The thin oxide film present on the surface of titanium generally does not react with cooling waters to form mineral deposits, as sometimes happens with some materials. When cleaning is necessary, conventional acid cleaning cycles may be used provided that proper inhibitors are present. For these reasons, heat exhangers tubed with titanium have been found to maintain high heat transfer efficiency indefinitely.

Galvanic Corrosion

The coupling of titanium with dissimilar metals usually does not accelerate the corrosion of the titanium. The exception is in reducing environments where titanium does not passivate. Under these conditions, titanium has a potential similar to aluminum and will undergo accelerated corrosion when coupled to more noble metals.

Titanium occupies a place high on the passive or noble end of the galvanic series in seawater, as shown in Table 8. In most environments, titanium will be the cathodic member of any galvanic couple. Accelerated corrosion of the other member of the couple may occur, but in most cases the titanium will be unaffected. Figure 2 shows the accelerating effect on the corrosion rate of various metals when they are galvanically connected to titanium in seawater. If the area of the titanium exposed is small in relation to the area of the other metal, the effect on corrosion rate of the other metal is negligible. If, however, the area of the titanium (cathode) greatly exceeds the area of the other metal (anode), severe corrosion may result on the anodic metal.

To avoid problems with galvanic corrosion, it is best to construct equipment of a single metal. If this is not practical, attempts should be made to select two metals that are close together in the galvanic series. Insulation of the dissimilar metal joint or cathodic protection of the less noble metal are other preventive measures that can be taken. If contact of dissimilar metals with titanium is necessary, the critical parts should be constructed of titanium, since it is usually not attacked. Use of large areas and heavy sections of the less noble metal, to allow for increased corrosion, also help to keep galvanic corrosion effects to a minimum.

Chlorides

Titanium is immune to all forms of corrosive attack in seawater and chloride salt solutions at ambient temperatures. It is also very resistant to attack in most chloride solutions at elevated temperatures [5, 6]. Under certain conditions in chloride solutions, titanium has been observed to undergo attack in the form of pitting and crevice corrosion. The temperature at which attack occurs is determined mainly by the pH of the solution, although chloride ion concentration has been observed to have a minor effect (more concentrated solutions, causing attack at lower temperatures). Figure 3 shows the relationship between the temperature and the pH of saturated brine, at which corrosive attack initiates on unalloyed titanium. This curve was drawn from laboratory data and is confirmed by service experience.

Figures 4 and 5 show the pH-temperature relationship for Ti-Pd (Grade 7) and the Ti-Code 12 (Grade 12) alloys. Note the greatly

Table 8. Galvanic Series in Flowing Seawater (13 ft/s at 75°F)

Metal	Potential,[a]
T 304 stainless steel (passive)	0.08
Monel	0.08
Hastelloy C	0.08
Unalloyed titanium	0.10
Silver	0.13
T 410 stainless steel (passive)	0.15
Nickel	0.20
T 430 stainless steel (passive)	0.22
70-30 cupronickel	0.25
90-10 cupronickel	0.28
Admiralty brass	0.29
G. bronze	0.31
Aluminum brass	0.32
Copper	0.36
Naval brass	0.40
T 410 stainless steel (active)	0.52
T 304 stainless steel (active)	0.53
T 430 stainless steel (active)	0.57
Carbon steel	0.61
Cast iron	0.61
Aluminum	0.79
Zinc	1.03

[a]Steady-state potential negative to saturated calomel half-cell.

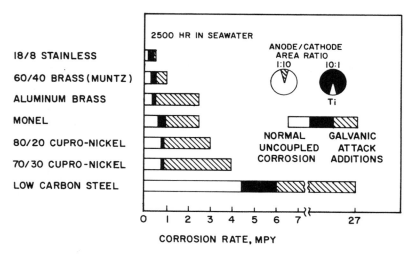

Figure 2.   Galvanic corrosion of titanium couples in seawater.

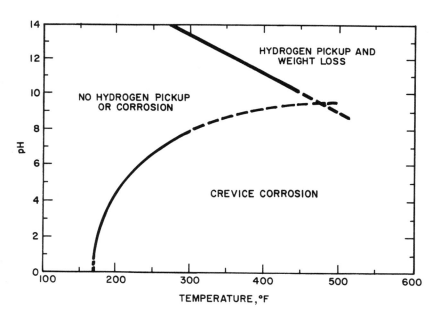

Figure 3.   Effect of temperature and pH on crevice corrosion of unalloyed titanium (Grade 2) in saturated brine.

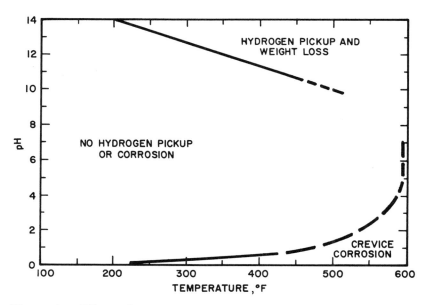

Figure 4. Effect of temperature and pH on crevice corrosion of Ti-Pd (Grade 7) in saturated NaCl brine.

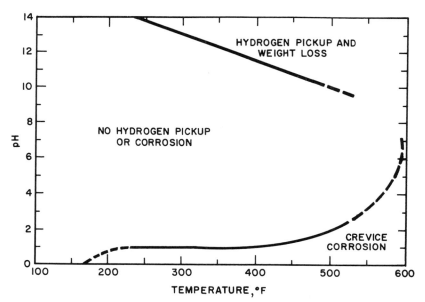

Figure 5. Effect of temperature and pH on crevice corrosion of Ti-Code 12 (Grade 12) in saturated NaCl brine.

Table 9. Crevice Corrosion of Titanium in Boiling Solutions

Environment	pH	Crevice test results[a]		
		Ti-50A	Ti-Code 12	Ti-Pd
$ZnCl_2$ (saturated)	3.0	F	R	–
10% $AlCl_3$	–	F	R	R
$MgCl_2$	4.2	F	R	R
10% NH_4Cl	4.1	F	R	R
NaCl (saturated)	1.0	F	R	R
10% Na_2SO_4	1.0	F	R	R
10% $FeCl_3$	0.6	F	F	R

[a]F, failed (samples showed crevice corrosion); R, resisted (samples showed no evidence of corrosion).

improved resistance to crevice corrosion of these alloys compared to unalloyed titanium over the pH range 9 to 0.5. The data in Table 9 further illustrate this point.

Crevice corrosion has not been observed when pH values are higher than 9. Failures have been observed at elevated temperatures in chloride solutions with pH 10 or higher, but these have been due to hydrogen embrittlement. It appears that the oxide film on titanium is not stable in highly basic solutions and is replaced by a hydride film that is protective at ambient temperatures. This may cause hydrogen embrittlement. For this reason, the use of titanium in pure caustic solutions should be limited to temperatures of 200°F or lower. The presence of chlorate or hypochlorite appears to inhibit the corrosion of titanium in caustic.

Chlorine, Hypochlorites, and Other Chlorine Chemicals

Titanium has excellent resistance to moist chlorine gas. For this reason it is used extensively in chlorine cells, heat exchangers, piping, reactors, pumps, valves, and other equipment in chlorine plants and in industrial processes that employ moist chlorine. Chlorine in solution tends to have a passivating effect on titanium by moving the corrosion potential in the positive direction. Thus titanium is widely employed to handle chlorinated brines and hypochlorites. Some corrosion rates for titanium in hypochlorite solutions are given in Table 10.

Table 10. Corrosion of Titanium in Hypochlorite Solutions

Environment	Temperature °F	Test duration, days	Corrosion rate, mpy	Pitting
17% hypochlorous acid, with free chlorine and chlorine monoxide	50	203	<0.1	—
16% sodium hypochlorite	70	170	<0.1	None
18-20% calcium hypochlorite	70-75	204	Nil	None
1.5-4% sodium hypochlorite, 12-15% sodium chloride, 1% sodium hydroxide	150-200	72	0.1	None

Source: Ref. 7.

Titanium is also used in bleach sections of pulp plants to handle corrosion caused by chlorine dioxide.

Titanium is not resistant to dry chlorine gas. It is attacked
rapidly and can ignite and burn if the moisture content is sufficiently
low. The amount of moisture required for passivation depends on
environmental conditions [8, 9]. About 1% H_2O is sufficient under
static conditions at room temperature. Somewhat less is required if
the chlorine is flowing. About 1.5% water is required at 200°C (392°F).

Acid Solutions

In general terms, titanium offers excellent resistance to oxidizing acids.
It has limited resistance to reducing acids. The presence of oxidizing
agents that induce passivity improves resistance in these reducing environments. Hydrofluoric acid, in very small amounts, will attack
titanium.

Oxidizing Acids

Titanium offers excellent resistance to oxidizing acids such as nitric
and chromic acids. It has been used for reactors, heat exchangers,
thermowells, and other equipment employed in producing nitric acid
[10] because of the low corrosion rates experienced over a wide range
of conditions. Solutions containing 20 to 70% HNO_3, at temperatures
from boiling to 315°C (600°F), have been handled in titanium equipment. Godard et al. [9] has cited an example of a titanium heat exchanger handling 60% HNO_3 at 195°C (380°F) and 300 psi which
showed no signs of corrosion after more than 2 years of operation.

Figure 6 shows the results of a Huey-type corrosion test for
unalloyed titanium (Grade 2), Ti-Pd (Grade 7), and Ti-Code 12 (Grade
12), in which the nitric acid solution is changed every 24 h. Note that
in this oxidizing environment, the resistance of the Ti-Pd alloy is no
better than that of unalloyed titanium. The Ti-Code 12 alloy, however,
shows a significantly lower corrosion rate.

In hot nitric acid, titanium may show a short period of relatively
high corrosion rate followed by almost complete passivation. It
appears that a small amount of titanium ion in solution acts to passivate titanium against further attack. This is illustrated by the data
in Table 11. Contrary to effects on stainless steel, the presence of
chromium ions also passivates titanium against attack by nitric acid as
shown in Table 12.

Degnan [12] has pointed out that titanium is considered to be an
outstanding material for equipment for heating nitric acid solutions.
Stainless steels, for instance, show accelerated attack in a hot wall
test or a continuous Huey test, due to chromium ions generated by
corrosion. Titanium, on the other hand, is passivated against further
attack by even the slightest degree of corrosion.

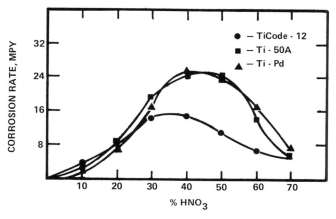

BOILING HNO$_3$ SOLUTION CHANGED EVERY
24 HR (TOTAL EXPOSURE 480 HR)

Figure 6. Boiling HNO$_3$ solution changed every 24 h—total exposure 480 h.

Table 11. Effect of Dissolved Titanium on the Corrosion Rate of Titanium in Boiling Nitric Acid Solutions[a]

Titanium ion added, mg/liter	Corrosion rate, mpy	
	40% HNO$_3$	68% HNO$_3$
0	29.5	31.8
10	—	0.8
20	8.6	2.4
40	1.9	0.4
80	0.8	0.4

[a]Duration of test, 24 h.
Source: Ref. 11.

Table 12. Effect of Chromium on Corrosion of Stainless Steel and
Titanium in Boiling 68%[a] (Nitric Acid)

Percent Cr	AISI 304L (annealed)	Unalloyed Ti
0.0	12-18	3.5-3.8
0.0005	12-20	—
0.005	60-90	0.9-1.6
0.05	980-1600	—
0.01	—	

[a]Exposed for three 48-h periods, acid changed each period.
Source: Ref. 12.

One word of caution: Titanium is not recommended for use in red
fuming nitric acid, particularly where the water content is less than
about 1.5% and the nitrogen dioxide content is above 2.5% [12].
Pyrophoric reactions have occurred in this environment.

Reducing Acids

Titanium is attacked by such reducing environments as hydrochloric,
sulfuric, and phosphoric acids. However, it is passivated by small
amounts of multivalent metal ions such as copper and iron. Many industrial acid process streams contain sufficient metal ion content to
permit excellent performance with titanium in fairly strong solutions
of these acids. The presence of dissolved oxygen, chlorine, nitrate,
chromate, or other oxidizing species also serves to passivate titanium
and reduce corrosion rate in acids.

Isocorrosion data for unalloyed titanium (Grade 2), Ti-Code 12
(Grade 12), and Ti-Pd (Grade 7) are shown in Fig. 7 for hydrochloric
acid without passivating species. The palladium-alloyed titanium
offers best resistance to HCl under these conditions.

The effect of ferric ion on passivating unalloyed titanium against
attack in 5 weight percent HCl is shown in Fig. 8. As little as 150 ppm
ferric ion can provide protection for unalloyed titanium in this concentration of acid, reducing corrosion rate from ore than 1000 mpy to less
than 10 mpy. Further effects of ferric ion on passivating unalloyed
titanium, Ti-Code 12, and Ti-Pd against attack by 3 and 4 weight percent hydrochloric acid are shown in Table 13. In general, these data
illustrate that if metal-ion concentration is sufficient to passivate
effectively, the corrosion rate for unalloyed titanium will be as low as
for the more expensive Ti-Pd alloy.

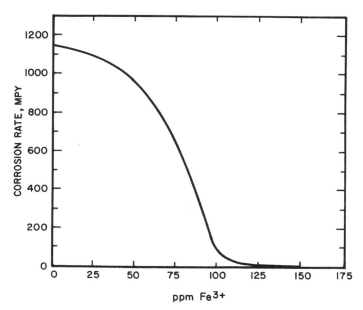

Figure 7.   Corrosion of Ti in HCl solutions.

Corrosion data in boiling sulfuric acid solutions are given in Table 14.
Although the Ti-Code 12 (Grade 12) alloy offers some improvement
over unalloyed titanium, the Ti-Pd alloy again offers best resistance.
Ferric, cupric, and other heavy metal ions will passivate titanium
against attack by sulfuric acids. The effect of cupric ion on corrosion by sulfuric acid of unalloyed titanium at 100°C (212°F) is shown
in Table 15. Similar effect of ferric ion in passivating titanium
against corrosion by 10 weight percent hydrochloric acid is also shown
by data in Table 15.
     The resistance of titanium to corrosion by phosphoric acid is similar
to resistance to sulfuric and hydrochloric acids. Unalloyed titanium
will handle up to 30 weight percent phosphoric acid at room temperature
and 1 weight percent acid at the boiling point with a corrosion rate of
10 mpy. The Ti-Pd alloy can handle up to 10 weight percent phosphoric acid at the boiling point, with a corrosion rate of about 5 mpy.
The presence of heavy metal ions and oxidizing agents in phosphoric
acid serves to passivate titanium in a manner similar to that demonstrated for the other reducing acids.

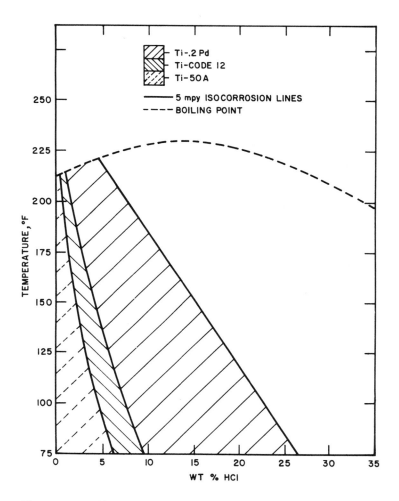

Figure 8. Effect of ferric ion on corrosion of unalloyed titanium in 5 weight percent boiling HCl.

Organic Chemicals

Titanium generally shows good corrosion resistance to organic compounds. In anhydrous environments where the temperature is high enough to cause dissociation of the organic compound, hydrogen embrittlement of the titanium is a consideration. Many organic compounds are absorbed on titanium surfaces and act as inhibitors.

Table 13. Effect of Ferric Chloride Inhibitor on Corrosion in Boiling HCl

		Corrosion Rate, mpy	
Wt % HCl	Alloy	No $FeCl_3$	2 g/liter $FeCl_3$
3.32	Ti-50A	740	0.2
3.32	Ti-Code 12	606	1.0
3.32	Ti-Pd	3	0.1
4.15	Ti-50A	843	0.4
4.15	Ti-Code 12	1083	2.3
4.15	Ti-Pd	6	0.3

Table 14. Corrosion Rates of Titanium Alloys in Boiling H_2SO_4

	Corrosion rate, mpy		
Wt % H_2SO_4	Ti-Pd	Ti-Code 12	Unalloyed, Ti-50A
0.54	0.1	0.6	252
1.08	0.2	35.4	440
1.62	2.1	578	713
2.16	3.8	759	945
2.70	14.9	1331	1197
5.40	29.8	2410	2047

Table 15. Effect of Inhibitors on Acid Corrosion of Titanium

Environment	Temperature, °F	Corrosion rate, mpy
10% HCl	Boiling	>2360
10% HCl + 16 g/liter $FeCl_3$	Boiling	8
20% H_2SO_4	212	>2360
20% H_2SO_4 + 2.5 g/liter $CuSO_4$	212	>2

Because of its resistance to corrosion in organic environments,
titanium is finding steadily increasing uses in equipment used to handle
these chemicals. Godard et al. [9] has pointed out that titanium is a
standard construction material for the Wacker Process, used to produce acetaldehyde by oxidation of ethylene in an aqueous solution of
metal chlorides.

Organic Acids

Titanium is, in general, quite resistant to organic acids. Its behavior
is dependent on whether the environment is reducing or oxidizing.
Only a few organic acids are known to attack titanium. Among these
are hot nonaearated formic acid, hot oxalic acid, concentrated trichlo-
racetic acid, and solutions of sulfamic acid. Aearation improves the
resistance of titanium in most of these nonoxidizing acid solutions. In
the case of formic acid, it reduces the corrosion rates to very low
values.

Titanium is resistant to acetic acid [9] over a wide range of concentrations and to temperatures well beyond the boiling point. It is being
used in terephthalic acid and adipic acid up to 215°C (400°F) and 67%
concentration. Good resistance is observed in citric, tartaric, car-
bolic, stearic, lactic, and tannic acids.

Some data for boiling organic acid solutions are given in Table 16.
As shown, the Ti-Code 12 (Grade 12) and Ti-Pd (Grade 7) alloys will
offer low corrosion rates in some cases where unalloyed titanium does
not.

Stress Corrosion Cracking

Titanium and its alloys have been relatively free of failures due to
stress corrosion cracking. One of the reasons for this is that environments which cause stress cracking have been recognized and use of
titanium has been restructed in these.

Unalloyed titanium with an oxygen content of less than 0.2% (ASTM
Grades 1 and 2) appears to be susceptible to cracking only in methanol and higher alcohols (liquid and vapor) in certain liquid metals such
as cadmium and possibly mercury, and in red fuming nitric acid. The
presence of halide ion in the alcohols accelerates cracking tendencies.
The presence of water (>2%) tends to inhibit stress cracking in alcohols
and in red fuming nitric acid.

Titanium alloys, including Ti-6Al-4V (Grade 5), are more susceptible to stress corrosion cracking. In addition to the environments that
affect unalloyed titanium, the alloys are also susceptible to hot salts,
as well as certain aqueous solutions and halogenated hydrocarbons [13].

Table 16. Resistance of Titanium to Boiling Nonaerated Organic Acids

Acid solution	Unalloyed titanium	Ti-Code 12	Ti-Pd
20% acetic	0	—	—
50% citric	14	0.5	0.6
10% sulfamic	538	455	14.6
45% formic	433	Nil	Nil
10% oxalic	3700	4100	1270

Hydrogen Embrittlement

Under most conditions, the oxide film on titanium acts as an effective barrier to penetration by hydrogen. However, under conditions that allow hydrogen to enter titanium and exceed the concentration needed to form a hydride phase (about 100 to 150 ppm), embrittlement can occur. Hydrogen absorption has been observed in alkaline solutions at temperatures above the boiling point. Acidic conditions that cause the oxide film to be unstable may also result in embrittlement under conditions where hydrogen is generated on the titanium surface. Coupling of titanium to a metal low in the galvanic series can result in generation of hydrogen on titanium as cathode, as can an impressed current from a cathodic protection system. In any case, it appears that embrittlement occurs only if the temperature is sufficiently high [i.e., above 75°C (170°F)] to allow hydrogen to diffuse into the titanium. Otherwise, if surface hydride films do form, they are not detrimental. However, they may reduce ability to resist erosion, resulting in higher corrosion rates.

Titanium tubing is being used effectively in a variety of environments where it is installed in a variety of tube sheet materials [14]. Hydrogen embrittlement of the titanium has been rare and has occurred in most cases where active corrosion was progressing on the dissimilar tube sheet and iron particles had been smeared into the titanium surface. The iron smears destroy the protective oxide surface, thereby allowing access of available hydrogen to the underlying titanium. Contact of titanium with soft iron that might result in pickup of iron particles should be guarded against. Anodizing is sometimes used to remove iron and build up oxide film thickness.

Embrittlement of titanium in gaseous hydrogen atmospheres has not been a problem. The presence of as little as 2% moisture effectively prevents the absorption of molecular hydrogen to temperatures as high as 315°C (600°F). Anhydrous environments and those lacking a source of oxygen to maintain the oxide film intact on titanium should be avoided, since absorption of hydrogen will occur at temperatures above 75°C (170°F).

FABRICATION

Titanium is routinely being fabricated into various types of process equipment by many fabricators. Techniques required to fabricate titanium are essentially the same as are required for other highly alloyed materials. These techniques, in most cases, simply represent good fabrication practice. Lack of attention to good practice for titanium, as with other metals, can lead to difficulties.

From a design standpoint, the properties of titanium must be kept in mind. Low modulus of elasticity, for instance, means greater springback on forming. Shrinkage and distortion of welds may be greater than is gnerally experienced with other metals. This, coupled with low strength at elevated temperatures, may require more careful attention to alignment during welding. Lower ductility might require more generous radii on bending or preheating prior to forming.

The welding of titanium requires special mention. Molten titanium absorbs oxygen, nitrogen, and hydrogen readily on exposure to air or moisture. To a lesser extent, solid titanium will also absorb these elements, at least down to temperatures in the range 315 to 425°C (600 to 800°F). Inert gas shielding during welding is, therefore, imperative down to temperatures in the range 315 to 425°C (600 to 800°F).

Good techniques for inert gas shielding of welds, which involve use of trailing shields as well as back shields, have been developed and are even being used in the field. These techniques, combined with careful procedures to exclude all moisture, dirt, and other foreign matter from the weld area, assure good welds.

The foregoing brief descriptions of fabrication details are well recognized by the experienced titanium fabricator and present little difficulty. Proper attention to these details results in weld properties equivalent to those of the base metal. Contamination of welds may mean loss of ductility, as well as degraded corrosion resistance.

REFERENCES

1. J. B. Cotton and B. P. Downing, Corrosion resistance of titanium to sea water, *Trans. Inst. Mar. Eng.* 69(8): 311, 1957.
2. G. J. Danek, Jr., The effect of sea-water velocity on the corrosion behavior of metals, *Nav. Eng. J.* 78(5): 763 1966.
3. D. W. Stough, F. W. Fink, and R. S. Peoples, *The Corrosion of Titanium*, Titanium Metallurgical Laboratory, Rep. No. 57, Battelle Memorial Institute, Columbus, Ohio, 1956, p. 60.

4. J. A. Davis, The effect of velocity on the sea water corrosion behavior of high performance ship material, Paper No. 78, NACE Corrosion/74, Chicago, March 4-8, 1974.

5. *Process Industries Corrosion*, National Association of Corrosion Engineers, Houston, 1975.

6. T. P. Oettinger, Austenitic stainless steels and titanium for wet air oxidation of sewage sludge, *Materials Performance* 15(11): 29, 1976.

7. P. J. Gegner, Corrosion resistance of materials in alkalies and hypochlorites, in *Process Industries Corrosion*, National Association of Corrosion Engineers (1975), p. 296.

8. E. E. Millaway and M. H. Klineman, Factors affecting water content needed to passivate titanium in chlorine, *Corrosion* 23(4): 88, 1967.

9. H. P. Godard, W. B. Jepson, M. R. Bothwell, and R. L. Kane, *The Corrosion of Light Metals*, Wiley, New York, 1967.

10. E. E. Millaway, Titanium: its corrosion behavior and passivation, *Mater. Perform.* 4(1): 16, 1965.

11. A. Takamura, K. Arakawa, and Y. Moriguchi, Corrosion resistance of titanium and titanium-5% tantalum alloy in hot concentrated nitric acid, *The Science, Technology and Applications of Titanium*, (R. I. Jaffee and N. E. Promisel, eds.), Pergamon Press, Oxford, 1970.

12. T. F. Degnan, Materials for handling hydrofluoric, nitric, and sulfuric acids, in *Process Corrosion Industries*, National Association of Corrosion Engineers, Houston, 1975, p. 229.

13. M. J. Blackburn, J. A. Feeney, and T. R. Beck, Stress-corrosion-cracking of titanium alloys, in *Advances in Corrosion Science and Technology*, Vol. 3, (M. G. Fontana and P. W. Staehle, eds.), Plenum Press, New York, 1973, pp. 67-292.

14. J. A. McMaster, Titanium: economical corrosion control tool for petroleum refineries, NACE, Corrosion/77, San Francisco, 1977.

3.7
Tantalum

PHILIP A. SCHWEITZER / Chem-Pro Corp., Fairfield, New Jersey

INTRODUCTION

Tantalum is not a new material. Its first commercial use at the turn of the century was as filaments in light bulbs. Later, when it became apparent that tantalum was practically inert to attack by most acids, application in the laboratory and in the chemical and medical industries were developed. The rise of the electronics industry accelerated the development of many new applications.

Much of this growth can be attributed to a broader range of tantalum powders and mill products available from the producers—high-melting-point ability to form a dielectric oxide film and chemical inertness. Encouraging these applications, new reduction, melting, and fabrication techniques have led to higher purities, higher reliabilities, and improved yields to finished products.

TANTALUM MANUFACTURE

Ingot Consolidations

The first production route for tantalum was by powder metallurgy.
Tantalum powder, produced by one of several reduction techniques,
is pressed into suitably sized bars and then sintered in vacuum at
temperatures in excess of 2100°C (3800°F). When completed, the
pressed and sintered bars are ready for processing into mill shapes.
Forging, rolling, swaging, and drawing of tantalum is performed at
room temperature on standard metalworking equipment with relatively
few modifications.

The powder metallurgy route, although still in use and adequate
for many applications, has two major limitations: (1) the size of the
bar capable of being pressed and sintered to a uniform density, limits
the size of the finished shape available, and (2) the amount of residual
interstitial impurities such as oxygen, carbon, and nitrogen remaining
after sintering adversely affects weldability.

The use of vacuum melting, either by the consumable arc or electron beam process, overcomes these limitations. Either melting technique is capable of producing ingots which are large enough in size
and high enough in purity to meet adequately most requirements of
product size and specifications provided that starting materials are
selected with care.

Quality Descriptions

The greatest volume of tantalum is supplied as powder for the manufacture of solid electolytic tantalum capacitors.

Since it is necessary to distinguish between capacitor-grade powder
and melting-grade powder, manufacturers of electronic components
tend to use the phrase "capacitor grade" when ordering forms such as
wire, foil, and sheet to identify end use and desired characteristics.

The phrase *capacitor grade* means that the material should have
the ability to form an anodic oxide film of certain characteristics.
Capacitor grade in itself does *not* mean an inherently higher purity,
cleaner surface, or different type of tantalum. It does mean that the
material should be tested using carefully standardized procedures for
electrical properties. If certain objective standards such as formation
voltage and leakage current are not available against which to test the
material, the use of the phrase *capacitor grade* is not definitive.

Metallurgical grade could be simply defined as noncapacitor grade.

PROPERTIES OF TANTALUM

Alloys Available

Pure tantalum has a body-centered-cubic crystal lattice. There is no
allotropic transformation to the melting point, which means that un-

alloyed tantalum cannot be hardened by heat treatment. Additions of oxygen, carbon, or nitrogen above normal levels, either purposefully or accidentally, are considered as alloying ingredients no matter what the concentration.

Tantalum-niobium alloys containing more than about 5 to 10% niobium are much less corrosion resistant than tantalum itself.

Tantalum-tungsten alloys containing more than 18% tungsten are inert to 20% hydrofluoric acid at room temperature. Few data are available on the 90Ta-10W alloy. It is known to be somewhat more oxidation resistant (e.g., to air at higher temperatures) than tantalum. The indication is that it has about the same corrosion resistance to acids as tantalum itself.

Mechanical Properties

The room temperature mechanical properties of tantalum are dependent on chemical purity, amount of reduction in cross-sectional area, and temperature of final annealing. Annealing time does not appear to be critical. Close control over the many parameters that affect mechanical properties are mandatory to ensure reproducible mechanical behavior. Typical mechanical properties for tantalum are shown in Table 1.

Tantalum can be strengthened only by cold work with resulting loss in ductility. As certain residual impurities have pronounced effects on ductility levels and metallurigical behavior, the purpose of most consolidation techniques is to make the material as pure as possible. Cold-working methods are used almost without exception to preclude the possibility of embrittlement by exposure to oxygen, carbon, nitrogent, and hydrogen at even moderate temperatures. Temperatures in excess of 425°C (800°F) should be avoided.

There are basically three structures which can be ordered: (1) unannealed, (2) stress relieved, and (3) annealed.

In the unannealed condition, the structure will be typically wrought fibrous. Yield and tensile strength will be increased with corresponding decreases in elongation, as shown in Table 2. The amount of work hardening will be dependent on the amount of cold reduction since the last anneal. The rate of work hardening is rapid for the first 30% of reduction. The rate then diminishes so that there will be no appreciable strengthening until reduction of over 90% are taken. There is actually no limit to the amount of cold work that the metal can take: there are only equipment limitations or mechanical limitations, such as poor shape control in rolling or excessive thinning when forming, which dictate periodic heat treatment in vacuum to soften the metal.

Unannealed tantalum may be preferred for machinability. Corrosion behavior is not affected nor is the susceptability to interstitial contamination changed. Unalloyed sheet 0.030 in. thick and under can make a 1-in.-thickness bend, but the annealed condition is preferred when bending since the metal is not as stiff or springy.

Table 1. Typical Mechanical Properties: Annealed Tantalum Sheet

Thickness	0.2% yield strength, psi	Ultimate tensile strength, psi	Elongation, %	Rockwell hardness 15T	B
0.005 deep draw	29,000	41,000	22	–	–
0.005 regular	44,000	55,000	18	–	–
0.010 regular	40,000	52,000	32	–	–
0.030 regular	35,000	45,000	40	75	–
0.060 regular	35,000	45,000	42	78	48

Stress relieving at 1010°C (1850°F) in vacuum reduces yield and tensile strength and raises elongation levels. These properties will be

Table 2. Typical Mechanical Properties: Tantalum Sheet with Increasing Cold Work

Percent cold work	0.2% yield strength, psi	Ultimate tensile strength, psi	Elongation, %	Hardness, VHN
30	70,600	74,200	18	189
50	82,200	86,000	9	192
80	100,500	109,200	4	235
90	117,800	123,400	2	239
95	127,000	135,500	1	265
98	–	135,500	1	280

Table 3.   ASTM Specification Limits for Tensile Properties: Test Procedure ASTM E8-61T

	Ultimate psi		Maximum yield, psi	Minimum elongatation, %	Maximum hardness
	Minimum	Maximum			
Cold worked	75,000			2	
Stress-relieved					
Any section >0.021 in.	55,000			10	
Any section <0.021 in.	55,000			7.5	
Annealed					
Any section >0.021 in.		55,000	45,000	25	80 + 5T
Any section <0.021 in.		55,000	45,000	15	
To 0.005 in. minimum					

intermediate between annealed and unannealed. Stress relieving has been used more as a matter of expediency than design. Fabricators need some ductility to allow them to roll tubes into tube sheets. Until recently the only tubular heat-treating vacuum furnaces were limited to 1010°C (1850°F) maximum. This equipment limitation dictated the use of stress relieving. As newer furnaces allowing full annealing in vacuum are now on-stream, stress relieving may gradually fall into disuse.

Tantalum specified in the annealed condition is in its softest, most ductile condition. The usual objective of the procedure is too choose an annealing temperature that will result in complete recrystallization but avoid excessive grain growth. This temperature will be about 1175 ± 32°C (2150 ± 75°F). The temperature for recrystallization is, however, considerably affected by the purity and amount of cold work prior to annealing.

When the amount of reduction is limited, complete recrystallization is very difficult unless the annealing temperature is substantially increased. Purity is at the core of this problem. Unless sufficient work (about 75% reduction) is put into the material, tantalum does not have the impurities present to act as nucleation sites for grain growth unless an inordinate amount of energy in the form of annealing heat is added. This will result in recrystallization, but to a very large grain size. The larger the cross section, the more severe is this problem. Recrystallization to a finer grain size becomes more readily obtainable. The tensile test is normally used to determine the state of anneal. American Society for Testing and Materials (ASTM) specifications for annealed tantalum are shown in Table 3.

This test is backed by hardness tests, Olsen cups, and grain size determinations to ensure product quality. The Olsen test is particularly effective for thin sheet as tensile test elongations decrease as a function of material thickness. The Olsen cup has the added advantage of detecting any strong directionality tendencies or to show "orange peel," indicating a coarse grain (see Table 4).

Tantalum sheet develops directionality if a producer does not choose his rolling and annealing schedules with care. Directionality is a term used to describe nonuniform sheet properties in the rolling direction and transverse to the rolling direction. Sheet with directionality has reduced elongations in the transverse direction which may affect performance in spinning and deep-drawing operations. When these metalworking functions are to be performed, it is helpful for the user to so specify when ordering the material.

Physical Properties

Tantalum, the seventy-third element in the periodic table is a member of the group 5 elements. Physical properties for tantalum are shown in Table 5.

Table 4. Annealed Tantalum Sheet: Olsen Cup Data, 7/8-in. Ball

	Typical	Minimum depth specification	Typical force, lb
0.030 regular	0.450	—	3920
0.020 regular	0.425	—	3200
0.010 regular	0.350	0.320	1540
0.005 deep draw	0.380	0.320	820
0.005 regular	0.250	—	504

Table 5. Physical Properties of Tantalum

Atomic weight	180.9
Density	16.6 g/cm^3,. 0.601 lb/in.^{3}
Melting point	2996°C, 5432°F
Vapor pressure at 1727°C	9.525 x 10^{-11} mm Hg
Linear coefficient of expansion	1135 K; 5.76 x 10^{-6}/°C
	1641 K; 9.53 x 10^{-6}/°C
	2030 K; 12.9 x 10^{-6}/°C
	2495 K; 16.7 x 10^{-6}/°C
Thermal conductivity	20°C; 0.130 cal/cm-s °C
	100°C; 0.131 cal/cm-s °C
	1430°C; 0.174 cal/cm-s °C
	1630°C; 0.186 cal/cm-s °C
	1830°C; 0.198 cal/cm-s °C
Specific heat	100°C; 0.03364 cal/g
Electrical conductivity	13.9% IACS
Electrical resistivity	-73°C; 9.0 μΩ/cm
	75°C; 12.4 μΩ/cm
	127°C; 18.0 μΩ/cm
	1000°C; 54.0 μΩ/cm
	1500°C; 71.0 μΩ/cm
	2000°C; 87.0 μΩ/cm

CORROSION RESISTANCE

Tantalum is inert to practically all organic and inorganic compounds at temperatures under 150°C (302°F). The only exceptions to this are hydrofluoric acid and fuming sulfuric acids. At temperatures under 150°C (302°F) it is inert to all concentrations of hydrochloric acid, to all concentrations of nitric acid (including fuming), to 98% sulfuric acid, to 85% phosphoric acid, and to aqua regia (see Table 6). Corro-

Table 6. Materials to Which Tantalum Is Completely Inert Up to at Least 150°C (302°F)

Acetic acid	Hydrogen
Acetic anhydride	Hydrogen chloride
Acetone	Hydrogen iodide
Acids, mineral (except HF)	Hydrogen peroxide
Acid salts	Hydrogen sulfide
Air	Iodine
Alcohols	Hypochlorous acid
Aluminum chloride	Lactic acid
Aluminum sulfate	Magnesium chloride
Amines	Magnesium sulfate
Ammonium chloride	Mercury salts
Ammonium hydroxide	Methyl sulfuric acid
Ammonium phosphate	Milk
Ammonium sulfate	Mineral oils
Amyl acetate	Motor fuels
Amyl chloride	Nitric acid, industrial fuming
Aqua regia	Nitric oxides
Barium hydroxide	Nitrogen
Body fluids	Nitrosyl chloride
Bromine, wet or dry	Nitrous oxides
Butyric acid	Organic chlorides
Calcium bisulfate	Oxalic acid

Table 6. (Continued)

Calcium chloride	Oxygen
Calcium hydroxide	Perchloric acid
Calcium hypochlorite	Petroleum products
Carbon tetrachloride	Phenols
Carbonic acid	Phosphoric acid, < 4 ppm F
Carbon dioxide	Phosphorus
Chloric acid	Phosphorus chlorides
Chlorinated hydrocarbons	Phosphorus oxychloride
Chlorine oxides	Phthalic anhydride
Chlorine water and brine	Potassium chloride
Chlorine wet or dry	Potassium dichromate
Chloroacetic acid	Potassium iodide, iodine
Chrome plating solutions	Potassium nitrate
Chromic acid	Refrigerants
Citric acid	Silver nitrate
Cleaning solutions	Sodium bisulfate, aqueous
Copper salts	Sodium bromide
Ethyl sulfate	Sodium chlorate
Ethylene dibromide	Sodium chloride
Fatty acids	Sodium hypochlorite
Ferric chloride	Sodium nitrate
Ferrous sulfate	Sodium sulfate
Foodstuffs	Sodium sulfite
Formaldehyde	Sugar
Formic acid	Sulfamic acid
Fruit products	Sulfur
Hydriodic acid	Sulfur dioxide
Hydrobromic acid	Sulfuric acid, under 98%
Hydrochloric acid	Water

sion is first noticed at about 190°C (375°F) for 70% nitric acid, at
about 175°C (345°F) for 98% sulfuric acid, and at about 180°C (355°F)
for 85% phosphoric acid. Fuming sulfuric acid attacks tantalum even
at room temperature. Similarly, hydrofluoric acid, anhydrous HF, or
any acid medium containing fluoride ion will rapidly attack the metal.
Commerical phosphoric acid may attack tantalum due to the presence
of small amounts of fluoride impurity. One exception to fluoride attack
appears to be in chromium plating baths. Hot oxalic acid is the only
organic acid known to attack tantalum. The corrosion rates of tantalum in various acid media are given in Table 7.

Fused sodium and potassium hydroxides and pyrosulfates dissolve
tantalum. It is attacked by concentrated alkaline solutions at room
temperature; it is fairly resistant to dilute solutions. Tantalum is
completely inert to body fluids.

Tantalum's resistance to oxidation by various gases is very good
at low temperatures, but it reacts rapidly at high temperatures. Only
HF and SO_3 attack the metal under 100°C (212°F); most gases begin to
react with it at 300 to 400°C (570 to 750°F). As the temperature and

Table 7. Corrosion Rates of Tantalum in Selected Media

Medium	Temperature		Corrosion rate, mpy
	°C	°F	
Acetic acid	100	212	Nil
$AlCl_3$ (10% Soln.)	100	212	Nil
NH4Cl (10% Soln.)	100	212	Nil
HCl 20%	21	70	Nil
	100	212	Nil
Conc.	21	70	Nil
	100	212	Nil
HNO_3 20%	100	212	Nil
70 %	100	212	Nil
65%	170	338	< 1
H_3PO_4, 85%	25	76	Nil
	100	212	Nil

Table 7. (Continued)

Medium	Temperature		Corrosion rate, mpy
	°C	°F	
H_2SO_4			
10%	25	76	Nil
40%	25	76	Nil
98%	25	76	Nil
98%	50	122	Nil
98%	100	212	Nil
98%	200	392	3
98%	250	482	Rapid
H_2SO_4, fuming (15% SO_3)	23	73	0.5
	70	158	Rapid
Aqua regia	25	78	Nil
Chlorine wet	75	167	Nil
H_2O			
Cl_2 sat.	25	76	Nil
Sea	25	76	Nil
Oxalic acid	21	70	Nil
	96	205	0.1
NaOH 5%	21	70	Nil
	100	212	0.7
10%	100	212	1
40%	80	176	Rapid
HF, 40%	25	76	Rapid

concentration of such gases as oxygen, nitrogen, chlorine, hydrogen chloride, and ammonia are increased, oxidation becomes more rapid; the usual temperature for rapid failure is 500 to 700°C (930 to 1290°F). The conditions under which tantalum is attacked are noted in Table 8.

Table 8. Temperatures at Which Various Media Attack Tantalum

Medium	State	Remarks
Air	Gas	At temperatures over 300°C (572°F)
Alkaline solutions	Aqueous	At pH > 9, moderate temperature, some corrosion
Ammonia	Gas	Pits at high temperature and pressures
Bromine	Gas	At temperatures over 300°C (572°F)
Chlorine, wet	Gas	At temperatures over 250°C (482°F)
Fluorides, acid media	Aqueous	All temperature and concentrations
Fluorine	Gas	At all temperatures
HBr, 25%	Aqueous	Begins to corrode at temperatures over 190°C (374°F)
Hydrocarbons	Gas	React at temperatures around 1500°C (2732°F)
HCL 25%	Aqueous	Begins to corrode at temperatures over 190°C (374°F)
HF	Aqueous	Corrodes at all temperatures and pressures
Hydrogen	Gas	Causes embrittlement, especially at temperatures over 400°C (752°F)
HBr	Gas	At temperatures over 400°C (752°F)
HCL	Gas	At temperatures over 350°C (662°F)
HF	Gas	At all temperatures
Iodine	Gas	At temperatures over 300°C (572°F)
Nitrogen	Gas	At temperatures over 300°C (572°F)
Oxalic acid, sat. soln.	Aqueous	At temperatures of about 100°C (212°F)
Oxygen	Gas	At temperatures over 350°C (662°F)
H_3PO_4, 85%	Aqueous	Corrodes at temperatures over 180°C (356°F), at higher temperatures for lower concentrations
Potassium carbonate	Aqueous	Corrodes at moderate temperatures depending on concentration

Table 8. (Continued)

Medium	State	Remarks
Sodium carbonate	Aqueous	Corrodes at moderate temperatures depending on concentration
NaOH, 10%	Aqueous	Corrodes at about 100°C (212°F)
NaOH	Molten	Dissolves metal rapidly (over 320°C) (608°F)
Sodium pyrosulfate	Molten	Dissolves metal rapidly (over 400°C) (752°F)
H_2SO_4, 98%	Aqueous	Begins to corrode at temperatures over 175°C (347°F); lower concentrations begin to corrode at higher temperatures
H_2SO_4 (oleum) (over 98% H_2SO_4)	Fuming	Corrodes at all temperatures
Sulfuric trioxide	Gas	At all temperatures
Water	Aqueous	Corrodes at pH > 9, reacts at high temperatures

FABRICATION

Welding

The weldability of tantalum is limited only by the ingenuity of the welder in reducing the time and temperature at which the metal may be exposed to moisture, carbon, oxygen, and nitrogen.

Tantalum is one of the *reactive metals*. Like titanium, zirconium, and columbium, tantalum will react with contaminants when exposed to them at temperatures as low as 315°C (600°F). It must be remembered, however, that contamination is a time-temperature relationship. Seconds at a high temperature as in spot welding or some minutes at 315°C (600°F) when extruding may not be harmful. But there is danger everywhere when heat is present. Once exposed to any atmosphere excepting inert gases under closely controlled conditions, the metal will in all probability become embrittled. Only hydrogen contamination can be removed by outgassing by heating to 650°C (1200°F) and cooling in vacuum.

Because of tantalum's affinity for *gettering*, any fusion weld must be performed in an atmosphere free of contamination. This means not only the weld puddle, but all hot metal, must be protected. Four different procedures are used to achieve this protection. Although one method may be preferable to another, each has its own limitations.

Familiarity with each is necessary to achieve a balance between special-
ization and flexibility.

Electron Beam Welding

Electron beam (EB) welding requires highly sophisticated expensive
equipment. Most welding is done in a vacuum chamber, although
efforts are being made to allow welding to be performed outside a
vacuum chamber using differential pumping. Unless this is done, the
size of the chamber limits the size of the work. EB welding produces
good-quality welds which are characterized by narrow weld zones and
good penetration. It is useful for intricate, hard-to-reach welds,
particularly fillets and tees of different cross sections. Equipment
tooling and setup are all expensive.

Flow-Purged Chamber

A flow-purged chamber is used when the work is too large to fit any
available chamber and the joints are too complex to permit open-air
welding. The enclosure is constructed of polyethylene sheet and
masking tape. Argon flowing through the "bag" displaces or mixes
with the entrapped air to a level where welding can be performed.
Argon must be allowed to flow until the work is cool. Enclosure of all
sides of the work is mandatory.

Dry Box or Vacuum Purge Chamber

The dry box provides the best inert gas atmosphere possible. The
parts to be welded are placed in the box; the chamber is sealed and a
vacuum pulled on the box. Normal practice is to pump down to
approximately 50 μm. The box is then back filled with high purity
tank argon to a slight positive pressure. Welding is done from the outside using rubber gloves inserted through the sides of the chamber.
    The major limitations of the dry box are the size of the box, which
predetermines the size of the work that can be handled, and the skill
of the operator necessary to manipulate the work. This is not to mention the initial investment of the box itself. High-quality butt welds
joining 0.010-in. tantalum using 65 A at 5 in./min with a 3/32 in. tungsten electrode are commonplace using this technique.

Open-Air Welding

Open-air welding with tantalum is possible provided that the most
stringent precautions are observed. Only relatively simple joints
allowing adequate shielding are possible.
    Protection must be given to the arc, to the heated metal in front,
to the sides, and to the cooling metal behind and underneath and weld

bead. The protective atmosphere is provided by using a gentle but
adequate gas flow from the maximum-size cup feasible consistent with
good visibility. A blanket of inert gas must be supplied by properly
constructed trailing shields, which provide a flow of gas until the
metal is cooled below the critical temperature. Backup shielding on
the underside from the weld bead must also be used to give protection
until the material is cooled.

Absolute cleanliness is essential for good tantalum welding using
any of the foregoing techniques. A generally accepted procedure
to remove the naturally occurring oxide film is to abrade slightly with
emery cloth the surfaces to be joined. Then etch with a nitric- hydrofluoric acid solution. Finally, rinse before welding with a solvent such
as acetone to be sure that all grease is removed. Handle the work only
with clean, lint-free nylon gloves.

Spot welding tantalum is fairly straightforward. Tip diameters,
typically 1/8 in., should be kept to a minimum. Tungsten-tipped
electrodes are used to prevent copper pickup. Typical settings for
welding 0.015-in.-thick annealed stock used in the manufacture of furnace elements are 8 to 10 cycles per second. Air pressure is set at
25 psi and the least amount of current possible for a good nugget to
form is permitted. The short cycle time helps preclude contamination.
Some discoloration on heavier gauges may result, but this can be removed by acid cleaning.

Suggestions for good tantalum brazing are: absolute cleanliness,
careful joint design and fit-up to assure complete brazing alloys, and
a brazing temperature well above the flow temperature [870°C (1600°F)
recommended].

Soldering is not used or recommended in tantalum fabrications.

Machining

Although tantalum cannot be classed as an easy material to machine, it
is not among the most difficult. The material is soft with low strength
and good ductility. Chips will not break cleanly. Instead, the material is gummy and will tear rather than break clean unless the
machinist follows a few simple procedures.

Tools

High-speed tool steels are preferred, as carbides are too brittle for
tantalum. Tools should be kept sharp. Breakdown will occur rapidly
after the first signs of dullness. Extreme cutting angles are recommended to keep the tool and chip well clear of the work.

Side rake   15°
Back rake   45°
Front rake   5°

Speeds and Feeds

Slow speeds and high feeds are the best procedure. Speeds should not exceed 75 ft^2/min, with the recommended speed 25 ft^2/min. Feed should be 0.015 in. per revolution for roughing and 0.005 in. per revolution for finishing. Depth of cut can vary from 1/32 in. to light finishing cuts as desired. The tool should be well supported with little overhang. Feed should be continued as long as tool and work are in contact. Do not allow the tool to dwell on the work.

Lubrication

Single-point turning is generally dry. For milling and drilling, a generous amount of cutting fluid helps cool the work and carries away chips. Chlorothane or 40% chlorothane in black oil is recommended. Conventional cutting fluids are not satisfactory.

Forming

Tantalum in the annealed condition is an extremely ductile material. Unusually high reductions without annealing are possible because of its low rate of work-hardening characteristics. Tubing 1-1/16 in. in diameter with a 1/8 in. wall by 16 in. long has been drawn from circular blanks without annealing in a series of seven draws. It is nevertheless recommended that the producer be advised when spinning or drawing is planned.

Spinning

Successful tantalum spinning can be accomplished using conventional spinning techniques. The slow work-hardening rate permits repeated drafts without in process anneals unless unusually severe formations are to be attempted. Spinning thinner material such as 0.110 or 0.020 in. thick may have to be annealed more often.

Mandrels should be made of steel or aluminum bronze. Hardwood and composition mandrels are usually too soft to permit sufficient ironing for good surface finish. If steel mandrels are used, bottoms should be faced with aluminum bronze to prevent galling of the blank. Steel roller wheels or yellow brass tools are used with generous amounts of yellow soap. A commercial compound, Warren's Spinning Compound 1, has been found to offer good lubricating characteristics for tantalum.

Deep Drawing

Although tantalum is a soft, ductile metal, certain precautions are suggested for optimum results. Conventional reductions are possible

provided that due allowance is made for tantalum's galling characteristics. The metal will have more of a tendency to seize on the punch
and/or draw ring. This friction may result in premature failure unless lubrication is generous. Aluminum bronze is recommended for the
draw ring when justified.

The metal can be stretched and ironing is feasible with aluminum
bronze dies. As in any deep-drawing operation, some experimentation
with hold-down, punch and draw ring radii, and clearance may be
necessary to prevent wrinkling. Initial reductions of 50% are possible.
Drawing may be continued, although gradually decreasing the amount
of reduction per draw is recommended.

SUGGESTED APPLICATIONS

Because of its relatively high price, tantalum can only be recommended
for use in extremely corrosive media, in areas where no corrosion of
the part can be tolerated, or where very high purity materials are
being processed. Although some plastics and even glass fill these requirements to a large extent, tantalum is a structurally sound material
of construction, can take considerable mechanical abuse, and has a
much higher heat transfer coefficient. Tantalum should be used as a
material of construction in locations and for equipment where hot concentrated hydrochloric, sulfuric, or phosphoric acid will be present.
Tantalum is used by the medical profession for instruments and for
metal implants in the body. In the manufacture of high-purity chemicals and pharmaceuticals, tantalum ensures that no impurities are introduced from the container or reactor.

ACKNOWLEDGMENT

Data for this chapter were furnished by Zane B. Laycock of NRC, Inc.

3.8
Zirconium

DONALD R. KNITTEL* / Teledyne Wah Chang Albany, Albany, Oregon

INTRODUCTION

The element zirconium, atomic number 40 and molecular weight 91, was discovered by Klaproth in 1789. A pure and ductile metal was not obtained until around 1920, when Van Arkel, DeBoar, and Fast produced some zirconium using the iodide decomposition process. Large quantities were not available until after the magnesium reduction technique was developed by W. J. Kroll at the Bureau of Mines in 1947. Zirconium became available for use in industrial process applications around 1958.

---

*Present address: Cabot Corporation, Kokomo, Indiana

Zirconium's primary ore source is zircon, a zirconium silicate sand found as heavy beach sand in most parts of the world. It ranks 18 in order of elemental relative abundance, comprising about 0.0165% of the earth's crust. It ranks sixth in order of abundance of engineering metals, behind aluminum, iron, magnesium, titanium, and manganese [1]. Zirconium's abundance, wide distribution, and corrosion-resistant properties makes it an increasingly important engineering material for use in a variety of severely corrosive environments [2]. Because of its low thermal neutron capture cross section (0.17 barn), excellent corrosion resistance, and good mechanical properties, zirconium's primary use has been fuel cladding for nuclear reactors. Considerable alloy development work has taken place over the past 30 years to develop alloys with higher strength and heat resistance while maintaining the corrosion resistance of pure zirconium metal [3, 4]. The results are the Zircaloys, a group of zirconium 1.5% tin alloys developed in the United States and zirconium 1 to 2.5% niobium alloys developed in Russia and Canada. These alloys have reduced corrosion resistance compared to pure zirconium in 350 to 400°C (660 to 750°F) steam and water. Their advantages are, they have greater yield strength and more predictable corrosion rates.

There will be no further discussion in this chapter on zirconium's application in the nuclear industry; however, zirconium's development for use in industrial related corrosive environments will be discussed in detail. Zirconium, for industrial uses, can contain up to 4.5% hafnium. Hafnium has no noticeable effect on the corrosion resistance; however, its high thermal neutron captive cross section (113 barns) makes its presence intolerable in nuclear service [5-7]. Hafnium occurs naturally with zirconium at levels of about 2 to 4%.

Zirconium for industrial uses is supplied according to the American Society for Testing Materials (ASTM) or the American Society for Mechanical Engineering (ASME) equivalent specifications. The industrial grades ASTM nomenclature are Grade UNS R60702 (Grade 702, Grade 11, commercial grade), usually specififed for general corrosion resistant service; Grade UNS R60704 (Grade 704), an alloy with 1.5% tin and 0.2% iron plus chromium, used in chemical service where added strength is desired; and Grade UNS R60705 (Grade 705), an alloy with 2.5% niobium used where high strength and ductility are required; and Grade UNS R60706 (Grade 706), also alloyed with niobium and developed for service requiring extreme ductility. The ASTM chemical requirements of these alloys are listed in Table 1. Both Grades 702 and 705 are ASME Boiler Code and Pressure Vessel qualified. The design limits for both grades are given in Table 2.

The alloys most important to industrial service applications are Grades 702 and 705. The rest of this chapter deals primarily with these two alloys, with occasional remarks about Grade 704.

Table 1. ASTM Chemical Requirements[a] for Zirconium Alloys

Element	Composition, % Grades				
	R60702	R60704	R60705	R60706
Zirconium + hafnium min.	99.2	97.5	95.5	95.5
Hafnium, max.	4.5	4.5	4.5	4.5
Iron + chromium	0.2 max.	0.2–0.4	0.2 max.	0.2 max.
Tin	–	1.0–2.0	–	–
Hydrogen, max.	0.005	0.005	0.005	0.005
Nitrogen, max.	0.025	0.025	0.025	0.025
Carbon, max.	0.05	0.05	0.05	0.05
Niobium	–	–	2.0–3.0	2.0–3.0
Oxygen, max.	0.16	0.18	0.18	0.16

[a]By agreement between the purchaser and the manufacturer, analysis may be required and limits established for elements and compounds not specified in the table of chemical composition.
Source: American Society for Testing and Materials, Philadelphia, B-551, 1981 Annual Book of ASTM Standards, Part 8.

Table 2. ASME Mechanical Requirements of Zirconium Grade 702 and Grade 705 Unfired Vessels

Material form and spec. no.	Grade	Condition	Specified tensile strength,[a] ksi	Minimum yield strength,[a] ksi	Notes	Maximum allowable stress values in tension for metal temperature not exceeding °C,[a] (ksi)						
						40	90	150	200	280	320	370
Flat-rolled products SB 551	702[b]		52.0	30.0		13.0	11.0	9.3	7.0	6.1	6.0	4.8
	705[c]		80.0	55.0		20.0	16.6	14.2	12.5	11.3	10.4	9.9
Tubing SB 523	702	Seamless	52.0	30.0		13.0	11.0	9.3	7.0	6.1	6.0	4.8
	705		80.0	55.0		20.0	16.6	14.2	12.5	11.3	10.4	9.9
Tubing SB 523	702	Welded	52.0	30.0	d, e	11.1	9.4	7.9	6.0	5.2	5.1	4.1
	705				d, e	17.0	14.1	12.1	10.6	9.6	8.8	8.4
Forgings SB 493	702		52.0	30.0		13.0	11.0	9.3	7.0	6.1	6.0	4.8
	705		80.0	55.0		20.0	16.6	14.2	12.5	11.3	10.4	9.9
Bar SB 550	702		52.0	30.0		13.0	11.0	9.3	7.0	6.1	6.0	4.8
	705		80.0	55.0		20.0	16.6	14.2	12.5	11.3	10.4	9.9

[a]Tabulated in English units because it appears this way in the ASME Code.

[b]Grade 702 (UNS R60702).

[c]Grade 705 (UNS R60705).

[d]85% joint efficiency has been used in determining the allowable stress value for welded tube.

[e]Filler metal shall not be used in the manufacture of welded tube.

Source: ASME Boiler and Pressure Vessel Code, 1980 Edition, Sec. VIII, Div. 1, American Society of Mechanical Engineers, New York.

Table 3. Physical Properties of Zirconium

Melting point	°C	1,845
Boiling point	°C	3,580
Density	g/cm^3	6.506
Allotropic transformation	°C	865
Thermal conductance	cal/(s) (cm^2) (°C/cm)	0.0505
Specific heat	cal/g at 25°C	0.066
Heat of fusion	kcal/g-atom	5.50
Heat of vaporization	kcal/g-atom	142
Atomic volume	W/D	14.1
Electronegativity	Pauling's	1.4
Covalent radius	Å	1.45
Vickers hardness		110
Thermal expansion	deg/cm x 10^6	5.78
Electrical resistivity	$\mu\Omega$-cm	41.4
Crystal structure	Below 865°C	Hexagonal
Crystal structure, space	865-1845°C	Body-centered cubic
Thermionic work function	eV	4.1
Compressibility	cm^2/kg x 10^{-6}	11.77
Magnetic susceptibility	emu/g-atom x 10^6	119
Poisson ratio		0.34
Young's modulus	kg/cm^2 x 10^{-4}	0.939
Grueneisen constant		128,900
Hall coefficient	V-cm/A-Oe x 10^{12}	0.18

PHYSICAL AND MECHANICAL PROPERTIES

A compilation of physical properties for zirconium is given in Table 3. Table 4 lists the ASTM minimum requirements for mechanical properties of zirconium alloys. Grade 704 has higher tensile strength and is less

Table 4. Minimum ASTM Requirements for the Mechanical Properties of Zirconium at Room Temperature (Cold Worked and Annealed)

Grade	Tensile strength, kg/cm^2	Yield strength (0.2% offset), kg/cm^2	Percent elongation (2-in. gage)	Bend test[a]
702	3900	2100	16	5T
704	4200	2500	14	5T
705	5600	3900	16	3T
706	5200	3500	20	2.5T

[a]T equals the thickness of the bend test specimen. Bend tests are not applicable to material over 0.187 in. (4.75 mm) in thickness.
Source: ASTM B-551, *1981 Annual Book of ASTM Standards*, Part 8, American Society for Testing and Materials, Philadelphia.

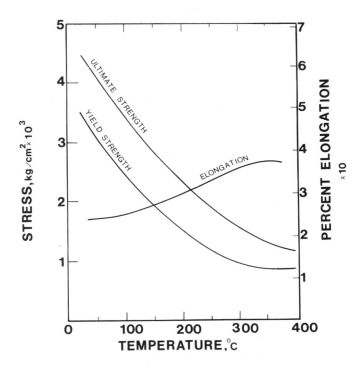

Figure 1. Typed tensile properties for the transverse direction of zirconium Grade 702. (From Ref. 74.)

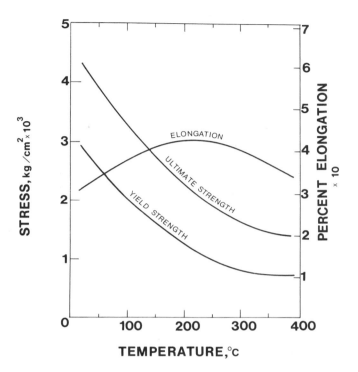

Figure 2. Typical tensile properties for the longitudinal direction of
zirconium Grade 702. (From Ref. 74.)

ductile than Grade 702. The fabrication properties, however, are very
similar. Grade 705 has greater tensile strength than Grade 704 and is
more ductile than Grade 702. Grade 705 can be cold bent around a
radius three times the metals thickness compared to five times the
thickness for Grades 702 and 704. Grade 706 has less tensile strength
than Grade 705, but has greater fabricability especially in die forming.
The dependence of the mechanical properties of Grade 702, with respect to temperature, are illustrated graphically for both transverse
and longitudinal direction in Fig. 1 and 2 (oxygen levels are about
1200 ppm). Mechanical properties of the different grades of zirconium
depend to a larger extent on the purity. The most notable impurities
affecting tensile strength and hardness are oxygen, nitrogen, and
iron.

   Zirconium and its alloys exhibit a fatigue limit behavior similar to
most ferrous alloys. Grade 702 fatigue characteristics are illustrated
in Fig. 3. The fatigue limit is increased by most alloy additions.
Oxygen has a very strong effect on fatigue properties.
   Zirconium has a strong tendency to gall. An oxide layer on the
metal can be generated which has a low coefficient of friction and reduces galling. The oxide can be formed by heating in air at about
560°C (1040°F) for 4 h. The integrity of the oxide coating can be
monitored using electrical contact resistance [8]. With this treatment
zirconium can be used for rotating and sliding parts (i.e., bolts,
bearing sleeves, rotating shaft, pump impellers, etc.). Care in handling the assembly is required because the thin film is easily damaged.

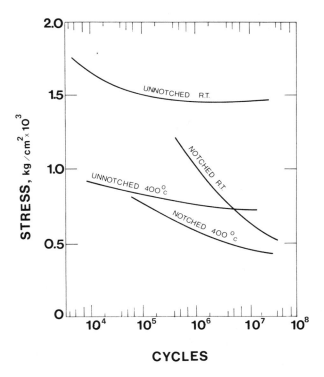

Figure 3. Flexible fatigue curves for zirconium Grade 702. (From
Ref. 74.)

CORROSION PROPERTIES

Zirconium is resistant to many chemical solutions. These include mineral acids, molten alkalies, alkaline solutions, and most organic and salt solutions [5, 6, 9-14]. It has excellent oxidation-resistant properties to 400°C (750°F) in air, steam, carbon dioxide, sulfur dioxide, nitrogen, and oxygen [3, 4, 10]. This excellent corrosion resistance depends on the integrity of the inert passive oxide film that forms on the surface when exposed to air. Zirconium is attacked in hydrofluoric acid, wet chlorine, concentrated sulfuric acid, aqua regia, ferric chloride, and cupric chloride solutions [11, 15-20].

Corrosion properties discussed in this section are for a zirconium alloy equivalent to Grade 702. The corrosion properties of the other Grades of zirconium, Grade 704 and Grade 705, are almost equivalent. Where differences have been observed they will be noted in the text.

Acid Solutions

Halogen Acids

Zirconium is vigorously attacked by hydrofluoric acid at all concentrations, even small amounts (1 ppm) in an acidic solution will cause accelerated corrosion [20]. The other halogen acids (hydrochloric, hydrobromic, and hydroiodic acids) do not attack zirconium [6, 15, 21, 22]. Corrosion information for hydrobromic and hydroiodic acids is based on tests conducted at Teledyne Wah Chang Albany (TWCA). Corrosion rates in boiling 20% and 45% HBr were less than 5 mils per year (mpy) and in 5% HI with 85% acetic acid at 80°C (176°F) the corrosion rate was less than 1 mpy. In hydrochloric acid zirconium has excellent resistance at all concentrations and temperatures even above boiling [15, 16, 21-23]. This corrosion information is illustrated in Fig. 4. Oxidizing impurities (e.g., nitrate, ferric, or cupric ions) in hydrochloric acid cause pitting and embrittlement [16, 18, 23, 24]. Intergranular attack has been observed in welds of zirconium with high carbon and iron levels exposed to 33% hydrochloric acid at 150°C (302°F) [25]. This selective weld attack could be alleviated by using higher purity zirconium or annealing the welds [25-28]. Selective weld attack of Grades 702, 704, and 705 was not observed at our laboratory in tests conducted in 32% hydrochloric acid at 80°C (176°F).

Nitric Acid

Zirconium's corrosion rate is less than 1.0 mpy in nitric acid at concentrations below 70% and temperatures up to 250°C (480°F), and at concentrations between 70% and 98% to boiling [11, 13, 15, 16, 19, 30].

Corrosion tests conducted on zirconium exposed to 35 to 65% nitric acids, with chloride ion impurities (20 ppm maximum) and up to 15% zirconyl nitrate had corrosion rates of less than 1 mpy in both liquid and vapor exposures [17]. It is reported to have corrosion rates less than 1 mpy at temperatures below 70°C (158°F) in white fuming nitric acid and less than 5 mpy in red fuming nitric acid at 25°C (77°F) [11, 19]. Corrosion tests conducted by M. G. Fontana at Ohio State University and by Titanium Alloy Manufacturing (TAM) show that zirconium's corrosion properties in fuming nitric acid were not as good as cited above [10]. Interest in these environments should be evaluated by individual cases.

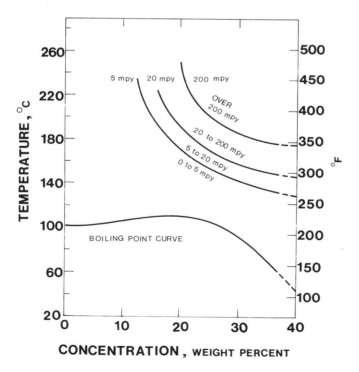

Figure 4. Corrosion curves of zirconium in hydrochloric acid. (Excerpted by special permission from *Chemical Engineering*, June 2, 1980. Copyright 1980 by McGraw-Hill, Inc., New York.)

Table 5. Corrosion Rates for Zirconium in Organic Solutions

Environment	Concentration, wt %	Temperature, °C	Corrosion rate, mpy
Acetic acid	5-99.5	35-boiling	<0.07
Acetic anhydride	99.5	Boiling	0.03
Aniline hydrochloride	5, 20	35-100	<0.01
Chloroacetic acid	100	Boiling	<0.01
Citric acid	10-50	35-100	<0.2
Dichloroacetic acid	100	Boiling	<20
Formic acid	10-90	35-boiling	<0.2
Lactic acid	10-85	35-boiling	<0.1
Oxalic acid	0.5-25	35-100	<0.5
Tartaric acid	10-50	35-100	<0.05
Tannic acid	25	35-100	<0.1
Trichloroacetic acid	100	Boiling	>50
Urea reactor	58% urea 17% NH_3 15% CO_2 10% H_2O	193	<0.1

Source: Excerpted by special permission from Chemical Engineering, June 2, 1980. Copyright 1980 by McGraw-Hill, Inc., New York.

Organic Acids

Zirconium has corrosion rates of less than 5 mpy in almost all organic acids except di and trichloroacetic acids at boiling [12]. These results are indicated in Table 5. If attack occurs in organic acid environments, it usually results from the presence of a catalysts (e.g., fluorides).

Phosphoric Acid

Corrosion-resistant properties of zirconium in phosphoric acid are illustrated in Fig. 5 [11, 15, 30, 31]. The addition up to 5% nitric acid to 88% phosphoric acid at room temperature does not cause accelerated attack [30]. Preferential weld attack was observed in 50% boiling solutions in tests conducted at TWCA. It was found that vacuum annealing welds at 750°C (1380°F) for 1/2 h significantly reduces the

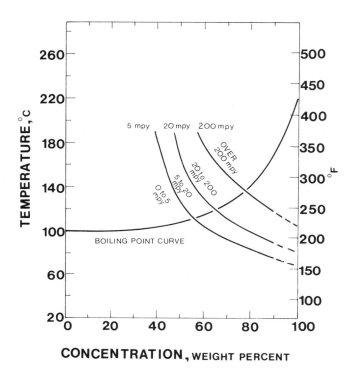

Figure 5. Corrosion curves of zirconium in phosphoric acid. (From
Ref. 16.)

selective weld attack. Fluoride ions (5 ppm) in 40 or 50% boiling phosphoric acid causes corrosion rates of zirconium in excess of 1000 mpy.

Sulfuric Acid

The isocorrosion curves in Fig. 6 illustrate zirconium's corrosion
characteristics in sulfuric acid [11, 15, 16, 30, 32]. This figure is
for Grade 702. Grade 705 is slightly less resistant (the 5-mpy line
intercepts the boiling curve at about 60% sulfuric acid). The dashed
weld limit line divides the figure into two regions. In the lower concentration region the weld corrodes at about the same rate as the
parent metal. In the higher concentration region welds are preferentially attacked and weld use is not recommended unless annealing
treatment is applied. A 750°C (1380°F) vacuum anneal for 30 min
significantly reduces selective weld attack [16, 33, 34]. After the
anneal, welds have similar corrosion-resistant properties compared to

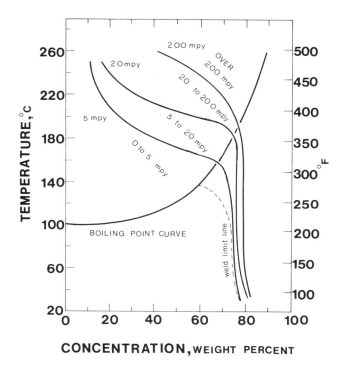

CONCENTRATION, WEIGHT PERCENT

Figure 6. Corrosion curves of zirconium in sulfuric acid. The reduction of corrosion due to welding is indicated by the weld limit line. (Excerpted by special permission from *Chemical Engineering*, June 2, 1980. Copyright 1980 by McGraw-Hill, Inc., New York.)

the parent metal. Impurities in sulfuric acid (e.g., Cu^{2+}, Fe^{3+}, NO_3^-, and Cl^- at 200-ppm levels) cause accelerated attack in the 65 to 75 % sulfuric acid region. However, below 65% and at temperatures to boiling, zirconium has corrosion rates less than 5 mpy [35]. Increasing the impurity concentration of any of the impurities listed above to 1% in 60% sulfuric acid does not increase the corrosion rate above 5 mpy [35]. Forty to 55% sulfuric acid saturated with chlorine gas accelerated attack and pitting of zirconium is observed; however, in higher sulfuric acid concentrations (55 to 70%) zirconium's corrosion properties are not effected by chlorine gas [17]. Corrosion of zirconium in 80% sulfuric acid can be reduced by adding 1% bichromate or 0.0082 to 0.0548 molar tetravalent zirconium ions [4, 30]. If a protective film is formed on the metal by oxidation in air for 2 h at 600°C (1110°F), the corrosion rate in 80% sulfuric acid is very low. This passive film, however, gradually dissolves and no new film takes its

Table 6. Corrosion Rates of Zirconium in Some Mixed Acids

Test solution,[a] Wt %		Temperature, °C	Corrosion rates, mpy
1% H_2SO_4	99% HNO_3	RT,[b] 100	0.06
10% H_2SO_4	90% HNO_3	RT, 100	WG[c]
14% H_2SO_4	14% HNO_3	Boiling	0.1
25% H_2SO_4	75% HNO_3	100	150
50% H_2SO_4	50% HNO_3	RT	0.63
68% H_2SO_4	5% HNO_3	Boiling	2000
68% H_2SO_4	1% HNO_3	Boiling	11
75% H_2SO_4	25% HNO_3	RT	260
88% H_3PO_4	0.5% HNO_3	RT	0.0
88% H_3PO_4	5% HNO_3	RT	WG[b]
Aqua Regia		RT	Dissolved
20% HCl	20% HNO_3	RT	Dissolved
10% HCl	10% HNO_3	RT	Dissolved
7.5% H_2SO_4	19% HCl	Boiling	0.5
34% H_2SO_4	17% HCl	Boiling	0.3
40% H_2SO_4	14% HCl	Boiling	0.2
56% H_2SO_4	10% HCl	Boiling	2.0
60% H_2SO_4	1.5% HCl	Boiling	1.0
69% H_2SO_4	1.5% HCl	Boiling	5.0
69% H_2SO_4	4% HCl	Boiling	15.0
72% H_2SO_4	1.5% HCl	Boiling	20.0

[a]Corrosion test with HCl/H_2SO_4 and HCl/HNO_3 solutions were conducted at TWCA.

[b]RT, room temperature.

[c]WG, weight gain.

Source: Refs. 7, 11, 30.

place. The dissolution rate rises to where it would be if no oxide film
had been present [4].

Other Acids

Zirconium has excellent corrosion resistance in up to 30% chromic acid
at temperatures to 100°C (212°F) [10]. It cannot be used in chrome
plating solutions because of the fluoride catalyst. In sulfurous acid
zirconium has a corrosion rate less than 0.2 mpy at 100°C (212°F).
   Zirconium is also resistant to some mixed acid systems. It can be
used in acid mixtures of sulfuric-nitric, sulfuric-hydrochloric, and
phosphoric-nitric. The sulfuric acid concentration must be below 70%
[11, 30]. Zirconium is aggressively attacked in 1:3 volume mixtures
of nitric acid and hydrochloric acids (aqua regia) [7, 23]. In 1:1
volume mixtures, zirconium is attacked but much slower than in the
1:3 mixture [7, 23]. In mixtures greater than 3:1 nitric acid to hydrochloric acid, zirconium is resistant. Some data for the mixed acid
systems are given in Table 6.

Alkaline Solutions

Zirconium is resistant to almost all alkaline environments to boiling
[10, 17, 36]. It is resistant in sodium hydroxide and potassium hydroxide solutions even under anhydrous conditions. It is also resistant
to calcium hydroxide and ammonium hydroxide environments at concentrations to 28% and temperatures to boiling. Zirconium is comparable
to nickel in caustic environments. Resistance of zirconium in these
environments distinguishes it from tantalum and titanium, which are
attacked by hot alkalies.

Organic Solutions

The corrosion resistance of zirconium in organic solutions is excellent
[23, 37]. It has been tested extensively in organic-cooled reactors
where the coolant consisted of mixtures of high-boiling aromatic hydrocarbons (e.g., terphenyls) [3]. These coolants are noncorrosive to
zirconium. Early experiments in the organic coolants indicated that
hydriding could be a major problem. It was found that chlorine impurity in the organic was the major cause of gross hydriding. Elimination of the chlorine and maintaining a good surface oxide film by ensuring the presence of adequate water (>50 ppm) alleviates the hydriding problem. Zirconium suffers stress corrosion cracking in alcohol solutions with halide impurities present. In some chlorine-containing carbon compounds (e.g., carbon tetrachloride and dichlorobenzene),
zirconium has excellent corrosion resistance at temperatures below
200°C (390°F).

Salt Solutions

Zirconium is resistant to most salt solutions [11]. These include halogen, nitrate, carborates, and sulfate salts. The metal has a corrosion rate of 0.01 mpy in seawater and salt spray environments [17, 32]. Zirconium is aggressively attacked in oxidizing chloride solutions. Examples are chloride solutions with cupric or ferric ions; however, chloride solutions with mercuric or stannic ions do not attack the metal. It is also attacked when the chloride ion concentration is greater than about 12 normal (70% $CaCl_2$) [17]. Since it is aggressively attacked in hydrofluoric acid, the general consensus is that zirconium is attacked in any fluoride solution. This is not altogether true [4, 20]. Zirconium has corrosion rates of less than 1 mpy in 20% solutions of NaF, KF, AlF_3, or CaF_2 at room temperature. Twenty percent ammonium fluoride solutions, however, aggressively attack zirconium. When solutions containing the fluorides above are heated to boiling,, only 20% AlF_3 and 20% CaF_2 solutions do not attack zirconium [38, 39].

Gases

Zirconium is not attacked by most gases (CO, CO_2, SO_2, C_3H_8, N_2, steam, and air) until temperatures exceed 300 to 400°C (570 to 750°F) [3, 10]. Zirconium reacts with hydrogen at 300°C (570°F) to form hydrides. The metal oxidizes readily in air when temperatures exceed 500°C (930°F). Zirconium's resistance to air, hot water, and steam is of special interest to nuclear power applications. The metal can be exposed for prolonged periods without pronounced attack at temperatures below 425°C (795°F). In hot water and steam applications at 360°C (680°F), zirconium can tolerate up to 350 ppm chloride and iodide ions, 100 ppm fluoride ions, and 10^4 ppm sulfate ions without influencing the corrosion resistance. Zirconium is stable in NH_3 up to about 1000°C (1830°F). It reacts readily with all halogens at moderate temperatures [200 to 400°C (390 to 750°F)] to form volatile halides. Zirconium is resistant to attack in dry Cl_2 gas at 50°C but is attacked at a rate of 200 mpy in wet chlorine gas [17]. In chlorine-saturated water zirconium corrodes at less than 1 mpy. Zirconium hydrated in hot [400°C (750°F)] gas halide acid environments [40].

Molten Salts

Zirconium is resistant to attack in many molten eutectic salts [3, 10]. An excellent set of references is found in an article by Janz et al. [41]. Zirconium is very resistant to corrosion by molten sodium hydroxide up to temperatures greater than 1000°C (1830°F) [36]. Other investigators find that in hot sodium hydroxide the resistance lasts only until the white oxide forms [42]. It is somewhat less resistant in

Table 7. Corrosion of Zirconium in Some Liquid Metals

Liquid metal	Melting temperature, °C	Temperature, °C		
		300	600	800
Bi	271.3	Unknown	Poor	Poor
Bi-In-Sn	60	Unknown	Poor	Unknown
Bi-Pb	125	Good	Limited	Unknown
Bi-Pb-In	70	Unknown	Poor	Unknown
Bi-Pb-Sn	97	Good	Limited	Unknown
Ga	298	Limited	Poor	Poor
Hg	-38.4	Poor	Poor	Unknown
Li	180.5	Good	Limited	Limited
Mg	650	—	Poor[a]	Unknown
Na, K, or NaK	12.3-97.9	Good	Good	Unknown
Pb	327.4	Good[a]	Limited	Limited
Sn	231.9	Good	Unknown	Unknown

[a]At its melting point.
Source: Ref. 43.

potassium hydroxide, although not enough to prohibit use for reasonable periods [10, 17]. In nitrate salts and eutectic mixtures of sodium nitrate, potassium nitrate, and nitrate, zirconium's oxidation properties are similar to those in air [3].

Molten Metals

Corrosion results for zirconium in molten metals should be considered very carefully because corrosion rates are very sensitive to trace impurities (e.g., oxygen, hydrogen, or nitrogen) in the liquid metal. Table 7 illustrates zirconium's corrosion resistance in several liquid metal systems. It has corrosion rates of less than 1 mpy in liquid lead to 600°C (1110°F), lithium to 800°C (1470°F), mercury to 100°C (212°F), and sodium to 600°C (1110°F) [10]. Zirconium is rapidly attacked in molten zinc, bismuth, and magnesium.

Special Corrosion Topics

Pitting

Zirconium has a tendency to pit in all halide solutions except fluoride
[44]. Zirconium's susceptibility to pitting is greatest in chloride ion
solutions and decreases as the halide ion becomes heavier [i.e., the
pitting potential in a 1 N solution of Cl^-, Br^-, and I^- is 150, 400, and
680 mV, respectively, normal hydrogen electrode (nhe)]. Zirconium,
in most halide environments, does not pit because the corrosion potential is less noble than the pitting potential. The presence of oxidizing
ions in halide solutions will increase the corrosion potential and when
the potential becomes greater than the pitting potential pits will initiate. Zirconium also pits in some hypochloride solutions [17, 45].
Nitrates will inhibit pitting of zirconium; however, it is not known
what minimum nitrate concentration is necessary for a given chloride
concentration [46, 47]. Equal concentrations of chloride and nitride
solutions show no tendency to pit below 1.2 V nhe [30]. Sulfate
ions also inhibit pitting; however, a considerable amount ($SO_4^{2-}/Cl^- >$
50) of these ions are required [47, 48]. An equation relating the
amount of sulfate ions necessary to inhibit chloride pitting is log
(SO_4^{2-}) = 0.935 log (Cl^-) + 1.9 [48].

Stress Corrosion Cracking

Stress corrosion cracking (SCC) of zirconium in iodine environments
encountered as a fission product on the fuel side in nuclear reactor
cladding has been studied extensively [3]. Cesium, a fission product,
may also be responsible for SCC of zirconium in nuclear environments.
Chemical environments known to cause SCC are listed in Table 8. SCC
of zirconium in methanol containing small amounts of HCl has been
studied extensively [49-52]. The greatest susceptibility is when 0.4
volume percent of aqueous hydrochloric acid is added to methanol. No
cracking is observed when sufficient water is available to maintain the
passive film (about 3.0% in 0.4 volume percent aqueous HCl). Iodine
in methanol will also cause stress corrosion cracking [53]. Zirconium
will crack in nitrate solutions. The optimum cracking conditions are
at pH 6 in $NaNO_3$ solutions and in 70 to 80% nitric acid [54, 55]. Aqueous chloride solutions polarized beyond the film breakdown (pitting)
potential either electrochemically or with oxidizing impurities are susceptible to stress corrosion cracking [56-59]. Cox [56] lists a number
of fused eutectic salts and liquid metals that he has studied and observed stress corrosion cracking of zirconium.

Hydriding

As noted earlier, zirconium hydrides readily in hydrogen and hydro-
halogen gases (HBr, HCl, HI) above 300°C (570°F) [59]. There is

Table 8. Environments Known to Cause Stress Corrosion Cracking of Zirconium

Environment	Technique	References
Methanol	U bend	49
Methanol + 0.4% HCl	U bend	49, 50
Ethanol + 0.4% HCl	U bend	49
Methanol + 0.4% HCl + 3% H_2O	U bend	49
Methanol + 1% H_2SO_4	U bend	49
Methanol + 1% HCOOH	U bend	49
Aqueous $NaNO_3$, pH 6	Slow strain rate	54
Methanol + 1% I_2	Tensile, U bend	53, 58
25% $FeCl_3$	Tensile	57
90% HNO_3	U bend	38, 55
80% HNO_3	U bend	38, 55
70% HNO_3	Slow strain rate	55

evidence that the metal absorbs some hydrogen at room temperature. Corrosion products in some oxidizing environments (e.g., concentrated H_2SO_4, $FeCl_3$, and $CuCl_3$) contain zirconium hydrides [20, 38]. Zirconium is not as susceptible to hydrogen embrittlement as some of the other reactive metals. To our knowledge hydrogen embrittlement has not been a problem in industrial applications. The accelerated test (solution of 10% HCl + 2% SeO_2) used to estimate the resistance of tantalum to hydrogen embrittlement was carried on both zirconium and tantalum using a current density of 500 mA/cm². Tantalum became brittle and crumbled into powder after 20 min, while zirconium showed no signs of embrittlement even after 10 h [24].

Pyrophoric Tendency

Zirconium, when exposed to certain oxidizing environments, forms a corrosion film that reacts in air to produce the oxide. This ignitable film is not the normal tenacious oxide layer formed on zirconium in most environments. The ignitable film is a much softer, rough, non-metallic layer on the metal surface. The film can be ignited by sparks, open flame, or frictional heat. One environment that produces this pyrophoric product layer is red fuming nitric acid saturated with NO_2

gas. Other environments that produce pyrophoric corrosion products on zirconium are ferric chloride solutions, cupric chloride solutions, and 70 to 80% sulfuric acid with small amounts (200 ppm) of ferric or cupric ions [38, 60]. X-ray diffraction analysis of the pyrophoric corrosion products on zirconium samples exposed to sulfuric acid show broad peaks indicative of small crystalline particles or an amorphous material. Chemical analysis indicates the presence of considerable amounts of hydrogen.

TYPICAL APPLICATIONS

As stated earlier, zirconium's primary use is to house uranium fuel pellets which expose the metal to fission products on the inside of the tubing and a cooling transfer fluid on the outside. The temperature at the metal surface is between 300 and 400°C (570 to 750°F). The transfer fluid can be water, organic solution, or eutectic salts. This demonstrates zirconium's applicability as a heat exchanger where the heat transfer media is similar to those used in the nuclear industry.

Zirconium finds extensive use in 40 to 70 weight percent sulfuric acid. As demonstrated in Fig. 5, the corrosion rate below 70% and to boiling is less than 5 mpy. Impurities (e.g., Cl^-, Fe^{3+}, Cu^{2+}, and NO_3^-) individually have only a light effect on the corrosion resistance properties. Zirconium has been used to replace impervious graphite heat exchangers and lead heating coils in an environment containing sulfuric acid, zinc and sodium sulfates, hydrogen sulfide, and carbon disulfide [61]. Zirconium has been used in this application for over 10 years without any corrosion problems. Some other examples of sulfuric acid applications include ester manufacturing, steel pickling alcohol stripping towers, and hydrogen peroxide manufacturing [62, 63].

Zirconium is one of the few metals that is useful at all concentrations of hydrochloric acid below boiling. Zirconium heat exchangers, pumps, and agitators have been used for over 15 years in an azo dye coupling reaction. Besides being very corrosion resistant in this environment, zirconium does not plate out undesirable salts which would change color and stability of the dyes [64]. Heat exchangers made of zirconium are used by Thiokol Chemical Corporation and E. I. de Pont de Nemours and Co., Inc. in the production of polymers [65]. Other hydrochloric applications where zirconium has been used are phthalic-hydrochloric acid and wet hydrogen chloride-chlorine atmospheres [66]. A very important application for zirconium is in processes that cycle between hydrochloric or sulfuric acid and alkaline solutions. Zirconium crucibles are used extensively in laboratories where caustic fusion is alternated with strong acid leaching.

Other applications include nuclear fuel reprocessing concentrated nitric acid. Zirconium has been considered for use in environments exposed to hot concentrated zinc chloride and ammonium chloride solutions [67]. It has been used for over four years, with no observation of deterioration, as trays to support and conduct current to zinc anode bars in a zinc sulfate electroplating solution adjusted to a pH of 3 to 4 with sulfuric acid [68].

Zirconium can be used in most organic environments. A urea-synthesis process is centered on the use of zirconium as the reactor liner. With zirconium lining, the reaction can take place at higher temperatures, permitting greater conversion of CO_2 to urea and avoiding the corrosive intermediate products [69, 70]. Extensive testing has been conducted in zirconium in acetic acid production plants, where acetic acid is syntheses from methanol and carbon monoxide. Zirconium has very little corrosion even in the presence of the catalyst cobalt acetate and potassium iodide [71, 72].

Zirconium is an excellent material for use in heat transfer applications. It can be used with very thin walls, and fouling is generally not a problem. Clean zirconium surfaces are not wet by water. This advantage of dropwise condensation enhances equipment performance.

Zirconium's use in industrial applications is increasing. This is due to material availability and an increased body of corrosion and design data.

FABRICATION PRACTICES

Zirconium and its alloys can be formed, machined, and welded into component parts for equipment using standard machines and techniques. Common problems encountered are weld contamination with air, spring back and cracks in bends, galling, and the inability to weld zirconium to other metals. Formability is dependent on the oxygen content of the metal. The lower the oxygen level, the easier the metal is to form. Proper fabrication practices are fully described in Ref. 5, 73, and 74.

Forming

Zirconium can be formed, which includes bending and punching, with conventional equipment used for stainless steel. Forming rates should be slower than for other metals. Bend radii five times the materials thickness can be obtained with Grade 702 and three times with Grade 705. Smaller bend radii can be obtained by bending the material hot [200 to 500°C (390 to 930°F)]. The punching operation requires very close punched die tolerances (1 to 2% of the metal thickness).

Machining

Machining zirconium is comparable to machining a good grade of aluminum. Caution must be exercised with fine machine chips which can ignite. The black oxide scale, if present, must be removed before any cutting operation is undertaken because the hard oxide will dull tools rapidly. The best metal removal rates are obtained by using slow speeds with heavy feeds and ample flooding of coolant.

Welding

We will spend more time with welding than with the other fabrication procedures. This is because good welding practices will help ensure that corrosion-resistant properties of the weld are optimized.

As with other reactive metals, welding zirconium requires shielding of the weld puddle and hot bead from air. The shielding must be made with very pure argon, helium, or a mixture of these two. Any carbon dioxide, nitrogen, oxygen, or hydrogen in the shielding gas mixture will dissolve into the molten metal and make the welds brittle. Inadequate shielding will produce blue or white deposits on the welds. Silver or straw color are signs of good shielding.

As in all welding techniques, clean practices are required in welding zirconium. Clean weld joints will help minimize contamination problems. If filler wire is used, it must be cleaned as thoroughly as the weld joints.

Standard welding equipment is generally used. Tungsten electrode inert gas (TIG) shielded arc welding technique is the most widely used. Any technique can be used as long as adequate shielding is provided. Shielding must be established before the weld starts and retained after welding has stopped. If filler metal is used, it must be zirconium or one of its alloys. One must be aware that zirconium should not be welded to other metals, except possibly titanium.

After obtaining a good weld, free from contamination, the next property important to corrosion resistance is the metallurgical structure. The weld bead and heat effect zone are subject to accelerated attack, as already mentioned. Welds can be made most resistant to corrosion by a full recrystallization anneal. This is accomplished by heating at 750 to 790°C (1380 to 1455°F) for about 1/2 h in an inert atmosphere. In situations where this is not possible, the best corrosion resistance is obtained when welds are allowed to cool very slowly.

Zirconium can be field welded using the same procedures outlined above with the additional requirement that wind shielding be used to prevent disruption of the protective gas cover.

Figure 7.   Zirconium 16-in. enclosed impeller for use in acetic acid
production.   This impeller is cast and machined by Oremet.   (Courtesy
of Oregon Metallurgical Corp., Albany, Oreg.)

ZIRCONIUM PRODUCTS

The variety of mill products available are comparable to that of stainless steel.   This includes sheet, plate, billet, wire, foil, tubing (both
seamless and welded), fasteners, and so on.   Zirconium can also be
obtained as cladding on steel plates and as castings.   Figure 7 shows
a zirconium casting.

Figure 8. (a) Zirconium heating coil for use in hydrochloric acid. Fabricated by Saffran Engineering. (Courtesy of Saffran Engineering, St. Clair Shores, Mich.) (b) Zirconium stripper column for use in hydrochloric acid. Fabricated by Futura. (Courtesy of Futura Metal Technology, Westlake Village, Calif.)

Zirconium can be fabricated into tanks, liners, strainers, agitators, exhausters, hubs, fans, valves, pipe fittings, nozzles, and pump parts to name a few. Figure 8 illustrates some components made from zirconium for industrial use. Figure 8a is a heating coil made for use in a hydrochloric acid application. Figure 8b is a zirconium stripper column made for use in a hydrochloric acid application.

REFERENCES

1. S. R. Taylor, Trace element abundances and the chondrites earth model, *Geochim. Cosmochim. Acta 28*: 1989, 1964.
2. A. Hurlick, Planet Earth's metal resources, *Met. Prog.*, Hl, October, 1977.
3. B. Cox, in *Advances in Corrosion Science and Technology* (M. G. Fontana and R. W. Staehle, eds.), Vol. 5, Plenum Press, New York, 1976, p. 173.
4. B. G. Parfenov, V. V. Gerasimov, and G. I. Venediktova, *Corrosion of Zirconium and Zirconium Alloys*, trans. from the Russian by Israel Program for Scientific Translation, Jerusalem, 1969.
5. J. H. Schemel, *ASTM Manual on Zirconium and Hafnium*, American Society for Testing and Materials, Philadelphia, 1977.
6. J. H. Schemel, in *Corrosion Resistance of Metals and Alloys* (F. L. LaQue and H. R. Copson, eds.), Reinhold, New York, 1963, p. 666.
7. R. J. Brumbaugh, Corrosion resistance of zirconium, *Ind. Eng. Chem. 43*: 2878, 1951.
8. P. L. Ko, Wear of zirconium alloys due to fretting and periodic impacting, *Wear Mater. 2*: 388, 1979.
9. J. B. Cotton, in *Corrosion* (L. L. Shreir, ed.), Vol. 1, Newnes-Butterworths, London, 1976, pp. 5-52.
10. S. M. Shelton, *Zirconium: Its Production and Properties*, U.S. Bur. Mines Bull. 561, 1956.
11. L. B. Golden, I. R. Lane, Jr., and W. L. Acherman, Corrosion resistance of titanium, zirconium and stainless steel, mineral acids, *Ind. Eng. Chem. 44*: 1930, 1952.
12. I. R. Lane, Jr., L. B. Golden, and W. L. Acherman, Corrosion resistance of titanium, zirconium and stainless steel in organic compounds, *Ind. Eng. Chem. 45*: 1967, 1953.
13. E. A. Gee, L. B. Golden, and W. E. Lusby, Jr., Titanium and zirconium corrosion studies, common mineral acids, *Ind. Eng. Chem. 41*: 1668, 1949.
14. H. H. Uhlig, *Corrosion Handbook*, Wiley, New York, 1948.
15. C. R. Bishop, Corrosion tests at elevated temperatures and pressures, *Corrosion 19*: 308t, 1963.
16. D. R. Knittel and R. T. Webster, "Corrosion resistance of

zirconium and zirconium alloys in inorganic acids and alkalies,"
Symp. Ind. Appl. Zirconium Titanium, ASTM B-10 Meet., New
Orleans, October 15-18, 1979.

17. P. J. Gegner and W. L. Wilson, Corrosion resistance of titanium
and zirconium in chemical plant exposures, *Corrosion 15*: 341t,
1959.

18. D. J. Stoops, M. D. Carver, and H. Kato, *Corrosion of Zirconium
in Cupric and Ferric Chloride*, U. S. Bur. Mines Rep. Invest.
5945, 1961.

19. R. S. Sheppard, D. R. Hise, P. J. Gegner, and W. L. Wilson,
Performance of titanium vs. other materials in chemical plant
engineering, *Corrosion 18*: 211t, 1962.

20. T. Smith and G. R. Hill, A reaction rate study of the corrosion
of low-hafnium zirconium in aqueous hydrofluoric acid solutions,
J. Electrochem. Soc. 105: 117, 1958.

21. W. E. Kuhn, The corrosion of zirconium in hydrochloric acid at
atmospheric pressure, *Corrosion 15*: 103t, 1959.

22. W. E. Kuhn, High temperature-high pressure corrosion of zirconium in hydrochloric acid, *Corrosion 16*: 136t, 1960.

23. V. V. Andreeva and A. I. Glukhova, Corrosion and electrochemical properties of zirconium, titanium and titanium-zirconium alloys
in solutions of hydrochloric acid and of hydrochloric acid with
oxidizing agents, *J. Appl. Chem. 11*: 390, 1961.

24. M. N. Fokin, R. L. Baru, and M. M. Kurtepov, *Corrosion and
Metal Protection* (J. L. Rosenfeld, ed.), Indian National Scientific Documentation Center, New Delhi, 1975, p. 113.

25. B. S. Payne and D. K. Priest, Intergranular corrosion of commercially pure zirconium, *Corrosion 17*: 196t, 1961.

26. W. E. Kuhn, Influence of impurities on the high temperature
corrosion and reaction kinetics of zirconium in hydrochloric acid,
Corrosion 18: 103t, 1962.

27. W. E. Kuhn, Higher purity boosts zirconium resistance, *Chem.
Eng.*, 154, February, 1960.

28. W. E. Kuhn, Effect of solution heat treatment on corrosion kinetics of zirconium in 20 percent hydrochloric acid at high temperatures and pressure, *Corrosion 19*: 169t, 1963.

29. W. K. Boyd and E. L. White, *Compatibility of Rocket Propellents
with Materials of Construction*, DMIC Memorandum 65 OTS PB
161215, Battelle Memorial Institute, Columbus, Ohio, September 15,
1960.

30. V. V. Andreeva and A. I. Glukhova, Corrosion and electrochemical properties of titanium, zirconium, and titanium-zirconiumm
alloys in acid solutions II, *J. Appl. Chem. 12*: 457, 1962.

31. T. Smith, The kinetics of the corrosion of low-hafnium zirconium
in aqueous sulfuric acid solutions, *J. Electrochem. Soc. 107*: 82,
1960.

32. L. B. Golden, in *Zirconium and Zirconium Alloys*, American

Society for Metals, Cleveland, Ohio, 1953, p. 305.

33. B. Frechem, J. Morrison, and R. T. Webster, Improving the corrosion resistance of zirconium weldments, Symp. Ind. Appl. Zirconium Titanium, ASTM B-10 Meet., New Orleans, October 15-18, 1979.

34. V. T. Günther, Interrelationship between structure and corrosion behavior of zirconium, *Werkst. Korros. 30*: 308, 1979.

35. D. R. Knittel, M. A. Maguire, R. E. Grammer, and R. T. Webster, Corrosion behavior of zirconium in H_2SO_4 and H_2SO_4 plus impurities, 12th Nat. SAMPE Tech. Conf., Seattle, October 7-9, 1980.

36. C. M. Graighead, L. A. Smith, and R. I. Jaffee, *Screening Tests on Metals and Alloys in Contact with Sodium Hydroxide at 1000 and 1500 F*, U.S. Energy Comm. Rep. No. BMI-706, prepared by Battelle Memorial Institute, November 6, 1951.

37. N. E. Hamner, *Corrosion Data Survey*, National Association of Corrosion Engineers, Houston, 1974.

38. Test conducted at Teledyne Wah Chang Albany.

39. G. Jangg, E. T. Baroch, R. Kieffer, and A. Watti, Research on the corrosion properties of zirconium alloys, *Werkst. Korros. 24*: 845, 1973.

40. F. Coen-Porisini and G. Imarisio, High temperature corrosion in the thermochemical hydrogen production from nuclear heat, *Rev. Int. Hautes Temp. Refract. 13*: 250, 1976.

41. G. J. Janz and R. P. T. Tomkins, Corrosion in molten salts; an annotated bibliograph, *Corrosion 35*: 485, 1979.

42. J. N. Gregory, N. Hodge, and J. V. G. Iredule, *The Static Corrosion of Nickel and Other Materials in Molten Caustic Soda*, AERE C/M 272, March 1956.

43. R. F. Koenig, *Corrosion of Zirconium and Its Alloys in Liquid Metals*, Rep. No. KAPL-982, prepared for U.S. Atomic Energy Commission by the General Electric Company, October 1, 1953.

44. Y. M. Kolotyrkin, Electrochemical behavior of metals during anodic and chemical passivation in electrolytic solutions, First Int. Cong. Met. Corros., London, 1961, p. 10.

45. V. A. Gilman and Y. M. Kolotyrkin, Pitting corrosion of zirconium in perchorate solutions, *Zashch. Met. 2*: 360, 1966.

46. G. Jangg, R. T. Webster, and M. Simon, Pitting behavior of zirconium alloys, *Werkst. Korros. 29*: 16, 1978.

47. M. Maraghini, G. B. Adams, Jr., and P. Van Rysselberghe, Studies on the anodic polarization of zirconium and zirconium alloys, *J. Electrochem. Soc. 101*: 400, 1954.

48. S. S. Chouthai, Pitting corrosion of Zirconium in chloride solutions, *Trans. Indian Inst. Met. 31*: 34, 1978.

49. K. Mori, A. Takamura, and T. Shimose, Stress corrosion cracking of Ti and Zr in HCl-methanol solutions, *Corrosion 22*: 29, 1966.

50. G. C. Palit, P. K. De, and K. Elayaperumal, Stress corrosion cracking of zirconium in methanol-aqueous HCl solution: effect of HCl concentration, *Proc. 14th Semin. Electrochem.*, Karaiduki, India, 1973 p. 439.

51. K. Elayaperumal, P. K. De, and J. Balachandra, Electrochemical factors of stress corrosion cracking of zirconium in CH_3OH + HCl, *Corros. Sci. 11*: 579, 1971.

52. G. C. Palit and K. Elayaperumal, Environmental factors in stress corrosion cracking and anodic polarization of zirconium in methanol-HCl solutions, *Corrosion 32*: 276, 1976.

53. K. Elayaperumal, P. K. De, and J. Balachandra, Stress Corrosion cracking of Zircaloy 2 in methanol-iodine solutions, *J. Nucl. Mater. 45*: 323, 1972/73.

54. T. C. Kasiuiswanathan, R. Chandrashekar, and K. I. Vasu, Dynamic stress corrosion cracking of zirconium in aqueous nitrate solutions, *Proc. 14th Semin. Electrochem.*, Karaiduki, India, 1973, p. 434.

55. J. A. beavers, J. C. Griess, and W. K. Boyd, Stress corrosion cracking of zirconium in nitric acid, NACE Corrosion/80, Chicago, March 3-7, 1980.

56. B. Cox, Environmentally induced cracking of zirconium alloys, *Corrosion 28*: 207, 1972.

57. J. T. Dunham and H. Kato, *Stress-Corrosion Cracking Susceptibility of Zirconium in Ferric Chloride Solution*, U.S. Bur. Mines Rep. Invest. 5784, 1961.

58. B. Cox, Stress corrosion cracking of Zircaloy-2 in neutral aqueous chloride solutions at 25°C, *Corrosion 29*: 157, 1973.

59. K. C. Thomas and R. J. Allio, The failure of stressed zircaloy in aqueous chloride solutions, *Nucl. Appl. 1*: 252, 1965.

60. R. I. Jaffee and I. E. Campbell, *The Properties of Zirconium*, Battelle Memorial Inst. AEC file NP-266, June 18, 1947.

61. L. B. Bowen, Use of zirconium heat exchangers in the viscose rayon process, Symp. Ind. Appl. Zirconium Titanium, ASTM B-10 Meeting, New Orleans, October 15-18, 1979.

62. R. T. Webster, The use of zirconium in plastic sheet, film and fiber production processes, NACE Corrosion/79, Atlanta, March 12-16, 1979.

63. W. G. Ashbaugh, Zr distillation trays boost production, save $125,000 in expansion cost, *Chem. Process.*, 68, March 1966.

64. Zirconium coils boost dye production 30%, *Chem. Process.*, 39, August 1967.

65. G. W. Armstrong, Zirconium reduces polymer plate down time, *Chem. Process.*, 52, December 1967.

66. J. H. McClain and R. W. Nelson, *Some Useful Applications of Zirconium*, U.S. Bur. Mines Inf. Circ. 7686, 1954.

67. R. E. Smallwood, The corrosion of reactive metals in zinc chloride solutions, Symp. Ind. Appl. Zirconium Titanium,

ASTM B-10 Meet., New Orleans, October 15-18, 1979.

68. J. Keck, Zirconium fails to corrode as an anode support in electroplating process, *Chem. Process.*, 110, mid-November 1979.

69. D. W. McDowell, Jr., Corrosion in urea-synthesis reactors, *Chem. Eng.*, 118, May 1974.

70. E. Guccione, New urea process boasts high yields, low costs, *Chem. Eng.*, 96, September 1966.

71. H. Togano and K. Ōsato, Corrosion test on the construction materials for acetic acid synthesis from methanol and carbon monoxide, *Boshoku Gijutsu 10:* 529, 1961.

72. T. Shimose, A. Takamura, and S. Segawa, Corrosion behaviors of various metals and alloys in acetic acid environments, *Boshoku Gijutsu 15:* 49, 1966.

73. R. W. Duhl, Zirconium as a reactor lining, *Safety Air Ammonia Plants 9:* 78, 1967.

74. Teledyne Wah Chang Albany, Sales brochure, December 1980.

75. D. R. Knittel, Zirconium: a corrosion resistant material for industrial applications, *Chem. Eng.*, 95, June 2, 1980.

3.9
Cast Alloys

GLENN W. GEORGE / The Duriron Company, Inc., Dayton, Ohio

INTRODUCTION

The purpose of this chapter is to make the reader aware of the broad range of cast alloys that are available for use in corrosive services. The degree of emphasis placed on each group of alloys is in proportion to its usage level and importance in corrosive applications. Therefore, the iron-carbon materials and the stainless steels receive the most emphasis since these alloys account for the major share of usage.

For practically every wrought alloy composition there is an equivalent grade available as a casting. In addition, there are many alloy compositions available only as castings. Because of the design freedom associated with the various casting processes, it is possible to use alloys with properties and characteristics not possible in the wrought grades because producing the various wrought products requires an alloy that can be subjected to many rolling and forming steps. These rolling and forming operations require an alloy that possesses reduced strength and considerable hot-working and cold-working ductility. In fact, the compositions of many of the cast grades are modified relative to their wrought equivalents to take advantage of the casting processes since little or no mechanical working of the cast component will be necessary. By being able to modify the alloy compositions of the cast grades, unique and improved properties can be imparted to some cast alloys as compared to their wrought equivalents. Since there are variations in cast compositions and the resulting physical and mechanical properties versus the recognized wrought equivalents, various alloy designation systems have been adopted to separate the two. For example, the Alloy Casting Institute (ACI) Division of the Steel Founders' Society of America (SFSA) developed a system for high alloy stainless steels. Under this system, CF-8M is the alloy designation for the cast equivalent of 316 stainless steel. Similar designations exist for cast carbon steel, nickel, copper, aluminum, and other alloy systems.

References will be made throughout this chapter to American Society for Testing and Materials (ASTM) standards. It is felt that these standards offer the best and most recognized basis for understanding between casting purchaser and supplier. There are certainly other standards used, but it would be impossible to list all of them and many of the other standards are referenced in the ASTM documents. Also, the cast grade designations will be cross-referenced to their wrought equivalents since the wrought designations are more readily recognized by many individuals. There are many modifications of the standard cast alloy grades available under various trade names. Since these modifications are quite numerous and to avoid offending any producer through exclusion, only generic alloy designations will be used.

Before presenting the specific alloy groups and their applications, a few words concerning castings and recent developments relating to casting processes and alloys seem appropriate. Castings offer the designer several distinct advantages in producing a component, and this certainly applies to the design of components for corrosive services. Of course, castings are not the answer to all design problems and the limitations of the casting processes and cast components should be realized. Some of the advantages of castings include literally unlimited freedom in design configuration, minimization or elimination of machining and material waste, broad range of alloy choice, mechanical property isotropy, and production economies. The most serious limitation is variations in quality from casting to casting and foundry to foundry. Possible quality shortcomings involve surface finish, internal integrity, compositional purity, and dimensional control. All these limitations can be overcome through judicious application of sound foundry principles.

Producers of quality castings are constantly striving to provide an improved product at an economical price. Two recent developments in ferrous and high alloy casting production serve to illustrate this point. The first deals with melt refining and involves the introduction of the argon-oxygen decarburization (AOD) process into foundries. This process has been utilized for some time in the production of wrought stainless steels, but it is just beginning to find widespread use in casting production. AOD refining will have a significant influence on the foundry processing and alloy properties of not only high alloy materials but also carbon and low alloy steels. The second development deals with the binders used in mold and core making. It has been recognized for some time that these binders could lead to deleterious contamination of cast alloys. Most of these binders are organic materials and when contacted by molten metal the binders break down and certain elements, the most notable being carbon, are picked up by the alloy. New binders that are inorganic and form a ceramic-type bond have been developed, thus alleviating the pickup problem. This is very obviously important to producers of low carbon alloys, but these ceramic systems offer advantages to carbon and low alloy steel producers because they prevent the pickup of elements such as phosphorus and sulfur. These systems also provide improved dimensional stability, better surface finish, cleaner castings, and enhanced working environments. Both the AOD process and ceramic binders are covered in more detail in the section of this chapter dealing with casting production.

IRON-CARBON ALLOYS

The iron-carbon alloys comprise the most widely used group of cast materials. Included in this group of alloys are the cast irons, the carbon and low alloy steels, and some specialty alloys. Many of these alloys find use in corrosion-type applications and perform quite satisfactorily; however, the primary reason for choosing from this group is generally not corrosion resistance, but low cost, good mechanical properties, and in some cases, ease of production. It would be impossible to list and discuss all the various compositions belonging to the iron-carbon alloy group. Instead, the more common and widely used materials are presented.

Table 1. Carbon and Low Alloy Steels

ASTM Standard	Cast alloy designation	Wrought alloy type[a]	Composition								
			C^b	Mn^b	Si^b	P^b	S^b	Cr	Ni	Fe	Other elements
A 148	80-40	—	e	e	e	0.05	0.06	e	e	Bal.	e
A 148	80-50	—	e	e	e	0.05	0.06	e	e	Bal.	e
A 148	90-60	—	e	e	e	0.05	0.06	e	e	Bal.	e
A 148	105-85	—	e	e	e	0.05	0.06	e	e	Bal.	e
A 148	120-95	—	e	e	e	0.05	0.06	e	e	Bal.	e
A 148	150-125	—	e	e	e	0.05	0.06	e	e	Bal.	e
A 148	175-145	—	e	e	e	0.05	0.06	e	e	Bal.	e
A 216	WCA	1020	0.25	0.70	0.60	0.04	0.045	0.40^b	0.50^b	Bal.	$0.50\ Cu^b$ $0.25\ Mo^b$ $0.03\ V^b$
A 216	WCB	1025	0.30	1.00	0.60	0.04	0.045	0.40^b	0.50^b	Bal.	$0.50\ Cu^b$ $0.25\ Mo^b$ $0.03\ V^b$
A 216	WCC	1020	0.25	1.20	0.60	0.04	0.045	0.40^b	0.50^b	Bal.	$0.50\ Cu^b$ $0.25\ Mo^b$ $0.03\ V^b$
A 352	LCA	1020	0.25	0.70	0.60	0.04	0.045	—	—	Bal.	—
A 352	LCB	1025	0.30	1.00	0.60	0.04	0.045	—	—	Bal.	—
A 352	LCC	—	0.25	1.20	0.60	0.04	0.045	—	—	Bal.	—
A 352	LC1	—	0.25	0.50/ 0.80	0.60	0.04	0.045	—	—	Bal.	0.45-0.65 Mo
A 352	LC2	—	0.25	0.50/ 0.80	0.60	0.04	0.045	—	2.00-3.00	Bal.	—
A 352	LC2-1		0.22	0.55/ 0.75	0.60	0.04	0.045	1.35-1.85	2.50-3.00	Bal.	0.30-0.60 Mo
A 352	LC3	2310	0.15	0.50/ 0.80	0.60	0.04	0.045	—	3.00-4.00	Bal.	—
A 352	LC4	—	0.15	0.50/ 0.80	0.60	0.04	0.045	—	4.00-5.00	Bal.	—

[a]Wrought grade most closely corresponding to cast alloy. Composition and properties may vary between wrought and cast.

[b]Maximum unless otherwise noted.

[c]Minimum unless otherwise noted.

[d]Typical.

[e]Composition shall be selected by manufacturer to achieve desired mechanical properties.

[f]Castings may be annealed, normalized, normalized and tempered, or quenched and tempered.

[g]Castings may be annealed, normalized, or normalized and tempered.

[h]Castings will be normalized and tempered or quenched and tempered.

Heat-treated condition	Tensile strength		Yield strength		Elongation in 50 mm (2 in.)[c]	Reduction of area,[c]	Brinell hardness number[d]	Charpy impact			
										Test temp.	
	MPa[c]	ksi[c]	MPa[c]	ksi[c]				J[c]	ft-lb[c]	°C	°F
f	552	80	276	40	18	30	180	–	–	–	–
f	552	80	345	50	22	35	180	–	–	–	–
f	621	90	414	60	20	40	190	–	–	–	–
f	724	105	586	85	17	35	230	–	–	–	–
f	827	120	655	95	14	30	260	–	–	–	–
f	1034	150	826	125	9	22	330	–	–	–	–
f	1207	175	1000	145	6	12	–	–	–	–	–
g	415	60	205	30	24	35	140	–	–	–	–
g	485	70	250	36	22	35	150	–	–	–	–
g	485	70	275	40	22	35	160	–	–	–	–
h	415	60	205	30	24	35	150	18	13	- 32	- 25
h	450	65	240	35	24	35	160	18	13	- 46	- 50
h	485	70	275	40	22	35	160	20	15	- 46	- 50
h	450	65	240	35	24	35	–	18	13	- 59	- 75
h	485	70	275	40	24	35	160	20	15	- 73	-100
h	725	105	550	80	18	30	–	41	30	- 73	-100
h	485	70	275	40	24	35	170	20	15	-101	-150
h	485	70	275	40	24	35	–	20	15	-115	-175

Carbon and Low Alloy Steels

There are many cast grades of carbon and low alloy steels available.
Table 1 lists a representative group of ASTM alloys that finds use in
many applications where mild corrosives are handled. ASTM A148
deals primarily with tensile properties and, in fact, leaves the choice
of composition up to the casting producer. ASTM A216 is much more
specific concerning composition so that uniform welding characteristics
can be expected. ASTM A352 introduces alloying, but again the major
concern is mechanical properties, specifically low temperature impact
toughness.

At the levels of alloying present in carbon and low alloy steels, no
element or combination of elements can significantly influence the
corrosion rate of iron under immersed conditions [1]. For this reason
it is not necessary to distinguish among the many grades of carbon
and low alloy steel when considering corrosion resistance under
immersed conditions; rather, they will be assembled together and
discussed as one group. Carbon and low alloy steels find application
in many mildly corrosive media, such as neutral to alkaline waters
and brines, alkalies at ambient temperatures, neutral organics, and
concentrated sulfuric acid. This group of materials is used extensively for handling concentrated sulfuric acid under ambient temperature
conditions, but there are a few precautions that should be taken.
First, concentrated sulfuric acid is hygroscopic. It is possible for
the acid to pick up sufficient moisture from the atmosphere to dilute
the surface layer to the point that the steel will be rapidly attacked.
Second, velocity conditions can remove the protective ferrous sulfate
film that is formed and serious corrosion will result. Finally, hydrogen grooving can occur, particularly in pipelines and storage tanks,
making catastrophic failure a possibility. Small additions of copper
will make steels more resistant to hydrogen grooving.

Although small additions of alloying elements to steels do not
appreciably affect the corrosion resistance under immersed conditions,
the resistance to atmospheric corrosion is significantly improved in
this manner. Small additions of elements such as copper, phosphorus,
chromium, nickel, and molybdenum can improve atmospheric corrosion
resistance as much as five to eight times over unalloyed steels [1].
This factor could be important when purchasing steel castings for
structural applications where improved resistance to atmospheric
corrosion is a factor.

Even though Table 1 presents a representative group of carbon
and low alloy steel cast grades, it is by no means exhaustive. As
mentioned previously, there is not an appreciable difference among
all the carbon and low alloy steels with respect to corrosion resistance
under immersed conditions. Therefore, little would be gained with

regard to assisting the reader in selecting an alloy for corrosion
resistance by presenting additional details on specific alloys.
Generally, other ASTM standards deal with alloys to meet specific
mechanical property needs or are written around various casting
processes. To make the reader aware of the various materials available, Table 2 presents a fairly complete list of ASTM standards dealing with cast carbon and low alloy steels.

Cast Irons

The general term cast iron encompasses a great number of materials.
Included in this group of materials is gray iron, white iron, malleable
iron, nodular or ductile iron, and many grades of specialty irons. The
major factor distinguishing cast irons from steel is the amount of carbon
present. Generally, in cast irons there is more carbon present than
can be held in solution in austenite at the eutectic temperature. This
excess carbon precipitates and at room temperature the microstructure
of these alloys will show free carbon or graphite present as flakes or
nodules. The presence of this free graphite has a very significant
effect on the mechanical properties compared to steel and it also affects
the corrosion resistance.

The most common of the cast irons is gray iron. This is a material
containing from 1.7 to 4.5% carbon and 1 to 3% silicon [2]. The carbon
is present as graphite flakes and when the material is fractured it has
a gray appearance, thus the name *gray iron*. Gray iron is the least
expensive of all cast metals and this factor, together with its properties, has lead to it being the most widely used cast material on a
weight basis. As with steels, minor alloying additions have little influence on the corrosion resistance of gray iron. In this regard, gray
iron would find use in the same environments as carbon and low alloy
steels. It should be noted that gray iron is not suitable for oleum, as
it has been known to rupture in this service with explosive violence.
In certain environments, gray iron may show an apparent advantage
over carbon steel due to the nature of the porous graphite-iron corrosion product film that forms on its surface. This film gives a particular advantage under velocity conditions, and this is evident in the
widespread use of underground gray iron water pipes. Gray iron is
subject to a form of corrosion known as graphitization [3]. This involves a selective leaching of the iron matrix, leaving only a graphite
network. Even though no apparent dimensional changes occur, there
can be sufficient loss of section and strength to lead to failure. Gray
iron usually performs adequately under atmospheric corrosion conditions due to the heavy sections used. With these heavy sections, the
degree of corrosion that occurs is not sufficient to present a problem.

Table 2.   ASTM Carbon and Low Alloy Steel Casting Standards

ASTM standard	Title
A27	Mild- to Medium-Strength Carbon-Steel Castings for General Application
A148	High-Strength Steel Castings for Structural Purposes
A216	Carbon-Steel Castings Suitable for Fusion Welding for High Temperature Service
A352	Ferritic Steel Castings for Pressure-Containing Parts Suitable for Low-Temperature Service
A356	Heavy-Walled Carbon and Low-Alloy Steel Castings for Steam Turbines
A389	Alloy Steel Castings Specially Heat-Treated for Pressure-Containing Parts Suitable for High-Temperature Service
A487	Steel Castings Suitable for Pressure Service
A643	Steel Castings, Heavy-Walled, Carbon and Alloy, for Pressure Vessels
A732	Carbon and Low-Alloy Steel Investment Castings for General Application

There are a number of ASTM standards dealing with gray iron castings, but A48 is probably the one referred to most commonly. This standard classifies the alloy based on tensile strength, with values ranging from 138 MPa (20,000 psi) to 415 MPa (60,000 psi). No limits are placed on composition. Other ASTM standards dealing with gray iron include A126, A278, and A319.

Mechanical properties are a major drawback to more widespread use of gray iron. Presence of the graphite flakes in the microstructure results in a brittle alloy with very poor shock resistance, particularly at low temperatures. Through carefully controlled manufacturing procedures, it is possible to cause the graphite to be present in a nodular form, thus creating fewer discontinuities in the metal structure, resulting in a stronger and much more ductile material [4]. If the graphite is caused to nodularize by treating the alloy while it is liquid with small additions of magnesium or cerium, the resulting materials is called *ductile* or *nodular cast iron*. If the graphite is caused to nodularize by means of an appropriate heat treatment, the

Table 3. Malleable and Ductile Iron Properties

ASTM standard	Alloy grade	Tensile strength		Yield strength		Elongation in 50 mm (2 in.)	Brinell hardness number
		MPa	ksi	Mpa	Ksi		
A47	32510	345	50	224	32	10	—
A47	35018	365	53	241	35	18	—
A220	40010	414	60	276	40	10	—
A220	45008	448	65	310	45	8	—
A220	45006	448	65	310	45	6	—
A220	50005	483	70	345	50	5	—
A220	60004	552	80	414	60	4	—
A220	70003	586	85	483	70	3	—
A220	80002	655	95	552	80	2	—
A220	90001	724	105	621	90	1	—
A395	—	414	60	276	40	18	143-187
A536	60-40-18	414	60	276	40	18	—
A536	65-45-12	448	65	310	45	12	—
A536	80-55-06	552	80	379	55	6	—
A536	100-70-03	689	100	483	70	3	—
A536	120-90-02	827	120	621	90	2	—

Table 4. Ni-Resist Cast Irons

ASTM specification	Cast alloy designation	Composition							
		C	Mn	Si	S^a	Cr	Ni	Fe	Other elements
A436	Type 1	3.00	0.5/1.5	1.00/2.80	0.12	1.5-2.5	13.5-17.5	Bal.	5.5-7.5 Cu
A436	Type 1b	3.00	0.5/1.5	1.00/2.80	0.12	2.5-3.5	13.5-17.5	Bal.	5.5-7.5 Cu
A436	Type 2	3.00	0.5/1.5	1.00/2.80	0.12	1.5-2.5	18.0-22.0	Bal.	$0.5\ Cu^a$
A436	Type 2b	3.00	0.5/1.5	1.00/2.80	0.12	3.0-6.0	18.0-22.0	Bal.	$0.5\ Cu^a$
A436	Type 3	2.60	0.5/1.5	1.00/2.00	0.12	2.5-3.5	28.0-32.0	Bal.	$0.5\ Cu^a$
A436	Type 4	2.60	0.5/1.5	5.00/6.00	0.12	4.5-5.5	29.0-32.0	Bal.	$0.5\ Cu^a$
A436	Type 5	2.40	0.5/1.5	1.00/2.00	0.12	0.10^a	34.0-36.0	Bal.	$0.5\ Cu^a$
A436	Type 6	3.00	0.5/1.5	1.50/2.50	0.12	1.0-2.0	18.0-22.0	Bal.	3.5-3.5 Cu $1.00\ Mo^a$
A571	—	2.2/ 2.7	3.75/4.5	1.50/2.5	0.08	0.20^a	21.0-24.0	Bal.	—

[a]Maximum unless otherwise noted.

[b]Minimum unless otherwise noted.

[c]RT, room temperature.

[d]Typical, unnotched.

[e]Annealed.

material is called *malleable iron*. The corrosion resistance of these alloys is comparable to gray iron except that under velocity condition they may be slightly less resistant because they do not form the same sort of film as is present on gray iron. The major ASTM standards covering ductile iron are A395 and A536, and for malleable iron they are A47 and A220. Table 3 presents the mechanical properties for these alloys.

There are two groups of highly alloyed cast irons. One group is found in ASTM A436 and is referred to as the *Ni-Resist alloys*. This group consists of high nickel austenitic cast irons used primarily for their corrosion resistance [2]. The Ni-Resist alloys find use in mildly oxidizing acids, alkalies, salts, seawater, foods, plastics, and synthetic fiber manufacturing. These alloys tend to maintain their corrosion resistance under high-velocity conditions and show good wear resistance. Although the alloys show relatively low tensile strengths in the range 138 to 276 MPa (20,000 to 40,000 psi), they have much improved toughness over unalloyed gray iron. It is possible to obtain ductile or nodular Ni-Resist alloys. The corrosion resistance is

Heat-treated condition	Tensile strength		Yield strength		Elongation in 50 mm (2 in.)[b]	Brinell hardness number	Charpy impact		Test temp.[c]	
	MPa[b]	ksi[b]	MPa[b]	ksi[b]			J	ft-lb	°C	°F
–	172	25	–	–	–	131-183	136[d]	100[d]	RT	RT
–	207	30	–	–	–	149-212	109[d]	80[d]	RT	RT
–	172	25	–	–	–	118-174	136[d]	100[d]	RT	RT
–	207	30	–	–	–	171-248	82[d]	60[d]	RT	RT
–	172	25	–	–	–	118-159	204[d]	150[d]	RT	RT
–	172	25	–	–	–	149-212	109[d]	80[d]	RT	RT
–	138	20	–	–	–	99-124	204[d]	150[d]	RT	RT
–	172	25	–	–	–	124-174	–	–	–	–
e	448	65	207	30	30	121-171	21[b]	15[b]	-195	-320

essentially the same as that for the conventional flake graphite Ni-Resist alloys, but the mechanical properties, particularly the low-temperature impact toughness, are much improved. ASTM A571 covers ductile Ni-Resist. Table 4 presents the composition and properties of the Ni-Resist alloys.

The second group of highly alloyed cast irons is the *white irons*. The carbon in these alloys is essentially all in solution and the fracture surface appears white. There are two major subgroups of the white irons; the *Ni-Hards* and the *high chromium irons*. Nickel in the range 4 to 5% and chromium in the range 1.5 to 3.5% are the major alloying elements in the Ni-Hards [2]. Chromium levels up to 25% are found in the high chromium irons [2]. This group of alloys is intended primarily for abrasive wear applications. Some of these alloys are capable of Brinell hardness levels in excess of 700, with 600 being fairly typical. Areas where these alloys find use include mining, power generation, cement manufacture, pulp and paper, and similar industries where severe abrasive services are encountered. The composition and properties of these alloys can be found in ASTM A532.

The final cast iron to be discussed is high silicon cast iron. This
alloy contains nominally 14.5% silicon, 1% carbon, and the balance iron
[2]. This alloy has unexcelled corrosion resistance to a wide range
of chemicals and, in fact, it is probably the most universally corrosion resistant alloy that is commercially available. One of the major
uses for the alloy is in handling sulfuric acid. It is resistant to all
concentrations up to and including the normal boiling point. The alloy
will handle nitric acid above 30% to the boiling point. Below 30%, the
temperature is limited to about 180°F. By adding 4.5% chromium to
the alloy, it becomes resistant to environments such as severe chloride-
containing solutions and other strongly oxidizing services. The
chromium-bearing grade will handle hydrochloric acid up to 80°F, and
the presence of ferric chloride, which wrecks havoc on many alloys,
actually inhibits corrosion with this alloy when the acid is present at
less than 20%. Obviously, alloys with such outstanding corrosion resistance in very severe environments must have some shortcomings.
Unfortunately, the high silicon irons are very susceptible to thermal
and mechanical shock. The alloys cannot withstand any signficant
stressing or impact, and they cannot be subjected to sudden fluctuations in temperature. With proper precautions, it is possible to obtain
outstanding performance with these materials under some very severe
corrosion conditions. High silicon cast irons are covered by ASTM
A518.

STAINLESS STEELS

Stainless steels are iron-base alloys containing at least 11.5% chromium.
This level of chromium is required to produce passivity. The term
stainless steels covers a very complex and extensive series of alloys
that have been developed and tailored for many specific corrosive services. Cast stainless steels are available in all the grades comparable
to the wrought grades plus many additional grades for special end use
applications [5]. The standard wrought stainless steels are generally
identified by a three-digit American Iron and Steel Institute (AISI)
designation and the alloys are usually grouped for discussion by their
microstructure. The largest group is comprised of the 300 series
nickel-chromium austenitic grades. The 400 series alloys, which contain primarily chromium as an alloying element, are identified as either
martensitic or ferritic. The cast alloys are identified by a system
developed by the Alloy Casting Institute (ACI). To illustrate, the
case equivalent for 316 stainless steel is CF-8M. The C signifies that
the alloy is for corrosive services, the F represents the nominal
chromium-nickel range, the 8 indicates the maximum carbon level
allowed, and the M is used when molybdenum is present. Even though

the cast and wrought grades are discussed as being equivalent, it
should be noted that significant compositional and property differences
exist for some of the grades that result in their not being truly equivalent. Some of these differences will be presented when the specific
alloys are discussed. It is very desirable for users of stainless castings to reference the ACI designations when specifying alloys since
significant differences can exist relative to the wrought grades.
Grouping the cast alloys by microstructure is a convenient method for
presenting these materials, and this procedure is used in the following
discussion. Table 5 presents the alloys that will be discussed.

Martensitic Alloys

The martensitic alloys found in Table 5 are CA-6NM, CA-15, CA-15M,
and CA-40. The chromium in CA-15 is the minimum required to make
a rustproof alloy. Its resistance to atmospheric corrosion is good and
it finds use in mildly corrosive organic services. Since the alloy is
martensitic, it is used in some abrasive applications. Some specific
areas where the alloy has been used include alkaline liquors, ammonia
water, boiler feedwater, pulp, steam, and food products [6]. CA-15M
contains molybdenum, which provides improved elevated temperature
mechanical properties. CA-40 is a higher carbon version of CA-15.
The higher carbon content permits heat treatment to higher strength
and hardness levels. Corrosion resistance is comparable to CA-15
and this alloy finds use in similar applications. CA-6NM is an iron-
chromium-nickel-molybdenum alloy that is hardenable by heat treatment. Its corrosion resistance is comparable to CA-15, although the
addition of molybdenum does make CA-6NM more resistant to corrosion
in seawater. The major advantages of CA-6NM over CA-15 are its
much improved foundry characteristics and increased toughness. Some
specific applications for CA-6NM include boiler feedwater, seawater,
and other waters to 400°F. This alloy has found particular use in
very large turbine runners for power generation.

Ferritic Alloys

Included in Table 5 in this group of alloys are CB-30 and CC-50. CB-
30 is essentially all ferritic and, as such, is nonhardenable. Chromium
is present at a sufficient level to give this alloy much better corrosion
resistance in oxidizing environments. Areas where this alloy has been
used include food products, nitric acid, steam, sulfur atmospheres,
and other oxidizing atmospheres to 760°C (1400°F) [6]. A major drawback to this alloy is its very poor impact toughness, and for this reason CB-30 has been replaced in many applications by the CF grades.

Table 5. Stainless Steel Compositions and Properties

A351	A743	A744	Cast alloy designation	Wrought alloy type	C[a]	Mn[a]	Si[a]	P[a]	S[a]	Cr	Ni	Fe	Other elements
	•		CA-6NM	—	0.06	1.00	1.00	0.04	0.03	11.5-14.0	3.5-4.5	Bal.	0.40-1.0 Mo
	•		CA-15	410	0.15	1.00	1.50	0.04	0.04	11.5-14.0	1.00[a]	Bal.	0.50 Mo[a]
	•		CA-15M	—	0.15	1.00	0.65	0.040	0.040	11.5-14.0	1.0[a]	Bal.	0.15-1.0 Mo
	•		CA-40	420	0.20/ 0.40	1.00	1.50	0.04	0.04	11.5-14.0	1.0[a]	Bal.	0.5 Mo[a]
	•		CB-30	442	0.30	1.00	1.50	0.04	0.04	18.0-21.0	2.00[a]	Bal.	—
			CB-7Cu	326	0.07	1.00	1.00	0.04	0.04	15.5-17.0	3.6-4.6	Bal.	2.3-3.3 Cu
	•		CC-50	446	0.50	1.00	1.50	0.04	0.04	26.0-40.0	4.00[a]	Bal.	—
•	•	•	CD-4MCu	—	0.04	1.00	1.00	0.04	0.04	24.5-26.5	4.75-6.00	Bal.	1.75-2.25 Mo 2.75-3.25 Cu
	•		CE-30	312	0.30	1.50	2.00	0.04	0.04	26.0-30.0	8.0-11.0	Bal.	—
•	•	•	CF-3	304L	0.03	1.50	2.00	0.04	0.04	17.0-21.0	8.0-12.0	Bal.	—
•			CF-3A	—	0.03	1.50	2.00	0.040	0.040	17.0-21.0	8.0-12.0	Bal.	—
•	•	•	CF-8	304	0.08	1.50	2.00	0.04	0.04	18.0-21.0	8.0-11.0	Bal.	—
•			CF-8A	—	0.08	1.50	2.00	0.040	0.040	18.0-21.0	8.0-11.0	Bal.	—
	•		CF-20	—	0.20	1.50	2.00	0.04	0.04	18.0-21.0	8.0-11.0	Bal.	—
•	•	•	CF-3M	316L	0.03	1.50	1.50	0.04	0.04	17.0-21.0	9.0-13.0	Bal.	2.0-3.0 Mo
•			CF-3MA	—	0.03	1.50	1.50	0.040	0.040	17.0-21.0	9.0-13.0	Bal.	2.0-3.0 Mo
•	•	•	CF-8M	316	0.08	1.50	2.00	0.04	0.04	18.0-21.0	9.0-12.0	Bal.	2.0-3.0 Mo
•	•	•	CF-8C	347	0.08	1.50	2.00	0.04	0.04	18.0-21.0	9.0-12.0	Bal.	8 x C Cb[c] 1.0 Cb[a]
•			CF-10MC	—	0.10	1.50	1.50	0.040	0.040	15.0-18.0	13.0-16.0	Bal.	1.75-2.25 Mo 10 x C Cb[c] 1.20 Cb[a]
	•		CF-16F	303	0.16	1.50	2.00	0.17	0.04	18.0-21.0	9.0-12.0	Bal.	0.20-0.25 Se 1.50 Mo[a]
	•	•	CG-8M	317	0.08	1.50	1.50	0.04	0.04	18.0-21.0	9.0-13.0	Bal.	3.0-4.0 Mo
	•		CG-12	—	0.12	1.50	2.00	0.04	0.04	20.0-23.0	10.0-13.0	Bal.	—
•	•		CH-20	309	0.20	1.50	2.00	0.04	0.04	22.0-26.0	12.0-15.0	Bal.	—
•	•		CK-20	310	0.20	2.00	2.00	0.04	0.04	23.0-27.0	19.0-22.0	Bal.	—
•	•	•	CN-7M	332	0.07	1.50	1.50	0.04	0.04	19.0-22.0	27.5-30.5	Bal.	2.0-3.0 Mo 3.0-4.0 Cu

Composition

Applicable ASTM Standard

[a]Maximum unless otherwise noted.

[b]NT, normalize and temper; A, anneal; SA, solution anneal; AH, age harden.

[c]Minimum unless otherwise noted.

[d]Typical.

Heat treat condition[b]	Tensile strength		Yield strength		Elongation in 50 mm (2 in.)[c]	Reduction of area,[c]	Brinell hardness number[d]	Impact toughness[d]			
										Test temp.	
	MPa[c]	ksi[c]	MPa[c]	ksi[c]				J	ft-lb	°C[e]	°F[e]
NT	755	110	550	80	15	35	240	95	70	RT	RT
NT	620	90	450	65	18	30	210	27	20	RT	RT
NT	620	90	450	65	18	30	220	–	–	–	–
NT	690	100	485	70	15	25	240	3	2	RT	RT
A	450	65	205	30	–	–	180	3	2	RT	RT
SA AH	892	130	755	110	10	–	320	34	25	RT	RT
A	380	55	–	–	–	–	200	–	–	–	–
SA	690	100	485	70	16	–	240	75	55	RT	RT
SA	550	80	275	40	10	–	190	10	7	RT	RT
SA	485	70	205	30	35	–	150	149	110	RT	RT
SA	530	77	240	35	35	–	160	136	100	RT	RT
SA	485	70	205	30	35	–	150	100	74	RT	RT
SA	530	77	240	35	35	–	160	95	70	RT	RT
SA	485	70	205	30	30	–	150	81	60	RT	RT
SA	485	70	205	30	30	–	150	163	120	RT	RT
SA	550	80	255	37	30	–	170	136	100	RT	RT
SA	485	70	205	30	30	–	160	95	70	RT	RT
SA	485	70	205	30	30	–	150	41	30	RT	RT
SA	485	70	205	30	20	–	150	–	–	–	–
SA	485	70	205	30	25	–	150	102	75	RT	RT
SA	520	75	240	35	25	–	160	108	80	RT	RT
SA	485	70	195	28	35	–	160	–	–	–	–
SA	485	70	205	30	30	–	160	41	30	RT	RT
SA	450	65	195	28	30	–	140	68	50	RT	RT
SA	425	62	170	25	35	–	130	95	70	RT	RT

CC-50 has a higher chromium content than CB-30, resulting in improved corrosion resistance in oxidizing media. Also, at least 2% nickel and 0.15% nitrogen are usually added to CC-50, resulting in improved toughness. This alloy has found use in acid mine waters, sulfuric and nitric acid mixtures, alkaline liquors, and sulfurous liquors [6].

Precipitation Hardening Alloys

There are two alloys in Table 5 belonging to this group, CB-7Cu and CD-4MCu. Alloy CB-7Cu is equivalent to wrought alloy type 17-4PH. This is a martensitic alloy with minor amounts of retained austenite present in the microstructure. By aging the alloy at 496°C (925°F) after solution annealing, submicroscopic copper particles precipitate and significantly strengthen the alloy. In the age-hardened condition, the alloy possesses corrosion resistance superior to the straight martensitic and ferritic grades, but inferior to the CF and higher alloy grades. This alloy finds use where moderate corrosion resistance and high strength are required. Typical application areas include aircraft parts, food processing equipment, and pump shafting.

CD-4MCu is one of the cast alloys that has not wrought equivalent. It is a chromium-nickel-molybdenum-copper stainless steel developed to meet the demand for a high strength alloy possessing good ductility and hardness together with corrosion resistance equal or superior to the popular 19% chromium-9% nickel stainless steels [6]. The alloy is two phased, having a microstructure of ferrite containing large austenite pools. Typically, these constituents are present in a ratio of about 60% ferrite-40% austenite. Because the alloy contains such a high percentage of ferrite, it is strongly magnetic. The alloy develops yield strength and hardness levels about twice that of the cast 304 and 316 equivalents. This higher hardness has lead to greatly improved resistance to erosion and velocity conditions. Also, due to the duplex microstructure, the alloy is resistant to weld cracking and shows exceptional resistance to chloride stress cracking. The alloy can be precipitation hardened to even higher strength levels, but it is recommended that the alloy be used in the solution annealed condition because the strength levels are high in this condition and aging can lead to embrittlement and significant lowering of the impact strength.

CD-4MCu has found application in a very wide variety of services. Becuase of its high chromium level, it is outstanding in oxidizing environments such as nitric acid, and the presence of solids in these environments presents little problem since the alloy possesses good hardness. Reducing environments are also handled quite well. CD-4MCu is widely used in dilute sulfuric acid services up to fairly high temperatures. The alloy has excelled in fertilizer production and the wet process method for producing phosphoric acid where phosphate rock containing fluorides is encountered. Even though the alloy is low in nickel, it has outstanding resistance to sodium hydroxide—in fact, almost as good as the much higher alloyed CN-7M. Other services that have been successfully handled by CD-4MCu include seawater, concentrated brines, fatty acids, pulp liquors, hot oils, scrubber solutions containing alumina and hydrofluoric acid, and dye slurries [6].

Austenitic-Ferritic Alloys

Of the alloys listed in Table 5, these comprise the largest group both in terms of the number of compositions and the quantity of material produced. This group of alloys also illustrates most vividly the differences that can exist between so-called equivalent cast and wrought grades. Included in this group is CE-30, CF-3, CF-3A, CF-8, CF-8A, CF-20, CF-3M, CF-3MA, CF-8M, CF-8C, CF-10MC, CF-16F, and CG-8M.

The austenitic-ferritic alloys comprise the 19% chromium-9% nickel cast equivalents of the wrought 300 series stainless steels. In the wrought form, the 300 series alloys are fully austenitic. This structure is required to permit the necessary hot- and cold-forming operations to be performed to produce the various wrought shapes. Since castings are produced essentially to finished shape, it is not necessary for the alloys to be fully austenitic. The cast compositions of the 300 series equivalents can be balanced such that the microstructure contains from 5 to 40% ferrite. The amount of ferrite present can be estimated from the composition using Fig. 1 or from the magnetic response achieved with calibrated measuring instruments. It is important to note that the cast equivalents of the 300 series alloys can display a magnetic response ranging from none to quite strong. The wrought 300 series alloys contain no ferrite and in the annealed condition are nonmagnetic.

The presence of ferrite in the cast alloys offers a number of advantages. Weldability, particularly with regard to resisting fissuring, is significantly enhanced. The mechanical strength increases and this increase is proportional to the amount of ferrite present (see Fig. 2). Finally, the presence of ferrite raises the resistance of these alloys to stress corrosion cracking compared to the fully austenitic wrought equivalents. It is possible to specify castings with specific ferrite levels and, in fact, some of the alloys discussed below are designated as controlled ferrite grades.

CE-30 is a high carbon cast stainless steel. The alloy has a duplex microstructure of ferrite in austenite, and carbide precipitates are present as-cast. There is sufficient chromium present so that resistance in the as-cast condition to intergranular corrosion is not serious-

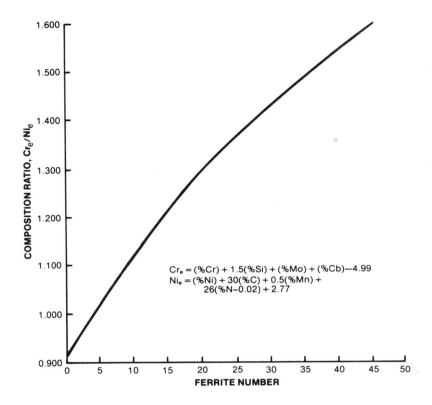

Figure 1. Diagram for estimating ferrite content in cast stainless
steels. (Adapted from Ref. 7.)

ly impaired. Since the alloy does retain good corrosion resistance as-
cast, it is useful where heat treatment is not possible or where heat
treatment following welding cannot be performed. Of course, corrosion resistance and ductility can be significantly enhanced by solution annealing. The alloy is resistant to sulfurous acid, sulfites, mixtures of sulfurous and sulfuric acids, and sulfuric and nitric acids
[6]. Also, the controlled ferrite grade CE-30A is resistant to stress
corrosion cracking in polythionic acid and chlorides. Other major
uses include pulp and paper manufacture, acid mine water, caustic
soda, and organic acids.

The CF alloys constitute the major segment of the corrosion-resistant casting alloys. These are the 19% chromium-9% nickel materials,
which possess very good processing characteristics and mechanical

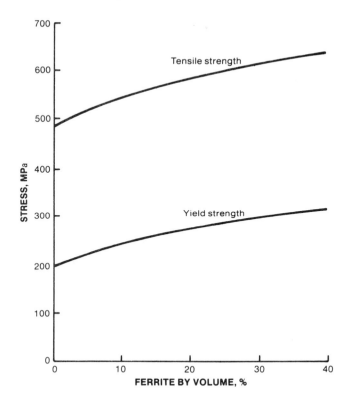

Figure 2. Relationship of yield strength and tensile strength to ferrite content for CF-8 and CF-8M. (Adapted from Ref. 7.)

properties and are resistant to a very wide range of corrosive enviro·.-ments. Microstructurally, the alloys are generally ferrite in austenite, with the composition balanced to provide 5 to 25% ferrite. For applications where nonmagnetic castings are required, a fully austenitic structure can be provided; however, because of the many desirable benefits derived from the presence of ferrite, it is usually present in the structure. To provide optimum corrosion resistance, these alloys are provided in the solution-annealed condition. The CF alloys are not hardenable by heat treatment. In all cases, the corrosion resistance of the cast alloy is at least equal to its wrought equivalent. For many applications, the CF alloys are overspecified; that is, they are applied where a lesser alloy would be adequate from a corrosion standpoint. This happens because the CF alloys have such outstanding

foundry characteristics compared to the 12% chromium materials that
the total cost to produce a casting can be less with the CF grades.
This factor should be kept in mind when selecting an alloy for
moderately corrosive services.

CF-8 is the base composition for the CF alloy group. Its wrought
equivalent is 304. It contains nominally 19% chromium-9% nickel-0.08
maximum carbon. The alloy has good strength and ductility and its
impact toughness is maintained to below -240°C (-400°F). CF-8 is
particularly good in strongly oxidizing media such as boiling nitric
acid. Other typical applications include adipic acid, copper sulfate,
fatty acids, organic liquids and acids, sewage, sodium carbonate,
sodium sulfite, vinegar, and white liquor [6]. CF-3 is the low carbon
(0.03 maximum) version of CF-8. It is generally specified where welding will be performed and postweld heat treatment is not possible. Because the carbon is so low, castings have been used in the as-cast
condition for some applications, although to achieve the best corrosion
resistance and mechanical properties, solution annealing is performed.
The overall corrosion resistance of CF-3 is somewhat better than
CF-8, but in general, they find use in the same applications. CF-8C
is the stabilized grade of the CF-8 alloy. Controlled amounts of
columbium (niobium) or columbium plus tantalum are added to the
alloy to tie up the carbon to prevent the formation of chromium carbides. This alloy finds use where field welding without subsequent
heat treatment will be performed. CF-8C provides essentially equivalent corrosion resistance to CF-8. With the introduction of AOD refining, the stabilized grades will probably find less use since the low
carbon grades will become less costly and more available. CF-8A and
CF-3A are the controlled ferrite grades of the 19% chromium-9% nickel
alloys. By controlling the composition and thus the percent ferrite
present (see Fig. 1), it is possible to achieve a 35-MPa (5-ksi) in-
crase in the minimum yield strength and a 45-MPa (7-ksi) increase in
the tensile strength. This also results in increased resistance to
stress corrosion cracking, with no sacrifice in overall corrosion resistance. CF-3A has found extensive use in nuclear power plant construction. Because of the thermal instability of the higher ferrite
microstructures, CF-3A has an upper temperature limit of 340°C
(650°F) and CF-8A an upper limit of 425°C (800°F). CF-20 is a high
carbon version of CF-8. It somewhat fills the gap between CA-15 and
CF-8. This alloy performs satisfactorily in moderately oxidizing environments. Microstructurally, the alloy is fully austenitic and is
nonmagnetic. Typical services include caustic salts, food products,
sulfite liquor, and sulfurous acid [6].

CF-8M is the cast equivalent of 316 stainless steel. This is a 19%
chromium-9% nickel-2% molybdenum-0.08 maximum carbon alloy. The
nickel level in this alloy is usually a little higher than CF-8 to offset
the ferritizing influence of the molybdenum and thus maintain a comparable ferrite level in the microstructure. The alloy can be made

fully austenitic and thus nonmagnetic. Addition of molybdenum improves the general corrosion resistance, provides greater elevated temperature strength, and particularly, improves pitting resistance in chloride environments. Some resistance to strongly oxidizing environments such as boiling nitric acid is sacrificed by adding mylybdenum, but passivity is enhanced under weakly oxidizing conditions compared to CF-8. This alloy has good resistance to reducing media. Overall corrosion resistance is equal to or better than 316 stainless steel. Because of the versatility of this alloy, it is used predominatly for pumps and valves in corrosive services. Some typical services for CF-8M include acetone, acetic acid, black liquor, chloride solutions, hot dyes, fatty acids, phosphoric acid, sulfate and sulfite liquors, dilute and concentrated sulfuric acid, sulfurous acid, and vinyl alcohol [6]. The alloy is supplied in the solution-annealed condition. CF-3M is the low carbon grade intended for use where postweld heat treatment is not possible. The areas of application for CF-3M are essentially the same as for CF-8M. CF-3MA is the controlled ferrite grade. Increases in minimum yield strength of 50 MPa (7 ksi) and tensile strength of 65 MPa (10 ksi) can be achieved. There is an upper temperature limit of 425°C (800°F) for usage of CF-3MA. CF-10MC is the stabilized grade of CF-8M for field welding applications. Again, as AOD refining becomes more widespread, this alloy will probably find less application in favor of the low carbon grades.

    CF-16F is a free-machining 19% chromium-9% nickel stainless steel. Small additions of selenium and phosphorus form inclusions that act as chip breakers during machining. The microstructure in addition to the selenium inclusions contains about 15% ferrite. Corrosion resistance is slightly inferior to CF-20. This alloy and CF-20 find use in similar applications.

    CG-8M is a higher molybdenum version of CF-8M. It is particularly resistant to reducing media and performs very well in sulfuric and sulfurous acids; in addition, it resists the pitting action of halogen compounds. The alloy is not suitable for strongly oxidizing environments. Microstructurally, the alloy usually displays a high ferrite content and this results in very good stress corrosion cracking resistance. However, because of its high ferrite level, the alloy has an upper temperature limit of 650°C (1200°F).

Austenitic Alloys

The alloys in Table 5 comprising this group include CH-20, CK-20, and CN-7M. These alloys display an essentially single-phase structure of austenite, although depending on the compositional balance, there may be a slight amount of ferrite present.

    CH-20 is similar to CE-30 but has higher nickel and lower chromium levels. This alloy is more ductile than CE-30 and stronger than CF-8.

CH-20 is considerably more corrosion resistant than CF-3 and is less susceptible than CF-8 to intergranular corrosion after exposure to sensitization temperatures. Solution annealing is required to achieve maximum corrosion resistance and mechanical properties. CK-20 finds use in many of the same services as CH-20 but at higher temperatures [6].

CN-7M is the most highly alloyed of all the cast stainless steels. It is fully austenitic and must be solution annealed for maximum corrosion resistance. The alloy is resistant to a wide range of corrosives. It finds widespread use in sulfuric acid services and is suitable for all concentrations to 65°C (150°F) and even higher temperatures for most concentrations. Because of its high nickel content, CN-7M has excellent resistance to alkaline environments. It can be used in sodium hydroxide up to 73% and 149°C (300°F). This good resistance in caustic environments plus good chloride resistance makes this alloy a good choice for the hot alkaline brines found in caustic-chlorine plants. Since the alloy contains appreciable chromium, it is superior to the CF grades in nitric acid—even better than CF-3, which is generally considered the best alloy for this service. Hydrochloric acid, certain chlorides, and strong reducing agents such as hydrogen sulfide, carbon disulfide, and sulfur dioxide will accelerate corrosion as will deaeration of the solution. Additional services where CN-7M is used include hot acetic acid, dilute hydrofluoric and hydrofluosilicic acids, nitric-hydrofluoric pickling solutions, phosphoric acid, and some plating solutions [6].

NICKEL-BASE ALLOYS

Table 6 lists the major cast nickel-base alloys that find use in corrosive environments. These alloys generally are used only in specialty areas and very severe service because of their high cost compared to stainless steels. ACI designations have been adopted for these alloys since, like the stainless steels, their compositions and properties do in many instances vary significantly from the wrought equivalents. The cast nickel-base alloys are found in ASTM Standards A494 and A744.

Cast nickel, CZ-100, is the equivalent of commercially pure wrought nickel [8]. The carbon and silicon levels are necessarily higher in the cast grade to provide adequate castability. By treating the alloy with magnesium in the molten state, the carbon in the cast alloy is caused to nodularize, and this leads to enhancement of the mechanical properties much as with ductile cast iron. The properties of CZ-100 are not improved by heat treatment; therefore, the alloy is generally provided in the as-cast condition. CZ-100 finds its widest use in alkaline services. It has unsurpassed resistance to all bases except

ammonium hydroxide [9]. Above 1%, ammonium hydroxide causes
rapid attack. In sodium and potassium hydroxide, CZ-100 is resistant
to all concentrations and temperatures. The presence of chlorates and
oxidizable sulfur compounds will accelerate the corrosion rate in caustic. Food processing, where product purity is important, is an area
where CZ-100 is used. Also, the alloy is used in synthetic fiber manufacture.

Cast Inconel*, CY-40, is the equivalent of wrought Inconel Alloy
600. Again, the carbon and silicon in the cast grade are higher than
the wrought material to provide adequate castability. CY-40 is solution annealed to maximize corrosion resistance and mechanical properties. The alloy is used primarily for oxidation resistance and strength
retention at high temperatures [9]. In chloride environments, the
alloy resists stress corrosion cracking and it is frequently substituted
for CZ-100 in caustic soda containing halogens. Since it resists
chloride stress cracking and corrosion by high-purity water, CY-40
is widely used in nuclear reactor services. Also, some CY-40 is used
in food processing.

Cast Monel*, M-35, is the equivalent of wrought Monel 400. The
compositional variations between the wrought and cast grades are
again due to foundry characteristics. Since nickel and copper are
mutually soluble in each other at all concentrations, there is no need
to heat treat M-35, and therefore, it is used in the as-cast condition.
This alloy finds widespread use in water handling, particularly in sea
and brackish waters [8]. It handles water services very well under
high velocity conditions, and many pump and marine components are
cast in M-35. Air-free hydrofluoric acid is commonly handled by
M-35, as are chlorinated solvents. Oxidizing conditions or contaminants accelerate the corrosion rate of Monel in all services. Since
M-35 does possess good resistance to fluorides, it is finding widespread use in uranium enrichment. Chloride stress corrosion cracking
is not a factor with this alloy. M-35 has found some use in sulfuric
acid services where reducing conditions exist.

The final alloys in the nickel-base group are the cast equivalents
or modifications of the nickel-molybdenum and nickel-chromium-molybdenum alloys. The N-12M alloys in this group comprise the nickel-
molybdenum materials. These alloys are solution annealed and heat
treatment, together with alloy purity, is critical to producing a microstructure that will provide maximum corrosion resistance. If impurities such as carbon and silicon are not controlled to as low levels as
possible, a secondary phase will be present in the microstructure
that will adversely affect corrosion resistance. It should be noted
that the N-12M-2 alloy is much more ductile than N-12M-1. The N-12M

---

*Trademark of the International Nickel Company, Inc.

Table 6. Nickel-Base Alloys Compositions and Properties

A494	A744	Cast alloy designation	Wrought alloy type	C^a	Mn^a	Si^a	P^a	S^a	Cr	Ni	Fe	Other elements
•	•	CZ-100	Nickel 200	1.00	1.50	2.00	0.03	0.03	—	95.00^c	3.00^a	1.25 Cu^a
•	•	CY-40	Inconel[f] 600	0.40	1.50	3.00	0.03	0.03	14.0-17.0	Bal.	11.00^a	—
•	•	M-35	Monel[f] 400	0.35	1.50	2.00	0.03	0.03	—	Bal.	3.50^a	26.0-33.0 Cu
•	•	N-12M-1	Hastelloy[g] B	0.12	1.00	1.00	0.040	0.030	1.00^a	Bal.	4.0-6.0	26.0-30.0 Mo 0.20-0.60 V
•		N-12M-2	—	0.07	1.00	1.00	0.040	0.030	1.0^a	Bal.	3.0^a	30.0-33.0 Mo
•	•	CW-12M-1 Hastelloy[g] C	0.12	1.00	1.00	0.040	0.030	15.5-17.5	Bal.	4.50-7.50	16.0-18.0 Mo 0.20-0.60 V 3.75-5.25 W
•		CW-12M-2	—	0.07	1.00	1.00	0.040	0.030	17.0-20.0	Bal.	3.0^a	17.0-20.0 Mo

[a]Maximum unless otherwise noted.

[b]SA, solution annealed; AC, as cast.

[c]Minimum unless otherwise noted.

[d]Typical.

[e]RT, room temperature.

[f]Trademark of the International Nickel Company, Inc.

[g]Trademark of Cabot Corp.

grades are particularly recommended for handling hydrochloric acid at all concentrations and all temperatures, including boiling [10]. Oxidizing contaminants or conditions can lead to rapid failure. Cupric or ferric chloride, hypochlorites, nitric acid, and even aeration are common causes of accelerated corrosion with these alloys. At 26°C (78°F), the tolerable ferric ion concentration in 10% hydrochloric acid is 5000 ppm, at 66°C (150°F) it is less than 1000 ppm, and at boiling it is less than 75 ppm [11]. This illustrates the importance not only of the contaminant level but also of temperature. The N-12M alloys also perform well in hot sulfuric acid, again without the presence of oxidizing contaminants. Phosphoric acid in all concentrations can be handled up to 140°C (300°F) by these alloys.

The CW-12M alloys comprise the nickel-chromium-molybdenum group. CW-12M-1 is actually a dual-purpose alloy intended for both corrosion and high-temperature services. Tungsten and vanadium are specifically added for high-temperature oxidation resistance. CW-12M-2 is intended primarily for corrosive services. By so limiting the in-

Heat-treated condition[b]	Tensile strength		Yield strength		Elongation in 50 mm (2 in.)[c]	Brinell hardness number[d]	Impact toughness[e]			
									Test temp[e]	
	MPa[c]	ksi[c]	MPa[c]	ksi[c]			J	ft-lb	°C	°F
AC	340	50	125	18	10	120	–	–	–	–
SA	480	70	195	28	30	140	–	–	–	–
AC	450	65	205	30	25	130	–	–	–	–
SA	520	76	315	46	6	190	–	–	–	–
SA	520	76	315	46	20	190	27	20	RT	RT
SA	500	72	315	46	4	200	–	–	–	–
SA	500	72	315	46	25	200	27	20	RT	RT

tended use of this alloy, the tungsten and vanadium are removed and the chromium, molybdenum, and nickel levels are raised. This provides improved corrosion resistance and enhanced mechanical properties, particularly with respect to ductility. Heat treatment and compositional purity are again important, for the same reasons as those mentioned above for the Ni-Mo alloys. The essential difference between the N-12M and CW-12M alloys is that approximately one-half of the molybdenum in the Ni-Mo group has been replaced with chromium. The addition of chromium gives the CW-12M alloys oxidation resistance while maintaining good resistance to reducing acids. For example, these alloys can handle hydrochloric acid contaminated with ferric chloride at all concentrations up to 50°C (120°F). At lower concentrations, higher temperatures can be tolerated. CW-12M alloys are quite useful in acid brine solutions. Other typical applications include hot contaminated mineral acids, solvents, chlorine- and chloride-contaminated solutions, hot organic acids, hypochlorites, acetic anhydride, and seawater.

Table 7. Cast Copper-Alloy Families

UNS No.[a]	Description	Major constituents
C80100-C81100	Coppers	Cu
C81300-C82800	High-copper alloys	Cu plus Cr, Be, Co, Ni, Si
C83300-C83800	Red brasses, leaded red brasses	Cu, Sn, Zn, Pb
C84200-C84800	Semired brasses, leaded semired brasses	Cu, Sn, Zn, Pb
C85200-C85800	Yellow brasses, leaded yellow brasses	Cu, Sn, Zn, Pb
C86100-C86800	Manganese bronzes, leaded manganese bronzes	Cu, Zn, Al, Mn, Pb
C87200-C87900	Silicon bronzes and brasses	Cu, Zn, Si
C90200-C91700	Tin bronzes	Cu, Sn
C92200-C92900	Leaded tin bronzes	Cu, Sn, Pb
C93200-C94500	High-leaded tin bronzes	Cu, Sn, Pb
C94700-C94900	Nickel-tin bronzes	Cu, Sn, Ni
C95200-C95300	Aluminum bronzes	Cu, Al, Fe, Ni
C96200-C96600	Copper-nickels	Cu, Ni, Fe
C97300-C97800	Nickel-silvers	Cu, Ni, Zn
C98200-C98800	Leaded coppers	Cu, Pb
C99300-C99700	Special alloys	Cu, Ni, Fe, Al, Zn

[a]Unified numbering system.

COPPER-BASE ALLOYS

The copper-base alloys comprise a very large group of materials. Included are relatively pure copper, many different grades of brasses and bronzes, the copper-nickel alloys, and various special alloys. It would be impossible to present and discuss each cast alloy here, so a broad listing of the various cast copper-alloy families are presented in Table 7 and a few selected alloys are discussed in some detail below. There are a number of ASTM standards dealing with cast copper alloys. The major ones are listed in Table 8.

Copper and its alloys are generally not chosen for their corrosion resistance alone but rather for that characteristic plus one or more other properties [12]. In fact, for many applications, some other characteristic, such as thermal conductivity, may be the overriding consideration. Copper castings have some advantages over the wrought materials much the same as was found with other alloy systems in that the casting process permits greater latitude in alloying since hot- and cold-working properties are not important. This is particularly true with regard to the use of lead as an alloying element.

Before getting to specific alloys and applications, it is appropriate to point out a few precautions that are pertinent when considering copper alloys for corrosive applications [13]. Copper alloys containing 20 to 40% zinc are susceptible to a form of corrosion known as dezincification [3]. This form of corrosion is like the graphitization of cast iron in that overall component dimensions do not change but strength is seriously impaired. For this reason, these alloys should not be used in dilute or concentrated acids, organic and inorganic acid salts, dilute or concentrated alkalies, neutral solutions of chlorides and sulfates, and mild oxidizing agents such as calcium hypo-

Table 8. ASTM Copper and Copper-Alloy Casting Specifications

ASTM standard	Title
B 61	Steam or Valve Bronze Castings
B 148	Aluminum-Bronze Sand Castings
B 271	Copper-Base Alloy Centrifugal Castings
B 369	Copper-Nickel Alloy Castings
B 505	Copper-Base Alloy Continuous Castings
B 584	Copper Alloy Sand Castings for General Applications

Table 9. Copper-Base Alloys Compositions and Properties

ASTM Standard	Cast alloy designation	Nominal composition									Tensile strength		Yield strength		Elongation in 50 mm (2 in.)[a]
		Zn	Sn	Pb	Mn	Al	Cu	Fe	Si	Other elements	MPa[a]	ksi[a]	MPa[a]	ksi[a]	
B584	C83600	5	5	5	–	–	85	–	–	–	207	30	97	14	20
B584	C85200	24	1	3	–	–	72	–	–	–	241	35	83	12	25
B584	C86200	27	–	–	3	4	63	3	–	–	621	90	310	45	18
B584	C86300	27	–	–	3	6	61	3	–	–	758	110	414	60	12
B584	C90500	2	10	–	–	–	88	–	–	–	276	40	124	18	20
B584	C87200	5	1	–	1.5	1.5	89	2.5	3	–	310	45	124	18	20
B369	C96200	–	–	–	0.9	–	87.5	1.5	0.1	10 Ni	310	45	170	25	20
B369	C96400	–	–	–	0.8	–	67	0.7	0.5	30 Ni 1 Nb	415	60	220	32	20

[a]Minimum.

chlorite, hydrogen peroxide, and sodium nitrate. Copper alloys containing less than 15% zinc can withstand a wide range of environments, including acids, alkalies, and salt solutions, as long as the services do *not* contain dissolved air, oxidizing materials such as nitric acid, ferric salts, dichromates, and chlorine; compounds such as ammonia and cyanide that form soluble complex ions with copper; and compounds that react directly with copper, such as sulfur, hydrogen sulfide, silver salts, mercury and mercury compounds, and acetylene.

Relatively pure copper (99+%) finds use in many applications. Because of its resistance to industrial, marine, and rural atmospheres, this material finds use in many architectural applications [14]. It also is used in freshwater applications and some seawater services, although certain copper alloys are more desirable here. No ASTM standard exists for cast copper.

The brasses are probably the most useful of all the copper alloys. Table 9 lists some of the more common cast brass compositions and properties. Red brass is much superior to copper for handling hard water. The brasses also find use in seawater with the higher strength, higher hardness materials used under high velocity and turbulent conditions. Other services where the brasses might be used include boric acids, neutral salts such as magnesium chloride and barium chloride, organics such as ethylene glycol and formaldehyde, and organic acids.

The bronzes are the next major group of copper alloys. Table 9 lists some of the more common cast bronze compositions and properties. From a corrosion standpoint, the bronzes are quite similar to the brasses. The addition of aluminum to the bronzes does improve resistance to high temperature oxidation and also raises the tensile properties. The bronzes do find use in sulfate environments where brasses might not be used. Silicon bronzes can handle cold dilute hydrochloric acid, and cold and hot dilute and cold concentrated sulfuric acid.

The copper-nickels are the final major group of cast copper alloys. Table 9 shows the composition and properties of the two most common cast copper-nickel alloys. These materials have the highest corrosion resistance of all the copper alloys. The copper-nickels have outstanding resistance to seawater and probably find their most widespread use in the area of sea- and brackish-water handling. Dilute hydrochloric, phosphoric, and sulfuric acids cause no attack. The copper-nickels are almost as resistant as Monel to caustic soda.

ALUMINUM ALLOYS

As with the other systems, there are a very large number of different aluminum alloys. There is a major difference with regard to aluminum cast alloys for corrosive applications in that, in general, adding alloy-

Table 10. Aluminum Alloys Compositions and Properties

Cast alloy designation	Composition[a]							Heat-treated condition	Tensile strength		Elongation in 50 mm (2 in.)[b]
	Mn	Si	Zn	Ti	Cu	Mg	Fe		MPa[b]	ksi[b]	
A356	0.20	6.5/7.5	0.10	0.20	0.20	0.20–0.40	0.20	T6[c]	254	37	5.0
B443	0.35	4.5/6.0	0.35	0.25	0.15	0.05	0.8	F[d]	117	17	3.0
514	0.35	0.35	0.15	0.25	0.15	3.5 4.5	0.50	F[d]	151	22	6.0

[a]Maximum unless otherwise noted.
[b]Minimum unless otherwise noted.
[c]T6, solution heat treated, artificially aged.
[d]F, sand cast, no heat treatment.

ing elements to aluminum tends to decrease corrosion resistance but pure aluminum castings are rarely made due to poor casting quality and low mechanical properties [14]. Most aluminum casting applications can be handled by three alloys: 356, 443, and 514. These alloys can be found in ASTM B 26 and B 108 for sand castings and permanent mold castings, respectively. Aluminum die castings also find some use, particularly for atmospheric corrosion resistance, and the die cast alloys can be found in ASTM B 85. These three aluminum specifications appear in Part 7 of the ASTM standards.

Table 10 shows the compositions and properties for the three alloys mentioned above. The 356 alloy is used where a combination of good strength characteristics and atmospheric corrosion resistance is required. This alloy also finds use in some water applications. Alloy 443 possesses excellent resistance to domestic and marine atmospheric conditions. It is also very resistant to the mildly acidic solutions encountered in the textile industry and finds widespread use there. Alloy 514 possesses very good stress corrosion resistance even at very high stress levels. It has high resistance to general and pitting corrosion. Applications include seawater, salt spray, mild alkalies, and high octane gasolines. In addition to these services, aluminum castings find use in nitrogen fertilizer solutions, ammonium nitrate, urea, 98% nitric acid, concentrated sulfuric acid to avoid iron contamination, sulfite solutions, synthetic fiber manufacture, and sewage disposal.

TITANIUM

Titanium castings have found increasing use in corrosive services due to their resistance to some very severe environments. Titanium is much like aluminum in that alloying generally decreases the corrosion resistance. For that reason, only two cast titanium materials are generally used for corrosive services, the commercially pure grade and a grade stabilized with a small addition of palladium. Cast titanium alloys are covered by ASTM B 367.

Before discussing the applications in which titanium excels, it seems appropriate to enumerate some precautions relative to its use [15]. Titanium is particularly prone to hydrogen pickup even at room temperature. This pickup causes the protective film to become embrittled and it can then be eroded away, particularly under high velocity conditions. This can lead to rapid wastage of the titanium surface. Titanium can exhibit pyrophoric reactions in certain environments, such as red fuming nitric acid, high pressure oxygen, and dry chlorine. Stress corrosion cracking can be a problem in certain environments, including nitrogen tetroxide, fuming nitric acid, methyl alcohol, chlorinated solvents, hot dry salt, and hydrogen chloride. Finally, titanium is very susceptible to crevice corro-

Table 11. Titanium Compositions and Properties

ASTM Standard	Cast alloy designation	Composition[a]						
		C	H	N	O	Pd	Ti	Fe
B 367	C-3	0.10	0.0100	0.05	0.35		Bal.	0.30
B 367	C-8A	0.10	0.0100	0.05	0.35	0.12+	Bal.	0.30

[a]Maximum unless otherwise noted.

[b]AC, as-cast.

[c]Minimum unless otherwise noted.

sion in some environments, although the addition of palladium overcomes this problem to a great degree. Titanium does have some shortcomings and castings are costly, but for certain severe services this
can prove to be a wise and economical alloy choice.

The compositions and properties of two of the commonly used
corrosion resisting grades found in B 367 are shown in Table 11. One
of the outstanding characteristics of titanium is its resistance to inorganic chlorides which destroy stainless steels and other common
materials of construction. Titanium can easily handle hot or cold
chlorinated brines and marine environments. In fact, titanium handles
these services so well that it has replaced the copper-nickels in many
seawater applications. Titanium is widely used in the manufacture of
ferric and cupric chlorides, corrosives that play particular havoc on
stainless steels. Services containing wet chlorine gas and bleaching
solutions containing chlorites, hypochlorites, and chlorine dioxide are
very successfully handled by titanium. As little as 50 ppm water is
required to avoid vigorous attack by the chlorine. Titanium is used
in strong nitric acid up to 200°C (392°F), but the other mineral acids
tend to attack it. The presence of oxidizing species such as ferric
or cupric ions sufficiently inhibit attack to permit use in acidic services. Also, the palladium-stabilized grade shows much improved

Other elements	Heat-treated condition[b]	Tensile strength		Yield strength		Elongation in 50 mm (2 in.)[c]	Brinell hardness number[a]
		MPa[c]	ksi[c]	MPa[c]	ksi[c]		
0.10 each 0.40 total	AC	450	65	380	55	15	235
0.10 each 0.40 total	AC	450	65	380	55	15	235

behavior in acidic services. Since titanium tends to be nontoxic, it is being used more frequently in food service applications.

ZIRCONIUM

There are no ASTM specifications for cast zirconium, but Table 12 shows a typical composition and properties for cast material used in corrosive applications. The major use of zirconium is in nuclear applications, but it is coming into more widespread use for some very severe corrosive services. One of its outstanding properties is its resistance to hydrochloric acid [15]. Zirconium can handle boiling hydrochloric acid to 20% and can handle higher concentrations at somewhat lower temperatures. The presence of oxidizing species such as cupric or ferric chloride and wet chlorine render zirconium unacceptable. Sulfuric acid to 70% can be handled to the boiling point. Zirconium can handle all concentrations of nitric acid to boiling. It is superior to stainless steels, nickel alloys, and titanium in organic acids and is particularly useful at high temperatures. Severe caustic services can be handled very successfully. Zirconium can also handle all inorganic salts except ferric and cupric chloride.

Table 12.   Zirconium Composition and Properties

				Composition[a]				
C	H	N	O	P	Fe	Hf	Zr	Other elements
0.10	0.004	0.03	0.20	0.010	0.30	4.50	Bal.	0.40 total

[a]Maximum unless otherwise noted.

[b]AC, as cast.

[c]Minimum unless otherwise noted.

MAGNESIUM ALLOYS

There are a number of cast magnesium alloy grades produced and
these can be found in ASTM specifications B80, B94, B199, and B403
for sand castings, die castings, permanent mold castings, and investment castings, respectively. Magnesium is not generally chosen for
its corrosion resistance. In fact, since magnesium is anodic in comparison to most metals, it is frequently used as a sacrificial anode to
protect other alloys. Magnesium alloys find widespread use in aircraft and related applications where weight savings are important.
Magnesium alloys do resist attack by most alkalies, many organic
chemicals, concentrated hydrofluoric acid, pure fluorides, chromates,
and dichromate. These alloys are generally attacked by other acids
and aqueous salt solutions. The addition of alloying elements to magnesium does not appreciably influence corrosion resistance.

CASTING PRODUCTION

Casting Design

If a component is to be made as a casting, it should be designed as a
casting. This statement may seem axiomatic, but there are too many
instances where it is not adhered to. Casting design is certainly
much too broad a topic to cover in detail here and there are many
references [16-19] that cover the topic quite well. There are a few
general points concerning casting design that are worthy of mention.

Heat-treated condition[b]	Tensile strength		Yield strength		Elongation in 50 mm (2 in.)[c]	Reduction of area, [c] %	Brinell hardness number
	MPa[c]	ksi[c]	MPa[c]	ksi[c]			
AC	380	55	276	40	12	20	200

Casting design and casting process go hand in hand, and the inherent advantages and disadvantages of each process must be considered starting with the design step. There are a number of processes in use today for manufacturing castings for corrosive services. The major processes include sand casting, plaster casting, investment casting, die or permanent mold casting, and centrifugal casting. Sand casting is by far the most common of the processes. This process provides design flexibility and low cost. The largest castings are made by this process, but through proper design and good foundry practices, small pieces can also be produced. Surface finish is not generally as good as with the other processes, but with proper controls it can be quite adequate. Available molding equipment is capable of producing consistent molds at very good production rates. Pattern equipment costs can vary depending on production quantities and molding systems, but it is generally the lowest of all the casting processes. Since relatively inexpensive patterns can be obtained, this makes sand casting attractive for low volume and prototype work.

There are many variations of the basic sand casting process. The most prevalent is green sand casting. The process is so named because the molds, consisting of a sand-clay-water mixture, are poured soon after they are made or before any appreciable drying has occurred; thus they are in the green condition. Dry sand molds are those where the moisture is removed before pouring. Shell molding is a relatively new variation whereby a very thin mold is produced on a heated pattern using a resinous or organic binder. This procedure finds particular use for making cores that are placed in green sand molds. The shell molding process lends itself to higher production rates, better surface finish, and improved dimensional control but has the drawbacks of leading to contamination of low carbon alloys and presenting some environmental problems. Some progress has been made in reducing alloy contamination through the use of additives to the molding sands, and inorganic binder systems have been developed to eliminate totally problems encountered with resin systems. Practically all nonreactive metals and alloys are produced in some form of sand casting process.

Plaster casting involves making a mold by pouring a plaster slurry around a contained pattern or into a core box. It is particularly suited for the lower melting point alloys and finds widespread application in producing aluminum- and copper-base castings. Dimensional control and surface finish are good and relatively thin sections can be poured. There is some limitation on casting size.

Investment casting involves producing an expendable pattern, usually of wax or plastic, then forming a shell or solid mold around this pattern. The pattern is removed by melting, burning, or dissolution, leaving a cavity that is an exact replica of the pattern. Surface finish and dimensional control are excellent. Also, the process produces a ceramic mold that does not contaminate the alloy or present environmental problems. Since the molds can be poured hot, very thin sections can be produced. Cost of production dies to produce the patterns is relatively high, although lower cost prototype tooling can be made. Production rates vary depending on the degree of mechanization. Casting size is definitely limited. Since dimensional control is so good, it is possible to produce castings with a minimum of metal that require very little or no machining. Castings in all alloys are produced by this process.

Die or permanent mold casting involves introducing a molten alloy into a metal mold either through gravity pouring or under pressure. Dimensional tolerances and surface finish are excellent. This is a high production rate process with very high initial costs for dies and casting machines. Casting size is very limited. Some ferrous die casting is being performed, but on a very limited basis. Generally, the lower melting point alloys are cast by this process.

Centrifugal casting, as it is generally performed, can be considered a special form of permanent mold casting. A metal mold coated with a ceramic wash or a sand mixture is spun on its central axis while the molten metal is introduced. The centrifugal force produces a dense casting requiring no risering and no central core. This process is widely used to produce tubes and cylindrical configurations for corrosive services in many different alloys.

Production of castings in reactive metals such as titanium and zirconium requires special melting and casting techniques. Since the alloys are reactive, it is necessary to melt and pour under vacuum.

Some sort of centrifugal assist is commonly used to produce a sound casting. The molds are made from either a graphitic bonded or ceramic-type system. Since the castings must be poured in a vacuum chamber, there is a very definite size limitation. The necessity of vacuum melting and pouring is an obvious contributor to the high cost of cast titanium and zirconium components.

Since design is so critical to the final cost and quality of a cast component and since there are a number of casting processes available, the final choice of which is very dependent on design, it is imperative that casting expertise be input into the design function at a very early stage. Most good-quality casting producers can provide this input and will be of significant assistance in producing a design that will provide the needed performance at a reasonable cost.

Casting Quality

Cast components find use in many very critical corrosive applications. As operating conditions become more severe, as maintenance costs and the costs of shutdowns continue to rise, as product contamination and product loss attract increasing interest, and as safety and reliability assume more importance, the matter of component integrity becomes even more important. There are certain defects inherent in castings, and control of these defects is critical to the production of a quality part. The important quality aspects of any casting intended for a corrosive service can be assigned to one of three areas: surface integrity, internal integrity, and alloy purity.

Surface integrity deals with surface finish, surface defects, and dimensional accuracy. Surface finish has significance for a number of reasons. There is an aesthetic consideration that cannot be overlooked. Equipment manufacturers do not want unsightly castings as part of their product and this is understandable. For velocity applications, rough surfaces can disrupt flow patterns and reduce efficiency. Included in the area of surface defects are cracks, dirt and slag inclusions, and porosity. Again, the aesthetic value of a casting is reduced by surface defects. More important, surface defects can be serious enough to lead to component failure. This is particularly true of cracks, and these defects are often not discernible through visual inspection. A more searching technique, such as dye penetrant or magnetic particle inspection, is required to detect many linear-type defects. Finally, dimensional accuracy can influence machining operations, assembly, and final component performance. Factors influencing surface integrity include casting design, casting process, foundry practices, and alloy selection. Casting performance and cost are influenced by surface integrity requirements, and these must be given adequate consideration and be carefully defined for inclusion in the final purchasing specification.

Internal integrity deals with the internal soundness of a casting. Common defects leading to unsound castings include shrinkage, porosity, and inclusions. The presence of these defects can be determined only through such techniques as radiographic or ultrasonic inspection. These defects can certainly lead to component failure and their effect can be very insidious. Internal quality levels can be defined relative to ASTM reference standards. For example, ASTM E446 contains reference radiographs for steel casting section thicknesses up to 51 mm (2 in.), and ASTM E155 contains reference radiographs for aluminum and magnesium castings. Good casting design and very careful control of all foundry practices are required to produce sound castings. Required internal integrity must be defined in casting specifications.

Surface and internal integrity are certainly significant with regard to acceptable performance of cast components. However, in corrosive service, the casting alloy is of utmost importance. Proper alloy selection must receive careful consideration. It is essential that a sufficiently resistant alloy be chosen, but the selection must also be the most economical. Metallurgical expertise in alloy selection is as vital as with wrought alloys. Selection of the proper alloy for a service does not, however, ensure acceptable service because there is the matter of alloy purity. The same basic alloy grade can display variations in performance in the same service, and this variation is commonly due to differences in alloy purity. The two major factors determining alloy purity are melting practices and mold environment.

Many factors are included under the broad topic of melting practices. Those having the most influence on alloy purity are raw materials, melting method, and melt additives. Raw materials are obviously critical, but they have particular significance for those alloys made without benefit of melt refining. In this case, to achieve a high-purity alloy necessitates using high-purity raw materials. The melting method is also significant. Most high-melting-point metals are melted in electric furnaces. There are two major types, arc and induction. Induction melting is quite commonly found in foundries and is used to melt ferrous and nickel-base alloys in particular. Arc melting is generally used in the larger foundries and has the advantage of being able to refine alloys to some degree. Most of the lower melting point materials, such as aluminum, copper, and magnesium, are melted in fuel-fired furnaces, although induction melting is also used. Melt additives include such materials as degassers and inoculants. These materials are generally added to correct a problem introduced during melting, such as deoxidizing steels. The additives generally perform their intended purpose quite effectively, but by making these additions the overall alloy purity is frequently reduced. There has been a major development in the past few years that will have significant influence on melting practices and resulting alloy purity. This is the refining process called argon-oxygen decarburization (AOD).

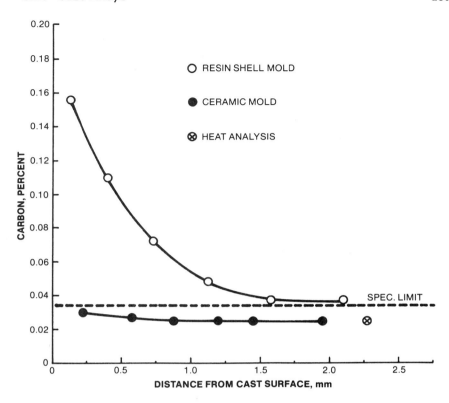

Figure 3. Carbon pickup by CF-3 casting poured in resin-bonded
and ceramic molds. (Adapted from Ref. 22.)

AOD refining is a duplex operation [20] utilized to improve the
compositional purity and properties of alloys. The process offers
several distinct advantages, including raw material cost savings, high
purity alloys, improved casting cleanliness, and enhanced corrosion
resistance and mechanical properties. Stainless steels can be routinely produced to low carbon specifications, thus eliminating the problem
of sensitization during welding. Desirable effects can also be achieved
with low alloy steels [21]. Reducing the sulfur levels in these alloys
significantly enhances the ductility and impact toughness of these
alloys. Lower hydrogen and nitrogen levels reduces the susceptibility
to embrittlement. Nickel-base alloys can also benefit from AOD refining. Significant improvements in the ductility of some nickel alloys
have been reported [21]. Sufficient field experience has not been

achieved to define adequately the influence of AOD refining on the corrosion resistance of cast alloys.

The mold environment encountered by molten metal can have a significant influence on the final composition of the alloy. It has been known for many years that the carbonaceous binders and additives used in producing molds and cores can lead to carbon pickup on the surface of castings. Very definitive work has been performed to show the degree of this pickup with stainless steels [22]. The pickup on the surface of the casting is quite substantial; in fact, for the low carbon grades, sufficient carbon is picked up by the bulk alloy as it flows through the gating system and into the mold cavity that the entire casting carbon level exceeds the specification maximum. Figure 3 shows the level of carbon pickup experienced by a CF-3 casting poured into a resin-bonded mold.

Contamination due to carbon pickup can lead to accelerated corrosion of cast components. In fact, premature failure has resulted, particularly in intergranular corrosion media [23]. For those alloys such as carbon steel where carbon pickup would not be of concern, certain other deleterious elements, such as sulfur and phosphorus, can be a problem.

Appreciable work has been done to eliminate, or at least ameliorate, contamination of alloys by mold and core binders. The resin-bonded systems have gained such wide acceptance that much effect has been directed toward reducing the contaminating effects of the mold environment. It has been determined that certain additives can reduce the level of contamination, but results are somewhat variable and not totally controllable. The only method found to date that totally eliminates the problem is to utilize inorganic, ceramic binders. Figure 3 shows a profile for a CF-3 casting poured in a ceramic mold.

REFERENCES

1. H. Leckie, Iron, carbon steel, and low alloy steels in the process industries, in *Process Industries Corrosion*, National Association of Corrosion Engineers, Houston, 1975, pp. 90-102.

2. Cast iron, in *Metals Handbook*, Vol. 1, American Society for Metals, Metals Park, Ohio, 1961, pp. 349-406.

3. M. Fontana and N. Greene, *Corrosion Engineering*, 2nd Ed., McGraw-Hill, New York, 1978.

4. F. LaQue, The corrosion resistance of ductile iron, *Corrosion* 14(10): 485t-492t, 1958.

5. E. Shoefer, The cast stainless steels, in *Handbook of Stainless Steels* (D. Peckner and I. Bernstein, eds.), McGraw-Hill, New York, 1977, pp. 2/1-2/18.

6. E. Shoefer, *High Alloy Data Sheets, Corrosion Series*, Alloy Casting Institute Division of Steel Founders' Society of America, Rocky River, Ohio, 1973.

7. W. Herrnstein, Cast iron-nickel-chromium alloys, in *Handbook of Stainless Steels* (D Peckner and I. Bernstein, eds.), McGraw-Hill, New York, 1977, pp. 10/1-10/18.

8. *Huntington Alloys Handbook*, 5th ed., International Nickel Company, Toronto, 1970.

9. D. Graver, Nickel and high nickel alloys, in *Process Industries Corrosion*, National Association of Corrosion Engineers, Houston, 1975, pp. 109-113.

10. R. Hughson, High nickel alloys for corrosion resistance, *Chem. Eng. 83*(24): 125-152, 1976.

11. T. Lee, III and F. Hodge, Resistance of "Hastelloy" alloys to corrosion by inorganic acids, *Mater. Perform. 15*(9): 29-36, 1976.

12. W. Lyman and A. Cohen, Engineering with copper alloys–I, *Chem. Eng. 85*(6): 99-102, 1978.

13. W. Lyman and A. Cohen, Engineeering with copper alloys–II, *Chem. Eng. 85*(9): 147-150, 1978.

14. R. Gackenbach, *Material Selection for Process Plants*, Reinhold, New York, 1960.

15. W. Boyd, The use of titanium, zirconium, tantalum, and columbium in chemical environments, in *Process Industries Corrosion*, National Association of Corrosion Engineers, Houston, 1975, pp. 207-212.

16. *Metals Handbook*, Vol. 5, American Society for Metals, Metals Park, Ohio, 1970.

17. H. Taylor, M. Flemings, and J. Wulff, *Foundry Engineering*, Wiley, New York 1959.

18. *Steel Casting Design, Engineering Data File No. 1*, Steel Founders' Society of America, Rocky River, Ohio.

19. *Steel Castings Handbook*, 4th ed. (C. W. Briggs, ed.), Steel Founders' Society of America, Rocky River, Ohio, 1970.

20. D. Hilty and T. Kaveny, Melting of stainless steels, in *Handbook of Stainless Steels* (D. Peckner and I. Bernstein, eds.), McGraw-Hill, New York, 1977, pp. 3/1-3/35.

21. L. Venne and C. Oldfather, Enhancement of physical properties by argon oxygen decarburization, *Mod. Cast. 68*(8): 70-71, 1978.

22. W. Luce, M. Fontana, and J. Cangi, Improved corrosion resistance, reliability, and integrity of cast stainless alloys, *Corrosion 28*(4): 115-128, 1972.

23. W. Herrnstein, J. Cangi, and M. Fontana, Effect of carbon pickup on the serviceability of stainless alloy castings, *Mater. Perform. 14*(10): 21-27, 1975.

4

Development and Application of Plastic Materials

JOHN H. MALLINSON / J. H. Mallinson, P.E. & Associates Inc.,
Front Royal, Virginia

INTRODUCTION AND HISTORY [1]

One of the great steps in human development was the discovery that
we could change the nature of materials. Slowly at first we adapted
materials to our use and then as the Industrial Revolution produced a
much more complex civilization, the problems of material preservation
changed toether with the emphasis. Economic competition is the driving force that compels engineers to provide the lowest cost unit over
the service life of equipment. Practically all engineering materials are
composites of some sort. Cannons made of wood were bound with
brass. Steel is painted to withstand corrosion. Today the corrosion
engineer is constantly looking for materials that are lighter, stiffer,
stronger, and more corrosion resistant than their predecessors.

Practically every engineering design is a balance of economic
choices, but economic factors act as the independent variable. Design
is for expected life, and reliability is an important part of the picture.
The performance demands of a vessel or process system are generally
involve a combination of many different materials. *Each part is the
engineer's optimum choice for the local function. Rarely, if ever
does a single material become a blanket prescription for success.*

Nature has natural polymers in abundance but at our level of technical learning we were unable to grasp the idea that trial and error
were not the keys to the great materials of the future. Only with new
instrumentation such as the electron microscope, ultracentrifuge, and
infrared analyzers has polymer development really pushed forward.

The plastics industries began about 1868, when Hyatt discovered
celluloid. This was followed by Backeland and his phenol-formaldehyde
resins (including Bakelite) in 1909, nylon in 1938, the polyesters and
polyethylene in 1942, and the epoxies and acrylonitrile-butadiene-
styrene (ABS) in 1947. In the first half of the century the thermosets
centered on the corrosion battle—at first on a small scale and then on a
much larger scale in the early 1950s. Some plastic and lined pipe were
available in the 1940s and the first phenolics were being used for
chemical industry corrosion problems.

During the last several decades polyvinyl chloride (PVC) and ABS
have become widely used in process equipment. The polyolefins are
moving into the forefront where low cost and chemical inertness are

required. They are available in many forms, such as duct, piping, and packing for scrubbers and columns. Polypropylene is applied increasingly in process equipment.

Fluorinated thermoplastic resins have found special uses in the chemical process industries becuase of their heat stability and inertness. For example, heat exchangers built of Teflon have carved out a unique position in many areas, particularly in crystallization, where fibrillation techniques require their use over rigid structures [1].

Of course, there are special reasons that make the use of other thermoplastics desirable. Nylons and acetals are excellent materials for small parts and gears; phenylene oxides provide high strength and have been made into bolts and nuts for corrosive service; and the urethanes provided a complete breakthrough in the insulation area. Some elastomers are used for lining process equipment where particular properties are desired, even for large earthen pits running into multimillion-gallon capacities. Great quantities of polyethylene and polypropylene tubing are used in process instrument applications because of their chemical inertness.

The basic push for other materials has constantly been prodded by economic considerations. Alkaline materials at moderate temperatures have never represented a troublesome corrosion area. The tough corrosion problems are found in oxidizing or reducing atmospheres or combinations. Steel pipe may last only a few days. Type 304 stainless steel may do little better. Type 316 stainless steel might last several months and in many cases only the highest-grade Cr-Ni-Mo is successful. The engineer's first attempt to get around this was to use lined steel equipment; in some cases ceramic ware, lead, rubber, glass, a few elastomers, and some phenolic linings were used, but engineers were faced with high-priced solutions. They have continually looked for a material as strong as steel, as light as aluminum, and with high chemical resistance to the service conditions at hand. When engineers found that they could tailor-make a material to suit particular corrosion problems, the field of possibilities broadened explosively.

In the chemical industry much has been achieved with the use of plastics to perform difficult assignments. Unfortunately, engineers are faced with a double standard of judgment and evaluation. The failure of steel, stainless steel, rubber-lined vessels, lead-lined steel, and the like, is looked upon in general as inevitable. Industrial management expects it and allows for it in the budget, but the failure of plastic materials calls for a task force investigation. Quite often the solution to the problem lies in improper selection of the plastic to be used, or not following recommended installation procedures. Lack of knowledge plus lack of the application of proper engineering are the most general causes of problems.

PVC is widely applied in piping such as drain, waste, vents, and rural irrigation systems, as well as electrical conduit, telephone duct, and wire and cable insulation. Filled phenolics are inherently flame

retardant with low smoke evolvement, and as such have potential use
in corrosive fume systems, particularly in ducts.

High-density polyethylene and cross-linked high-density polyethylene are becoming more widely used. Applications in sewers and chemical pipelines, with sizes up to 48 in. in diameter, are particularly note-
worthy. Usage is expected to increase 50-fold in the next decade because the chemical and abrasion resistance properties of plastics are
significantly better than those of steel.

Chemical process pumps of all descriptions are built from polypropylene, cast epoxies, reinforced epoxies, Teflon, and reinforced polyesters. These include magnetic pumps of polypropylene construction
and vinyl or Tygon peristaltic pumps.

The entire field of valves has been invaded by both the thermoplastics and the thermosets—ball valves particularly. Applications include
diaphragm valves, gate valves, butterflies, multiport valves, globe
valves, check valves, foot valves, diverters, pressure regulators,
solenoid valves, and needle valves.

Material blends combine some of the unique properties of each component. For example, ABS improves measurably the impact strength
of PVC. Inorganic additives such as fiberglass or graphite provide
considerably increased strength, and other additives produce fire retardancy. Process manufacturing innovations are continuing, such
as rotary molding and spray mold fluid-bed molding systems.

It is estimated that during the 1970s capital investment in plastic
processing equipment will progress at a rate three times that of the
1960s. Unfortunately, investment and discovery are not necessarily
synonymous. During the 1930s seven basic families of plastics were
introduced, but only four were introduced in the 1950s and three in
the 1960s [1]. During the 1970s only three new specialty materials
were discovered: thermoplastic polyester in 1970; polybutalene in
1973; and nitrile barrier resins for bottles in 1975.

The plastics industry no doubt is maturing. Up until 1972 the industry experienced double-digit growth, but by 1980 this had slowed
to about 8.7%. However, in all the chemical industries, plastics materials and resins are still the fastest-growing segment. Growth ma-
turity is occurring in phenolics and cellulosics. There are, however,
some plastics which are still growing at a double-digit rate: ABS,
about 13%; acrylics, 11%; high-density polyethylene, about 11%; and
the polyesters, a bit over 10%. Diverse growth is still occurring by
industries: transportation leads at 13%, followed by consumer uses,
9%; building and construction, 9%; and leisure items, about 8%. But
new product development has clearly reached a plateau. One should
be careful not to equate relative values with absolute values. In
absolute terms the industry is still growing at a very rapid rate and
probably will continue to do so for some time to come.

THE THERMOPLASTS [2]

It is difficult to say exactly when the field of synthetic polymers was born. In the late 1800s and early 1900s, some synthetic polymers were created in the laboratories, such as cellulose and Bakelite. But only since the beginning of the twentieth century, when advanced new instruments for investigation came into being, has polymer investigation really advanced. Some of the molecules involved, being composed of building blocks based on monomers, have extremely large molecular weights, running into the millions of units.

An entire series of thermoplastic resins were uncovered over a period of many years:

Chlorinated polyethers
Fluorinated hydrocarbons
Polyamides
Polyethylene
Polystyrene
Polyvinyl chlorides
Vinylidenes
Vinyls

Many of the early thermoplastics had relatively low tensile strength and a limited resistance to heat. The search went on.

THE THERMOSETS [2]

Another group of substances that was of interest were the thermoset compounds. These are initially a liquid at room temperature; then, by means of a catalyst or accelerator, they are changed into a rigid product which sets, or cures, into its final shape. Typical of the polymer thermoset resins are:

Diallyl phthalates	Phenolics
Epoxies	Polyesters
Furanes	Polyurethanes

The field of reinforced plastics as we know it today really began after World War II, at which time radomes were made. The boating industry, however, was the first large-scale user and producer of reinforced plastic material. Without the wide acceptance of this material by the boating industry, the field of reinforced plastics would not be where it is today. Originally, in the young industry, delamination was a severe

problem, and many of the first boats made of this material cam apart.
No published information existed because there was no information to
be had. Many of the design data on which the industry is based came
from boat builders.

SPECIFIC PLASTIC FAMILIES

Thermoplasts [3]

Polyethylene

One of the most interesting of all the thermoplasts is polyethylene.
Molecular weights of 120,000 to 220,000 are common. High-density
polyethylene (HDPE) has a wide range of chemical resistance to oxidizing chemicals and salts and is useful in continuous service up to 82°C
(180°F). Ultraviolet stabilizers are widely used, particularly carbon
black. It also enjoys good lubricity and has been used for bearings
and heavy machinery. The fact that some 600 different polyethylene
compounds are offered commercially indicates its wide acceptability.
About 30% of the total annual consumption of all plastics is polyethylene. Although polyethylene has exceptional abrasion resistance, it
is outranked by the more recently developed polybutylene, which will
handle coal or ores and chemical and abrasive aggressive slurries and
is used as liners for hopper cars and discharge gates. One step
higher is ultrahigh molecular weight polyethylene, commonly called
UHMWPE.

Polyphenylene Sulfide

Polyphenylene sulfide is one of the most economical thermoplasts.
Although initially expensive, it is useful up to temperatures as high
as 230°C (450°F). Strangely enough, as temperature increases, there
is a corresponding increase in toughness. Polyphenylene sulfide can
provide outstanding performance in aqueous inorganic salts and bases,
and it is inert to many organic solvents. The use of a glass filler enhances its physical properties. Fasteners have been made of this remarkable material for use under highly oxidizing conditions, and coatings exhibit good hardness and chemical inertness.

Polypropylene

Polypropylene is a relatively young plastic, being born in the mid-
1950s. It is unique in that it is used in all four primary fabrication
processes: molding, extrusion, fibers, and film. It also competes
with metals and natural fibers. Further, it can be reinforced with

glass fibers to materially increase its physical properties, and polypropylene fibers can be used to reinforce polyesters. In the chemical industry it compares very favorably with rubber in use as a liner. It is also being used as a composite with a fiberglass overlay in many items of ductwork, mixing chambers, and other chemical equipment. If exposed to sunlight, an ultraviolet (UV) absorber or screening agent should be used to protect it from degradation. Thermal oxidative degradation is also a major enemy, particularly where copper is involved. In this case, copper inhibitives should be added to prevent catalyzation. Fortunately, polypropylene is not affected by most inorganic chemicals except the halogens and severe oxidizing environments. Chlorinated hydrocarbons can cause swelling and softening at elevated temperatures. With a fiberglass overlay, resistance up to 150°C (300°F) is possible. Fillers are quite often used to improve impact resistance for increased stiffness.

Polyvinyl Polymers

PVC is by far the largest family in this group and can be found in many products, from rigid pipe to film and relatively heavy sheets. In general, high molecular weight PVC has better physical properties. Like other plastics, additives are used to further specific end uses, such as thermal stabilizers, lubricity, impact modifiers, and pigmentation. PVC can be either rigid or flexible. Both plastisols and organosols can be produced with PVC. Hard films can be produced as thin as 1 mil. The largest consumers of PVC are the phonograph records and footwear industries. PVC is used extensively in chemical piping with a service range of -18 to 60°C (0 to 140°F). Above 60°C the use of PVC is not recommended. However, polyvinyl dichloride (PVDC) is used extensively in chemical piping systems up to a temperature of 82°C (180°F). One of the problems with the PVC formula is that in fluctuating temperatures there is a tendency for the material to "grow" and not return to its original length when the temperature subsides. This produces growth in pipelines which can be troublesome. Another disadvantage is that under sudden shock the resistance to impact is limited. It has found extensive use in venting systems and gravity lines in corrosive chemicals. It has a very wide range of chemical stability and has been successfully overlaid with fiberglass to provide external reinforcing. In prolonged high-temperature use, thermal degradation is evident. Even with a fiberglass backing, PVC, PVDC, and chlorinated PVC (CPVC) are subject to checking, crazing, and cracking which ultimately destroys the structure. In terms of usage, however, it is a very important plastic and is widely used in the building and allied industries, including that of household appliances.

Thermoplastic alloys

The alloying of several thermoplastic systems often produces better physical qualities than any system can produce by itself. These include impact resistance, flame retardency, and thermal stability. Alloying is generally done with intensive mixing or screw extruders. ABS and PVC are frequently blended to provide rigidity, toughness, flame retardency, and chemical resistance.

Polycarbonate can be blended with ABS to provide better heat resistance and toughness. Polyurethane improves the abrasion resistance and toughness of ABS, while retaining the advantage of reduced cost. The impact strength of polypropylene is increased by alloying with polyisobutylene. Alloying is generally done on a relatively small scale, varying from as low as 0.01% up to as high as 9%.

Thermosets [2, 3]

Epoxies

In general, the high-temperature cured epoxies show much better performance than the room-temperature-cured epoxies. They are somewhat less resistant than the polyesters to the wide range of corrosive chemicals but show better resistance on the alkaline side. Many different epoxy resins provide outstanding service in chemical process equipment under severe conditions. For the purpose of illustration, the specific chemical resistance characteristics of one resin are described. This resin, Epon 828,* has been widely used in chemical service, and is backed by a large number of successful case histories. Other epoxy resins with equally successful case histories may be considered for similar applications.

Resins of this type generally require elevated-temperature postcures. This is common practice in the manufacture of piping, tanks, and structures made from epoxy resins. This type of resin is inherently more expensive, and the manufacturing process is considerably slower, so that even simple molds can be turned over only twice weekly. The epoxies are nearly always combined with filament winding to produce an extremely strong product. The absence of room temperature cures makes the use of a high-chemical-grade epoxy such as this difficult for "do-it-yourself" plant application—an area in which the polyester largely predominates; reinforced epoxies are generally confined to piping systems and tanks.

Piping systems in reinforced epoxy have been well engineered, and their use is prevalent. Flanged and adhesive assembled systems

---

*Registered trademark of Shell Chemical Co., New York.

are commonly used. The following services are applicable to 93°C (200°F) unless otherwise stated:

Acids	Salts
10% Acetic acid [to 68°C (150°F)]	Aluminum
Benzoic acid	Most ammonium salts

Furanes

The furanes, although relatively old in conception, suffer from fabrication difficulties. They are much more difficult to fabricate, for example, than the polyesters because their setting up depends on a condensation reaction rather than the rapid polymerization that occurs with polyesters. The curing of furanes, with the liberation of water, is a self-limiting reaction which if pushed too rapidly generally results in severe internal stress and ultimate cracking of the structure. However, furanes show remarkable solvent resistance that would easily defeat the polyesters. They also have an upside temperature limitation of 205°C (400°F) in continuous operation, which makes them uniquely useful in many areas. Perhpas best of all, with the furane molecule it is possible to fabricate difficult systems that will meet the National Fire Protection Association (NFPA) standards, with a fire spread rating of 25 or less and a smoke rating of 50 or less. In general, they carry about a 30% premium in price over the polyesters.

Phenolics

One of the oldest thermoplasts, phenolic compounds have been widely used in the chemical industry because of their relative inertness to acids and bases and good temperature stability. As such, they have been used for many machine processing parts, and particularly in Europe have been developed into various trade-named applications on ducts, pumps, scrubbers, heaters, and the like. They are one of the two compounds that will pass the NFPA regulations for ducts and fans, having a fire spread and smoke spread rating of 50 or less. They are used widely in molded products in the United States but have not met with the general acceptance of the polyesters, epoxies, and furanes for use in the chemical industries. They are much more widely used in the chemical industries. The reason for this is that in general they do not have the impact resistance possessed by the polyesters and the epoxies, even when glass filled.

Polycarbonates

Probably one of the most economical resins developed in the last 25 years is the polycarbonate resin system. Because of its extremely high impact resistance and good clarity, it is widely used for windows in chemical equipment. Its impact resistance is little short of astounding, so it is a good candidate for many types of windows. Polycarbonate is used extensively in the lighting market for lenses on shipboard, in aircraft, and at airports. Even the telephone has felt its impact. Polycarbonate foam is finding increased application for machine housings. The use of 10% glass fiber reinforcing will result in its use as a replacement for sheet metal, with better corrosion resistance.

Polyesters and Vinyl Esters

General Purpose Polyester Resins The upper temperature limit is generally considered to be 50°C (120°F) and then only to a very limited amount of corrosive media, although the general purpose resins perform excellently in distilled water and seawater.

Isophthalic Polyester Resins These have an upper temperature limitation of about 70°C (150°F). They are useful in a much wider area than the general purpose resins, and for mildly corrosive conditions have a wide area of acceptance. They usually cover the general field of salts, mild acids, water of various kinds, gasoline, and the petroleum distillates.

High-Performance Polyester Resins These generally are of the bisphenol type or chlorinated polyesters. Generally, they have an upper wet limit of 120°C (250°F) and a dry limitation of 150°C (300°F). They are resistant to a wide range of oxidizing acids (up to 70% H_2SO_4 salts) and have shown remarkable performance in the chemical industry over a span of 25 years or more. The polyesters are generally quite resistant to attack from saturated hydrocarbon molecules but are much less resistant to unsaturated hydrocarbon molecules. They are particularly vulnerable to solvents such as carbon disulfide.

Vinyl Esters In general, these are only slightly less corrosion resistant than the bisphenol resins, but in the field of bleaching mechanisms such as chlorine, sodium hypochlorite, hypochlorous acid, and chlorine dioxide they offer better resistance than the bisphenol polyesters or chlorinated polyesters. Further modifications of the vinyl ester molecule have produced very high temperature performance resin systems which will go up to 140°C (280°F) in continuous service.

General Purpose Polyesters General purpose polyester resins are normally not recommended for use in chemical process equipment. Their use in the finished fabrication represents a potential savings of 10 to 20%. Only purchasers can make the ultimate decision regarding the premium they are willing to pay for chemical resistance. These resins are generally adequate for use with nonoxidizing mineral acids and corrosives that are relatively mild. This is the resin that predominates in boat building, its resistance to water of all types, including seawater, being more than adequate. Test work has indicated satisfactory application in the following areas up to 52°C (125°F).

Acids

10% Acetic acid	Oleic acid
Citric acid	Benzoic acid
Fatty acids	Boric acid
1% Lactic acid	

Salts

Aluminum sulfate	Ferrous chloride
Ammonium chloride	Magnesium chloride
10% ammonium sulfate	Magnesium sulfate
Calcium chloride (saturated)	Nickel chloride
Calcium sulfate	Nickel nitrate
Copper sulfate (saturated)	Nickel sulfate
Ferric chloride	Potassium chloride
Ferric nitrate	Potassium sulfate
Ferric sulfate	10% Sodium chloride

General purpose resins have been found to be unsatisfactory in:

Oxidizing acids
Alkaline solutions, such as calcium hydroxide, sodium hydroxide, and sodium carbonate
Bleach solutions, such as 5% sodium hypochlorite
Solvents, including carbon disulfide, carbon tetrachloride, gasoline, and distilled water

Where the use of general purpose resin is contemplated, an environmental test program should be inaugurated to determine if the resin will be satisfactory. All contemplated applications above 52°C (125°F) should receive rigorous testing.

Isophthalic Polyesters

The isophthalic polyesters offer better chemical resistance than the general purpose resins in certain application areas, at a slightly higher cost. They show much better resistance to attack in the solvent areas, and are used extensively in the manufacture of underground gasoline tanks, where a satisfactory service life in the storage of gasoline and under the varied conditions of ground-soil corrosion are successfully met. The following general usage of the isophthalic resins is given as a guide in applications up to 54°C (150°F).

Acids
    10% acetic acid   Oleic acid
    Benzoic acid   25% Phosphoric acid
    Boric acid   Tartaric acid
    Citric acid   10% Sulfuric acid
    Fatty acids   25% Sulfuric acid

Salts
    Aluminum sulfate   Iron salts
    10% Ammonium carbonate   5% Hydrogen peroxide
    Ammonium chloride   Magnesium salts
    Ammonium nitrate   Nickel salts
    Ammonium sulfate   Sodium and potassium salts that
    Barium chloride   do not have a strong alkaline
    Calcium chloride (saturated)   reaction
    Copper chloride   Dilute bleach solutions
    Copper sulfate

Solvents
    Amyl alcohol   Gasoline
    Ethylene glycol   Kerosene
    Formaldehyde   Naphtha

Isophthalic resins have been found to be unsatisfactory in:

Acetone
Amyl acetate
Benzene
Carbon disulfide
Solutions of alkaline salts of potassium and sodium
Hot distilled water
Higher concentrations of oxidizing acids

High-performance polyester resins

The use of furane and polyester composites affords engineers another means of problem solving. The furanes themselves will not support combustion, having a tunnel test rating of less than 20; thus they are rated as nonflammable. These materials weigh approximately one-fifth that of conventional iron or steel fittings. Thermal conductivity is slightly higher than for the polyesters, but not appreciably so. Furane-lined piping, tanks, and reinforced-plastic (RP) structures are widely available from a variety of fabricators. The strong point of furane systems is their resistance to solvents, including acetone, ethyl alcohol, benzene, carbon tetrachloride, carbon disulfide, chloroform, the fatty acids, methyl ethyl ketone, toluene, and xylene, many of which would quickly impair the polyester or epoxy reins.

    Whereas the furanes excel in solvent resistance, there are some areas in which they are not as good as the high-chemical-resistance polyesters. A few of these areas are wet and dry chlorine gas; chromic acid plating solutions; hypochlorous acid; some of the nitrate solutions, such as lead, nickel, and zinc; and brine solutions saturated with chlorine, sodium hypochlorite solutions, or trichloracetic acid.

For the most part, however, furane-lined RP structures have a very broad spectrum of corrosion resistance in both acids and alkalis. The furanes themselves do not possess the rugged physical strength of a glass-reinforced structure, although corrosion-resistant piping in low-pressure supply and drain line work may be purchased built completely from a furane resin. In addition, some tanks and structures have been made solely from furane resins. However, using the furane as the inner lining of a composite structure, with the fibrous glass polyester overlay either laid up by hand or filament wound, effectively combines the best qualities of both systems in many areas of application.

Furane tanks up to 30,000 gal, ductwork to 24 in., piping, and equipment are available to meet dificult corrosive conditions involving organic solvents coupled with oxidizing chemicals. Flame spread ratings meet the fire retardant classification and low smoke evolvement competes with the best polyesters (about 400). Furane costs run 15 to 35% higher than those for the best polyesters. Accelerated heat-curing techniques for the furanes have reduced curing times considerably, but they are still considerably longer than the polyesters. High-temperature performance [92°C (200°F)] with the furanes is favorable, with case histories showing 10 years of service in 5% H_2SO_4 [1].

Vinyl Ester Resins

In the mid to late 1960s a new family of resins appeared, termed the vinyl esters. Since that time they have earned a solid place as one of the family of thermosets available for use to meet difficult corrosion problems. They possess good resistance to corrosion by many different chemicals, both acid and alkali, at both room and elevated temperatures. Some vinyl esters on the market possess two double bonds and others possess three double bonds. With minor additions of elastomers their percent elongation can be altered substantially and may vary in the elongation range 2 to 8%. This is highly critical in the field of cyclic temperature and pressure operation and with additions of various abrasion-enhancing additives becomes most important in increasing the abrasion resistance of laminated structures. Depending on the formulation, these may or may not interfere with the chemical resistance of the laminate. Of most importance is the fact that these resin systems use catalyst and promoter systems similar to those used with the polyesters. They show excellent resistance to severe oxidizing environments and generally exhibit better resistance than polyester and epoxies in fields such as bleach, chlorine, chlorine dioxide, chlorine-caustic systems, and calcium and sodium hypochlorite. In some areas they also show better solvent resistance, but not as good as that of the furanes.

Physical properties are in general very similar to those of the polyesters. Variations in formulations of the vinyl esters have been developed substantially beyond the initial work, so that formulations capable of operating at relatively high temperatures are available in addition to fire-retardant grades.

Carbon-to-carbon double-bond linkages occur both in vinyl ester and polyester molecules. These unreactive double bonds are subject to chemical attack through oxidation and halogenation. In the vinyl esters the double bonds are generally at the end of the molecule and thus give a more chemically resistant structure. In solvent attack the closer the double bond is to the carbon atoms, the greater the attack. That is why a molecule such as carbon disulfide is an extremely potent solvent and destroys the polyesters and vinyl esters through a rapid softening action and loss of strength, which is particularly pronounced within the first 30 days.

The vinyl ester molecule furthers reactions with hydroxyl groups on the surface of glass fibers and gives excellent wet out and good adhesion, which is further evidenced by the high strengths obtainable with vinyl ester resins. The vinyl ester resins may be safely used as structural components intended for repeated use in contact with food, subject to the limitations described in the appropriate regulations in the U.S. Food, Drug and Cosmetic Act.

Some of the vinyl ester resins are used with benzoyl peroxide dimethylaniline catalyst promoter systems in structures intended for bleach usage, where attention is paid to eliminating cobalt from the promoter system, the theory being that any heavy metal will tend to produce resin degradation. It should be noted that such catalyst systems are less forgiving than is the MEKPO cobalt naphthenate system. They can, however, be fabricated successfully and are specified by many users. The spectrum of chemical resistance of the vinyl esters is very wide. It is suggested that the reader consult the extensive literature on chemical environment and service conditions for the various vinyl ester systems.

REINFORCEMENT

Literally any plastic material provided with reinforcement will show a substantial strengthening of the plastic's physical properties. E glass is at present the dominant reinforcement and is used with many plastic systems to provide additional strength. Other types of reinforcement, such as graphite, carbon fibers, ceramic fibers, and sapphire whiskers, have also been used, particularly in the space program. Jute has been used in the Asiatic countries, and metallic wire has also been used.

Although polypropylene fibers do not develop the strength of glass, they have been useful in reinforced plastic tanks and structures, as has Kevlar 49.* Kevlar 49 has been used especially for localized reinforcement to provide extra strength at key points in a vessel. In general, glass reinforcement will approximately double the strength of a plastic.

COMPOSITE LAMINATES

It is quite common to use a lined polypropylene, furance, Armalon,* ABS, or PVC, backed with a fiberglass overlay. This permits the handling of a wide variety of materials that would not alone be satisfactory in a polyester resin system. Where we are able to tailor-make chemical resistance by a combination of thermoplast and reinforced thermoset materials, considerable equipment economies have been achieved. Dimensional stability and abrasion resistance can be increased similarly.

Further, most filament-wound tanks are provided with a hand-laid-up 100-mil barrier before the high-density filament winding begins. By techniques such as these it is possible to handle solvents, 93% sulfuric acid, hydrofluoric acid, and similar materials.

The evaluation of plastic construction material together with various reinforcements can produce some intriguing results which combine the optimum properties of filament winding with the corrosion resistance of hand-laid-up structures. Very typically the corrosion barrier would be followed by alternating layers of a 55° wind followed by a chopped mat followed by a hoop wind followed by a chopped mat, after which, the 55° wind, chopped layup, hoop wind, and so on, are repeated. This will produce a laminate which runs about 50% resin and 50% glass. A simple graph illustrates that at approximately this point the optimum combination of chemical and physical properties is obtained. A recent variation of this is a 0 to 90° wind with chopped strand interspersed.

Many different kinds of dual laminates are available, including ABS and polyester, bisphenol and isophthalate fibrous glass systems, epoxy and polyester, glass and reinforced polyester, Teflon and reinforced polyester, polypropylene-lined reinforced polyester, and PVC and polyester composite. With these dual laminates we are able to combat many of the most difficult corrosive systems. In addition, they offer additional strength and improved dimensional stability and abrasion resistance.

---

*Dupont trademark for their Aramid fiber.
*Armalon is Dupont's trademark for glass-fiber-backed fluorocarbon.

Table 1 compares the physical properties of metals and reinforced plastics.

CHEMICAL ATTACK OF PLASTICS–THE COUNTERPART OF METALLIC CORROSION [5]

If and when reinforced polyesters or epoxies are misapplied, chemical attack, which is a relatively complicated phenomenon for these matereials, may occur in several ways. These may be broadly classified as follow:

1. Disintegration or degradation of a physical nature due to absorption, permeation, solvent action, or other factors
2. Oxidation, where chemical bonds are attacked
3. Hydrolysis, where ester linkages are attacked
4. Dehydration (rather uncommon)
5. Radiation
6. Thermal degradation, involving depolymerization, and possibly repolymerization
7. Combinations of these mechanisms, and possibly others

As a result of such attacks, the material itself may be affected in one or more ways; for example, it may be embrittled, softened, charred, crazed, delaminated, discolored, dissolved, blistered, or swelled. Although all plastics are attacked in essentially the same manner, certain chemically resistant types suffer negligible attack, or exhibit significantly lower rates of attack, under a wide variety of severely corrosive conditions, due primarily to the unique molecular structure of the resins, including built-in steric protection of the ester groups. However, because of the complicated way in which attack occurs, knowledge of the chemical structure of the resin does not preclude actual testing in most environments to determine resistance.

Cure, of course, plays an important part in the chemical resistance developed by a thermoset, as does the construction of the laminate itself and the type of glass or reinforcing used. The degree and nature of the bond between the resin and the glass or other reinforcement also play an important role.

All these modes of attack affect the strength of the laminate in different ways, depending on the environment and other service conditions and the mechanism or combinations of mechanisms that are at work. For example, in certain instances absorption can weaken the polymer network by relieving internal strains or stresses. Certain environments may weaken primary and/or secondary polymer linkages, with resultant depolymerization; others may cause swelling or microcracking; and of course others may hydrolyze ester groupings or linkage. Extraction can occur in certain environments, and in still

Table 1. Comparative Physical Properties of Metals and Reinforced Plastics (Room Temperature)

	Carbon steel 1020	Stainless steel 316	Hastelloy C	Aluminum	Glass-mat laminate	Composite structure glass-mat woven roving	Glass-reinforced epoxy filament wound[a]
Density, lb/in.^{3}	0.283	0.286	0.324	0.098	0.050	0.065	0.065
Coefficient of thermal expansion, in./in. °F10^{-6}	6.5	9.2	6.3	13.2	17	13	9-12
Modulus of elasticity, psi x 10^6, in tension (Young's modulus)	30.0	28.0	26.0	10.0	0.7-1.0[b]	0.8-1.5	4.0-4.5
Tensile strength, psi x 10^3	66	85	80	12	9-15[b]	12-20	100
Yield strength, psi x 10^3	33	35	50	4	9-15[b]	12-20	100
Thermal conductivity, Btu/(h)(ft^2)(°F/ft)	28.0	9.4	6.5	135	1.5	1.5	1.5-2.0
Strength/weight ratio[c] x 10^3	230	300	250	122	300	308	1500

[a]Data on glass-reinforced filament-wound epoxy have been drawn from a variety of sources. Filament winding, in general, polyester or epoxy, will result in much higher physical strength.

[b]These data are from the Recommended Product Standard for Custom Contact Molded Reinforced Polyester Chemical Resistant Process Equipment, PS 15-69.

[c]The physical-strength figures used here for the glass-reinforced plastic laminates are very conservative. For example, some filament-wound epoxy tensiles will run to 300,000 lb, giving them phenomenal strength-to-weight ratios of 4500 x 10^3.
Source: Ref. 4.

other environments, repolymerization can occur, with a resultant
change in structure. Chain scission and a decrease in molecular
weight can occur under certain conditions. Simple solvent action may
also occur. Absorption or attack at the interface between the glass
and the resin will result in weakening.

All in all, the mechanisms involved are rather complicated and certainly not well understood. In addition to the combinations of mechanisms that can occur at the same time, it is believed that synergism
probably enters the picture in many instances.

Fortunately, chemical attack on plastics is a "go/no go" proposition.
Attack on reinforced plastics in many of the ways described above may
occur in a relatively short time in an improper environment. In reach-
ing the extreme upper limit of its temperature range the phenomenon
known as "blistering" may take up to a year to become evident.
However, this is a physical rather than a chemical phenomenon. Experience has indicated that if an installation has been soundly engineered and has operated satisfactorily for a year, the probability is
good that operation will continue in a completely satisfactory manner
for a substantial period of time. Fortunately, all this can be determined in advance with a well-conceived test program, so that the engineer's liability at any time can be suitably limited. We are dealing
here, of course, with only the high-grade chemically resistant plastics.
In severe chemical service these highly resistant resins willl normally
be the material of choice as opposed to either a general purpose polyester or an isophthalic polyester.

FAILURE FROM OTHER CAUSES

In a very useful review of the reasons for failure in the thermosets,
particularly in FRP chemical equipment, Webster showed in a survey
of industrial fabricators and consultants that the following type of
failures are most prevalent [6]:

1. Pressure piping
    a. Cemented joints generally caused by failure to follow recom-
        mended procedures
    b. Chemical attack caused by poor resin selection or changes
        in environmental conditions
    c. Impact damage—physically from the outside or by hammering
        from the inside

Inadequate supporting and hammering together with vibration are the
real enemies of plastic piping systems, even assuming that a pipe is
completely capable of handling the service conditions of the installation.

2. Tanks
   a. Chemical attack
   b. Internal pressure
   c. External pressure
   d. Secondary bond failure

These failures come under two general categories: inadequate design
and faulty fabrication. Much work has been done to increase the
strength of the secondary bond so that it actually becomes a chemical
bond rather than a mechanical bond. This is particularly true with
any hydroxyl group in the molecule.

   In general, the success of thermosets in cylindrical tanks is extremely high; the material is easy to handle and in general costs less
than rubber-lined steel or 316 stainless steel. There are probably as
many tanks that fail from the outside in as from the the inside out.
Plastic tanks have the advantage of being corrosion resistant both inside and out.

3. Scrubbers and absorbers
   a. Chemical attack
   b. Vacuum
   c. Impact damage

These translate further into poor definition of service conditions, inadequate design, faulty fabrication, and damage during transportation.

4. Ducts, fans, and stacks
   a. Joining failures
   b. Chemical attack
   c. Fire

   An analysis of 39 fires in plastic duct systems showed the following
breakdown [7]:

Occupancy	Number of incidents	Percent of total
Plating, pickling (and similar processes)	29	74.2
Pollution control	6	15.4
Epitaxial crystal formation	1	2.6
Clean room	1	2.6
Oven	1	2.6
Printing	1	2.6

In general, these fires were caused by [7]:

Cause	Number of incidents	Percent of total
Immersion heaters	17	43.6
Cutting and welding	6	15.4
Electrical	5	12.8
Chemical reaction	4	10.3
High-temperature flue gas	3	7.7
Overheated flammable liquid	2	5.1
Unknown	2	5.1

It is staggering to realize that in only 2 of the 39 losses did fire protection exist. In the first of the two sprinklered duct losses, the total loss was $840. *The fire was put out by one sprinkler head. In the second case, the duct fire was stopped by one sprinkler head.* In the second case, the fire continued to burn in the plastic plating tank and the plastic tape control consoles, with a loss of $30,000. In nonsprinklered ducts, the fire losses average approximately $75,000 each.

In general, the NFPA calls for fire protection of 0.2 gpm/ft^2 of water [8]. For plastic ducts the fire extinguishing system should be one in which sprinkler protection can be provided, but at the same time the sprinkler must be protected from the ravages of the exhaust gases. This is done by pyroxylin covers or plastic bags. All ducts, scrubbers, and fans should be built of fire-retardant material. It is heresy to build an exhaust handling system out of anything but fire-retardant materials. Experience with fires in duct systems is dramatic. The fires are generally brief but all-consuming. Unsprinklered stacks and ducts invite trouble. One 4-ft-diameter by 200-ft-high stack burned to the ground in 45 min. Another very large duct system was consumed in 1 h. Both of these losses could have been prevented with sprinkler protection.

The problems engineers must solve when dealing with scrubbers, fans, exhaust systems, and stacks are [9]:

1. Good corrosion resistance
2. Low fire spread rating
3. Low smoke evolvement
4. Good fire protection
5. Economic cost

Sometimes a compromise must be made to meet the hazard engendered by blistering chemicals traveling at high speed. The engineer then chooses the appropriate material of construction, such as fiber-reinforced plastic, stainless steel or brick or ceramic-lined concrete, PVC or rubber-lined steel, stainless steel, lead, or Plastisol-lined systems. It is easy to conceive of a system built completely of a fireproof material such as stainless steel or brick-lined concrete.

Actually, in view of the hard reality of past experience, no system is really fireproof, as deposits on the system may burn and cripple the exhaust system by knocking out the fans.

THERMOSETS—MECHANISMS OF FAILURE FROM ABRASION [10]

Failures by abrasion are more prone to occur in the gaseous state than in the liquid state. This is because in the gaseous state a particle can be easily hurtled along at speeds of 5500 to 6100 m/min (18,000 to 20,000 ft/min), whereas in the liquid state velocities may be 90 to 120 m/min (300 to 400 ft/min). The gaseous state produces a sandblasted effect and failures are much more common in this area.

Among other variables, failure by abrasion is a function of the particle size squared, and is particularly pronounced in particles that are larger than 100 mesh in diameter. In fact, as a rule of thumb, in liquid streams particles smaller than 100 mesh in diameter present no problem. However, when they are above 100 mesh, they may present a problem. For example, fly ash per se does not represent a conveying problem in a liquid stream but fly ash mixed with bottom ash, which represents a very large particle, can easily generate erosion. Polyethylene and polybutalene have shown exceptional resistance to erosive attack within the temperature limitations of their capabilities. Also, special formulations have been made of epoxies, polyesters, and polypropylene plastics in many chemical process forms which serve unique purposes.

It will be realized that in general terms, corrosion as it relates to metallic materials does not exist in plastics. Plastics do fail, but for a variety of reasons. Because plastics do not belong in the electromotive table, it is generally not possible to develop an electromotive force unless the particular plastic has been made properly conductive.

Combined particles carried in a gas stream at high velocities, operating at the uppermost temperature limits of the plastic's range, represent the worst possible conditions. For example, a highly corrosion resistant fiberglass line carrying a gaseous stream at 11,000 m/min (36,000 ft/min) can fail in less than a day's time, but the same line carrying the same velocities operating at 70°C (150°F) can easily last for four years. This is particularly true of the polyesters, epoxies, and furanes.

Table 2. Taber Test Material

Material	Taber abrasion
High molecular weight polyethylene	
Heavily filled high ceramic laminate inside a filament-wound pipe	1-2
Polyurethane (depending on formulation)	1-20
Soft rubber	3
Polyester rein laminate with small-particle high alumina filler	10-20
Steel pipe	15
Polyester resin with 10 to 30% silicon carbide	16-25
Polypropylene	18
PVC	19-22
Clay sewer pipe	19-22
Polyester resin with 10 to 30% tungsten carbide	22-25
Good parent polyester resin (no additive)	53-142

There is a lot that can be done to eliminate failure by abrasion. Basically, this can be summarized in the following steps:

1. Eliminate right-angled turns by using long-radius ells.
2. Keep the slurry flow parallel to the wall as much as possible.
3. Watch out for throttling valves or orifices, which may accelerate the velocity 10-fold in a localized area and impinge in an erratic manner against the pipe wall.
4. Protect small localized areas where impingement does occur by embedding in the laminate a 16-gage metal sheet such as type 316 stainless steel, Carpenter 20, or Hastelloy.
5. Cure completely all laminated structures if they are to be used in abrasive service. The difference between 27 Barcol and 38 Barcol may double the life of the plastic equipment. This may sound obvious, but unfortunately, some fabricated equipment is sold that has not developed a complete cure.
6. Modify the resin for a broad-scale increase (450%) in abrasion resistance extending over the entire laminated structure. At present, our standard specification for laminates where an increase in abrasion resistance is required calls for an exact resin mix

using an additive furnished by us and used in the fabrication as the prescribed thickness for the wear surface. There is a wide difference in abrasion capabilities among resin systems, up to 300%.

7. Select the best plastic—be it thermoform or thermoset—to meet the conditions, which may be a combination of corrosion, abrasion, and temperature.

8. The plastics themselves vary greatly in their abrasion-resistant capabilities. Table 2 presents an idea of the range of resistance to abrasion of various types of material.

Other Causes of Failure

Checking, Crazing, and Cracking

Quite often when one observes a thermoset in performance over long periods of time such as 5 to 10 years at elevated temperatures, one observes checking, cracking, and crazing that may occur to a depth of 50 to 70 mils and be profusely abundant. However, the core of the structure shows that the action has essentially stopped at this depth. It is particularly prevalent with high temperatures [above 80°C (175°F)] and in oxidizing atmospheres. Depending on service conditions, one observes this very frequently and quite often to a mild degree in service conditions that are not particularly severe. In general, it does not affect the performance of the vessel or system except with electrically conductive liners, for which the conductivity will literally be destroyed if the conductivity of the surface has become a series of islands so that a satisfactory ground cannot be established to the grounding points.

Delamination

The general cause of delamination is faulty fabrication, and it is seen especially in vessels that operate under a vacuum. Generally, the delamination occurs at the interface of the mat and roving. Roving is extremely difficult to wet out. Occasionally, however, it is seen between layers of mat where a secondary bond has been made. Bond intensifiers, whose sole purpose is to increase the strength of the secondary bond, appear to have been successful. It should be realized that secondary bonds are essentially mechanical unless bond intensifiers are used, at which time the secondary bond becomes a chemical bond.

Pyrolysis

The simple word for pyrolysis is *charring*. The molecule of all plastic materials is an organic molecule which can be destroyed by severe oxidation. This can occur by fire or it can occur, for example, with very strong sulfuric acid (e.g., 93%). This completely destroys the molecule and severe oxidation takes place. The resin turns black and loses all physical strength, so that the structure is destroyed. The polyesters, for example, can withstand up to 78% H_2SO_4 for short periods of time and 70% H_2SO_4 almost indefinitely, but above that, oxidation becomes so severe that the molecule is destroyed. This applies to many of the organic compounds, with exceptions such as some of the thermoplasts—polyethylene, polypropylene, and PVC—for which higher concentrations are acceptable.

PLASTICS—THEIR SERVICE AND USAGE

Table 3 is not intended to cover all the plastics presently in use in the chemical industry, but it covers the most prevalent ones. Exceptions include polyphenyline oxide where very high temperatures are to be dealt with, and polycarbonates where very high impact resistance is required. Nylon and the acetals are also used extensively for small gears.

The temperatures shown for the upper service limit are a function of service conditions, so they vary with the chemical environment. Many of the thermoplasts have a poor memory so that with temperature changes they tend to "grow" and not return to their original state. This is not true of the thermosets, which appear to have an excellent memory.

MAJOR TYPES OF FAILURES AND THEIR CAUSES

The major types of failures and their causes are shown in Table 4.

WHY ADHESIVE JOINTS FAIL

The reasons adhesive flanged systems fail can be summarized generally as follows [12]:

1. Poor surface preparation
2. Poor adhesive application
3. The wrong adhesive; for example, an epoxy in chlorine service
4. Surface contamination, particularly from moisture
5. Poor adhesive mixture
6. Low-temperature conditions
7. Overstressed joints
8. Inhibited cure. In the polyester area it is quite possible to

Table 3. Plastics—Their Service and Usage

Item	Type of plastic	Normal upper service limit		Typical usage areas
		°C	°F	
1.	Polyvinyl chloride	60	140	Piping—water, gas, drain, vent, conduit, oxidizing acids, bases, salts.
				Ducts (breaks down into HCl at high temperatures).
				Windows, plus accessory parts and machine equipment in chemical plants; liners with FRP overlay.
2.	Chlorinated PVC	82	180	Similar to above but upper temperature limit is increased.
3.	Polyethylene	60	140	Tubing, instrumentation (laboratory)—air, gases, potable water, utilities, irrigation pipe, natural gas.
4.	High-density polyethylene	82	180-220	Chemical plant sewers, sewer liners, resistant to wide variety of acids, bases, and salts. Generally carbon filled. Highly abrasion resistant. Can be overlaid with FRP for further strengthening.
5.	ABS	60	140	Pipe and fittings. Transportation, appliances, recreational. Pipe and fittings is mostly drain waste vent. Also electrical conduit. Also resistant to aliphatic hydrocarbons but not resistant to aromatic and chlorinated hydrocarbons. Recently, formulations with heat resistance up to 105°C (220°F) have been introduced.

Table 3. (Continued)

Item	Type of plastic	Normal upper service limit		Typical usage areas
		°C	°F	
6.	Polypropylene	82-104	180-220	Piping and as a composite material overlaid with FRP in duct systems. Useful in most inorganic acids other than halogens. Fuming nitric and other highly oxidizing environments. Chlorinated hydrocarbons cause softening at high temperatures. Resistant to stress cracking and excellent with detergents. Flame-retardant formulations make it useful in duct systems. It can be further reinforced with glass fibers. For stiffening, increase the flex modulus to 106 and deformation up to 148°C (300°F).
7.	General purpose polyesters	50	120	Made in a wide variety of formulations to suit the end product requirements. General purpose resins are used in the boat industry, automobiles, tub showers, aircraft, and gasoline storage tanks.
8.	Isophthalic polyesters	70	150	Isophthalic polyesters have increased chemical resistance and are used extensively in chemical plant waste and cooling tower systems.
9.	Chemical-resistant (Wet) (Dry)	120-150	250-300	This includes the families of bisphenol, hydrogenated bisphenols, and chlorendic types. A wide range of chemical resistance, predominantly to oxidizing environments. It is not resistant to H_2SO_4 above 78%. Can operate continually in gas streams at 148°C (300°F). End uses include scrubbers, ducts, stacks, tanks, and hoods.

		(Wet)	(Dry)	
10.	Epoxies	120-150	250-300	More difficult to formulate than the polyesters. Better alkaline resistant than the polyesters, with less oxidizing resistance. Highly resistant to solvents, especially when post cured. Often used in filament-wound structures for piping. Not commonly used in ducts. More expensive than polyesters.
11.	Vinyl esters	93-140	200-280	Especially resistant to bleaching compounds in chlorine-plus-alkaline environments. Wide range of resistance to chemicals similar to the polyesters. Used extensively in piping, tanks, and scrubbers. Modifications have been developed to operate continuously up to 140°C (280°F).
12.	Furanes	150-200	300-400	Excels in solvent resistance and combinations of solvents with oxidizing chemicals. One of the two resins that pass the 50 smoke rating and 25 fire spread rating. Carries about a 30 to 50% cost premium over the polyesters. Excellent for piping, tanks, and special chemical equipment. Does not possess the impact resistance of polyesters or epoxies.

Source: Ref. 11.

Table 4. Major Types of Failures and Their Causes

System		
Piping	Cemented joint	Failure to follow recommended practices.
	Chemical attack	Poor resin selection, or changes in environmental conditions.
	Impact damage	Various (generally handling damage).
Tanks and process vessels	Chemical attack	Poor resin selection or environmental data wrong.
	Internal pressure	Operational errors and poor process design.
	External pressure (vacuum)	Poor equipment design.
	Secondary bond failure	Faulty fabrication.
	Impact	Handling damage.
Scrubbers and absorbers	Chemical attack	Service conditions wrong or changes in process specifications.
	Vacuum damage	Poor equipment design.
	Impact	Handling damage.
	Fire	Lack of maintenance, lack of adequate safeguards, or lack of interlocks.
		Possibly lack of employee training, or no scrubbing liquid.
Ducts, fans, and stacks	Joint failures	Generally at the glue line.
	Chemical attack	Poor resin selection; incorrect or changed service conditions.
	Fire	80% of fires originate from an internal or process source; 20% originate externally. Lack of sprinkler protection is by far the largest single reason for large losses.

Source: Ref. 6.

produce a wrap or repair with an inhibited cure. Briefly, this means that the cure occurred only partially or not at all. In short, polymerization did not occur, either becuase of an air-inhibited cure or a very wet (humid or steam) atmosphere. The repaired section or wrapped joint is washed away with the first rush of liquid on startup. (Even water or dilute acid will wash it out.)

The solution is to "hot coat" the wrap or repaired section with a polyester resin containing 2% paraffin wax. All commercial enterprises finish up with a hot-coat system; but often, smaller industrial plants doing their own repairs do not. If conditions are favorable, the repair or wrap is a success. If not, failure occurs, sometimes with discouraging consequences—process hold-ups followed by the time required for a proper repair.

9. Wet mat. If a mat or roving gets wet, it will cause a white or milky joint. When in doubt, wipe the joint with acetone solvent to absorb moisture.

10. Temperatures. Do not attempt thermoset (polyester) cures below 10°C (50°F). Temperatures below this will slow the cure or prevent it completely unless excess catalyst, promoter, or dimethylaniline is used (the dimethylaniline acts as an accelerator). An alternative is to use an air heat gun to stimulate the resin cure.

RECOMMENDED DESIGN PRACTICES [2]

The Popular Thermosets

The successful application of any corrosion-resistant material begins with a complete and accurate definition of the problem to be solved. In order to consider any material as a potential solution to the problem, engineers must be intimately familiar with the material to be applied. They need to know its strong points and its weaknesses, and above all, the limits on its successful application. For the general classes of reinforced-plastic material available, we must therefore determine, first, when each can be used successfully and when it should be avoided. Many different types of polyester and epoxy chemical-resistant resins (and furanes and vinyl esters) are being manufactured today, both in the United States and abroad.

Generally, we are dealing with families of resins that have the following characteristics:

An upper wet temperature limit of about the boiling point of water or perhaps a little higher [120°C (250°F)]

A dry limit of perhaps 175°C (350°F)

Remarkable resistance to many of the oxidizing acids, up to fairly strong concentrations

Potentially superior alkaline resistance

Good solvent resistance in some areas but limited resistance in others

In variegated shapes, most useful in the areas of low pressure or
vacuum

Can provide complete piping systems at relatively high operating
pressures and in most of the common sizes up to 150 psig

Can be tailor-made for ablative conditions where service under extended temperatures for short periods of time is desired

Low thermal and electrical conductivity, so that use can be made of
these properties

The reinforcement, which is generally glass, asbestos, or a synthetic
fiber, may be furnished by other suppliers.

Design Checkpoints for Successful Plastic Piping Systems

1. Do not use a metal expansion joint with plastic pipe or wire-
    reinforced rubber expansion joints. Stick to a bellows type,
    where action forces are low.
2. A Teflon bellows joint will also serve as a vibration isolator and
    flexible connector.
3. Try to use the natural changes in direction of the pipe to take
    care of expansion, without creating additional expansion loops.
4. Do not use a U bolt as an anchor; it violates the principle of
    point loading.
5. Loose-fitting U bolts used in pairs can serve as a guide.
6. Guides and supports are not synonymous. Use each for its own
    purpose.
7. When heat-tracing a line, take care to prevent it from freezing
    Spiral wrapping of the trace material should be covered with
    light insulation to contain the heat. Watch carefully the upper-
    temperature limitations of the plastic pipe.
8. The operating bending moment should not exceed 25% of the
    allowable moment.
9. When plastic piping passes through wall openings, protect it from
    hammering against the wall.
10. Be careful of rotational moments in plastic piping. Do not wring
    it out like a dishrag.
11. Support valves independently.
12. The handling of plastic piping is extremely important. Do not
    ruin it before you get it installed.
13. Most plastic piping system failures are physical, due to violations
    of good design principles.

Design Checkpoints for Plastic Tanks

1. Be absolutely certain of the pressure specifications. Except for pressure vessels, most plastic tanks are used for gravity storage. Make sure that venting arrangements are provided so that negative pressure will not collapse the tank.
2. Select the plastic carefully. Review the service conditions and tailor the selection to meet these conditions.
3. If valves are used with tank outlets, make sure that the valve is supported independently.
4. Proper drainage is essential. Make sure that the tank will drain completely and that the drain size is adequate to empty the tank with a reasonable period.
5. Eliminate vibration and hammering. The worst enemy of a plastic tank is vibration. It can reduce the life of a tank by as much as 80%, particularly from internal heating with a steam sparger.
6. Strengthen the knuckle area. Too often overlooked, the point at which the bottom of the tank joins the side wall can develop flexing strength as the tank is rapidly filled and emptied. This is particularly true of batch vessels. For example, a tank processing 10 batches a day will have processed over 36,500 batches in 10 years.
7. Design for a minimum expected life of 15 years. Many plastic tanks have been in service this long.
8. Look closely at the bottom support. For a plastic tank, a continuous support is generally used, but intermittent supports can be used if the bottom of the tank is made sufficiently strong.
9. Make tank lids strong. If there is any possibility that personnel will walk on top of the vessel, the lid should be capable of supporting four or five 200-lb persons. If this is not intended, a notice to this effect should be posted on the tank. Serious accidents have occurred in which workers have fallen into shallow tanks of hot acid purely because the lid was not designed to support personnel. As a minimum the lidding system should be capable of supporting 1000 lb evenly distributed plus a 250-lb load concentrated in a 16-in.-diameter circle.
10. Establish good communications between yourself and the tank vendor. Write exact specifications; do not, for example, confuse vacuum with positive pressure.
11. Look at agitators and platforms. The evolution of design in this area has progressed markedly in the past 10 years, so that agitators can now be mounted on tanks with suitable support. Further accessories, such as platforms, ladders, and level gages, can be purchased from any reputable vendor.
12. Watch the freezing conditions in plastic tanks. Expansion forces in freezing are large and some tanks can be severely damaged if freezing occurs.

13. Take care of the inner surface. Quite often the life of a plastic
     tank is determined by its inner corrosion barrier. Treat it with
     care; do not damage it mechanically. Wear rubbers or rubber-
     soled shoes inside a plastic tank.

RELATIVE COSTS

Piping Costs [13]

Tables 5A and 5B are taken from a Dow Chemical Company study of
installed prices of various materials of construction.

Table 5A. 500-FT Complex System

Piping material	Cost ratio
Carbon steel, Schedule 40	1.00
Polyester	1.09
Glass	1.58
Aluminum, Schedule 40	1.63
Rubber-lined steel	1.67
304 stainless steel, Schedule 5	1.70
316 stainless steel, Schedule 5	1.99
Kynar-lined steel	2.14
Teflon-lined steel	2.75
Alloy 20, Schedule 5	2.94
Titanium, Schedule 5	3.81
Hastelloy B, Schedule 5	5.05
Titanium, Schedule 40	5.96

Source: Ref. 13.

Table 5B. 1000-FT Straight-Run System

Piping material	Cost ratio
Carbon steel, Schedule 40	1.00
Glass	1.18
Aluminum, Schedule 40	1.19
Polyester	1.26
Rubber-lined steel	1.28
304 stainless steel, Schedule 5	1.51
316 stainless steel, Schedule 5	1.78
Kynar-lined steel	1.89
Teflon-lined steel	2.38
Nickel, Schedule 5	2.68
Alloy 20, Schedule 5	3.32
Titanium, Schedule 5	3.77
Zirconium, Schedule 5	6.97
Tantalum, Schedule 5	12.08

Source: Ref. 13.

Comparative Costs of Fiber-Reinforced Plastic Pipe with a Lined Pipe*

Many intensive studies have been done in this area. The one by Dow Chemical is very comprehensive and ranks a great number of materials of constructions by cost. There are, however, a number of items in both initial cost and final construction which should be emphasized. For example, look at the list below, which gives the cost of 20-ft sections of 10-in.-diameter pipe flanged at both ends and constructed as indicated.

Fiber-reinforced plastic	$ 560.00
Rubber-lined steel pipe	1160.00
Polypropylene-lined steel pipe	1260.00
Teflon-lined steel pipe	5000.00

*Actual costs, 1979; from a study by the author.

Fiberglass pipe can be installed very quickly, as a 20-ft section can be carried to the job site by two people and installed without mechanical assistance. All of the lined metal pipe, however, must be put up with a crane and then maneuvered into place by additional personnel. In addition, a gin pole must be erected to bring it into the final installation. Provided that it is suitable for the service conditions, fiber-reinforced plastic pipe enjoys a marked economic advantage over most of its competitors because:

1. The basic cost is less.
2. The installation cost is probably only 10 to 20% of that required for a lined steel pipe.

Pump Costs [2]

Practically every major pump manufacturer in the country makes a line of molded plastic pumps. These are of molded polyester or epoxy, in some cases further strengthened with a filament winding overlay. Smaller pumps are made of match-molded polyester, PVC, or polypropylene. Complete lines have been established by prominent pump manufacturers to meet the latest standards on pumps so that they will be interchangeable. On a relative cost basis for 300 gpm at a 90-ft-head pump, handling 1.3 specific gravity liquid at 60°C (120°F), the following relative quotes were obtained:

Glass-reinforced polyester	1.0
316 stainless steel	1.2
Worthite	1.5
Armored polypropylene	1.6
Carpenter 20	2.0
Hastelloy C	2.2

Tank Costs

Table 6 presents comparative pricing data on the cost of tanks.

LININGS [15]

Many plastics—PVC, the polyesters, epoxies, furanes, polypropylene, and the fluoroplastics—are used as lining materials for pipe, tanks, and vessels in a manner similar to rubber lining. It is difficult to generalize on liners but some items are worthy of note regarding the less exotic variety commonly employed in the chemical industry:

Table 6. Tank Costs[a] (7000-Gal Vertical Tank[b])

Material of construction	Wall thickness, in.	Weight, lb.	Estimated delivery, weeks	Price quoted
Carbon steel	3/16	5901	10-12	$ 6,324
304 stainless steel	3/16	6441	10-12	14,219
316 stainless steel	3/16	6441	10-12	16,985
Rubber-lined steel	3/16	6300	4-6	9,455
Heresite-lined steel	4-6 mils	5901+	4-7	8,125
Bis-a polyester	1/4	–	6-10	5,850[c]
QuaCorr furan	1/4-1/2	2450	8-10	7,753[c]
Haveg 61	–	–	30+	15,800

[a]Quotes obtained by a major engineering company.
[b]Dished head, flat bottom, two 4-in. nozzles, one 12-in. manway.
[c]Average of four to seven quotes.
Source: Ref. 14.

1. The liners in themselves generally run 60 to 150 mils depending on service, whereas a rubber liner would go 125 to 250 mils.
2. Quite often liners are filled or reinforced to enhance physical properties such as strength, conductivity, and permeability.
3. Permeability is largely a function of temperature (the driving force), so that with the polyesters the addition of flake glass does a great deal to reduce permeability.

Large steel vessels 30 to 50 ft in diameter and 60 to 100 ft high, serving as scrubbers, are quite often built with formulated flake glass polyester liners of 60 to 80 mils, with operating temperatures up to 150°C (300°F) successfully achieved.

4. Lining failures generally occur by a number of mechanisms:
    a. *Inadequate surface preparation.* A blast with a 3- to 5-mil profile is essential.
    b. *Inadequate priming coat.* Always measure thickness from the peak of the valley.
    c. *Poor ambient conditions.* Do not line in wet weather or when the ambient is near the dew point.
    d. *Pinholes.* Improper coverage; this would show up in a defective spark test. The solution is to grind out the area and begin again.
    e. *Temperature excusions.* Each lining has its own temperature limitations both under wet and dry conditions. Flake glass and fiberglass linings have a good wet limit of 60°C (140°F). In the range 60 to 82°C (140 to 180°F), the lining may or may not be successful. Above 82°C (180°F) many linings are not suitable, except fluorocarbons, which may go to 96°C (240°F) wet, but this varies widely with chemicals so it should be checked in each case. With temperature the driving force failure occurs by migration of the water molecule in the solution through the lining to the substrate and ultimate debonding. The same phenomena occur in all plastic vessels, but since the water molecule rapidly evaporates from the surface of a solid plastic vessel, it is not a problem. In *gases* successful applicatons go to 150°C (300°F). Excursions above that, depending on temperature, produce lining failure by pyrolysis.
    f. *Debonding.* The most prevalent form of failure observed in linings, debonding can be caused by permeation, poor preparation, or temperature excusions above the useful upper limit of the plastic.
    g. *Solvent action.* Softening of the lining with a pronounced decrease in hardness is often the result of solvent action, which can be fairly rapid.

Resin Selection

As more and more plastic material enters the market, the problem of suitable selection becomes increasingly difficult. Many thermoplastics are available as well as many furanes, epoxies, polyesters, and vinylesters. Even trace contaminants or the same contaminants at different temperatures will change the material selection. A classic example of this was a cascade system built to circulate a 5% sulfuric acid solution with traces of CS_2 at 95°C (200°F). Unfortunately, this same solution, when circulated at ambient temperature, condensed CS_2 from the textile tow. The CS_2 immediately attacked the polyester and through solvent action quickly destroyed it, although similar systems running at 95°C (200°F) on a continuous basis have performed satisfactorily for years. Using epoxies or furanes, the same system under either condition would run indefinitely.

REINFORCED-PLASTIC PRESSURE VESSELS [16]

First, let us review the application of plastics to pressure vessels. Section X of the *ASME Pressure Vessel Code*, covering fiberglass-reinforced plastic (FRP) pressure vessels, was issued in 1969. In general, we can say that under this code practically all the vessels currently being made are filament wound, bag molded, or matched metal molded. This does not mean, however, that hand-laid-up FRP pressure vessels are not constructed. Indeed, they are, but in the strictest sense they do not qualify under the code. The reason for this is that one-of-a-kind vessels are not covered in Section X of the code. At one time, there was a committee set up to accomplish this in conjunction with the Society of the Plastics Industry (SPI), but unfortunately, most of the personnel have changed over time. The issue has not been pushed aggressively and to put it in a more homely phrase, most of the effort has "fallen through the cracks." I am sure that most people realize that in the hand-laid-up end of the FRP business it is by and large a one-of-a-kind operation, whereas in the coded end, particularly in the generation of large numbers of relatively small vessels, it is another matter to qualify a vessel and then proceed with the rest of them according to the code.

Section X has been adopted by 26 states. Presently, two code vessel manufacturing stamps have been issued, but only one is being actively used. In short, one pressure vessel design has been qualified and registered with the national board. The adoption and use of Section X has been a grievous disappointment.

Economically, it would appear than when it is to the advantage of fabricators or manufacturers to cooperate in the development of standards of this sort, they will do so, but when they judge that it is not to their financial advantage, they are usually reluctant to put a lot of money into a standard. A standard of this type takes a lot of money, time, and effort to resolve (this one took six years, 1963 to 1969). Thus the corrosion-resistant group of the SPI has done very little about it. There has been virtually no cooperation from this group because they have nothing to gain from it, although in the strictest sense of the word they are making vessels that fall under the definition of a pressure vessel.

Perhaps we should stop and identify what a pressure vessel is. As a matter of convenience, the Factory Mutual system identifies a pressure vessel as a vessel over 5 ft^3 in volume, but this is not used by many states. For example, in Ohio there are a number of reactors in use of less than 1 ft^3. I am deliberately a bit elastic in my definition, as some of the case histories I cite later fall under the Factory Mutual definition, whereas others are much smaller.

The limitations set on Section X by the ASME are:

1. It does not apply to vessels subject to federal control.
2. It does not apply to vessels containing 120 gal or less of water.
3. It does not apply to any vessel equipped with an air cushion.
4. It does not apply to hot water supply storage tank 120 gal or less in capacity or having a heat input of 200,000 Btu/h or less.
5. It does not apply to vessels having an internal or external operating pressure not exceeding 15 psi, regardless of size.
6. It does not apply to vessels less than 6 in. in inside diameter.

Many states have adopted Section X as part of their pressure vessel code and through ANSI (the American National Standards Institute), the Occupational Safety and Health Administration (OSHA) has adopted it. Section X-coded pressure vessels have gone into the following types of applications:

1. Softener regeneration tanks for household use and in small industries (this is by far the largest application)
2. Jet engine starters
3. Torpedo launchers

The largest manufacturer of fiberglass pressure vessels in the world manufactures pressure vessels according to Section X standards. Thousands of units are built daily in a bag-molded operation. Vessels are commonly designed with a safety factor of 4. Millions of these vessels are in operation in water softener systems in North America, Europe, and Japan and are sold in massive quantities by large chain stores. They carry a 10-year guarantee and are the top of the product line. Nominally, vessels are built in sizes ranging from 1.3 gal up to 120 gal, which meets Section X specifications.

Design qualification tests require a minimum burst of six times design pressure. Maximum design is limited to about 150 psi for bag-molded pressure vessels and 1500 psi for bag-molded vessels. Temperatures must not exceed 65°C (150°F) or be lower than 18°C (65°F). Such vessels should not be used to store, handle, or precess lethal liquids. Code rules apply to the vessel and piping as far as the first flanged, threaded, adhesive, or welded joint. The vessel to be tested must be pressure cycled 100,000 times from atmospheric to the design pressure. After this the vessel must stand a hydrostatic burst test of not less than six times the design pressure.

FORMS AND PRODUCTS COMMERCIALLY AVAILABLE

Most thermoplasts are available in various forms of molded products in almost every conceivable shape and size within the practical limits of the material. PVC, ABS, polyethylene, and poypropylene are available in piping, ductwork, and small tanks.

The thermoplasts can also be combined with the thermosets to form composite structures and provide the best properties of each to meet some specific need. As such, they serve as linings for piping and duct systems, stacks, scrubbers, and tanks plus process vessels. Much work has gone into this field and it provides a very unique area of reference.

The thermosets have been developed widely into piping and duct systems and in small to very large stacks, tanks, process vessels, reactors, mixing chambers, scrubbers, and hoods. These predominantly cover the families of polyesters, epoxies, furanes, vinyl esters, and some phenolics, where the performance of a high nickel alloy is available for less than the cost of stainless steel. Much work has been done to develop minimum fire spread ratings (i.e., less than 25) and low smoke ratings (i.e., less than 50). The field of the thermosets is growing at a rate of at least 15% per year.

The development of very high strength reinforcing materials such as graphite and carbon have provided laminates of superior stiffness which have come into extensive use by the military as ablative shields on the nose cones of reentry missiles and have provided unique stiffness in special areas in the automotive and aerospace industries. Scrubbers have been built of plastic that are 30 ft in diameter and 100 ft tall. Stacks have been built up to 600 ft high. Piping 9 to 12 ft in diameter is commonplace. Ducts of the same size have been successfully installed. Plastic tanks of up to 750,000 gal have also been built.

Successful case histories of crystallization vehicles of 10,000-gal capacity have been in continuous service for as long as 13 years. Plastic piping has made large inroads into the field of lined piping because it is lighter and less expensive, more easily installed, and carries a reduced downtime liability.

The entire field of household drain piping has been invaded by PVC and ABS.

SANDWICH CONSTRUCTION

The use of fiberglass sandwich construction is common. Experience initially gained in the boat industry spread into the chemical industry. A typical sandwich is an end-grain Ecuadorian balsa with fiberglass skins. The advantage is a tremendous increase in strength at a relatively low cost. With such construction, economies of 20% over solid FRP laminates are easily obtained in vessel construction. Very large tanks, vapor heads, and rectangular vessels have been built using this construction technique. Other sandwiches consist of honey-comb or foam cores and metallic or FRP skins. The use of urethane as a sandwich material with FRP skins on both sides is a real advance in building an FRP process vessel or tank operating at very low temperatures. In addition, the vessel—if suitably specified—is literally corrosion proof inside and out, as it is immune to fumes, drippage, spillage, and accidental overflows that would wreak havoc on a con-ventional insulation system. One such typical agitated vessel serving as a crystallizer has been in service 12 years. It is 12 ft in diameter and 12 ft high, holds 9500 gal, and operates near 0°C (32°F). A 10-mm (3/8-in.) inner shell is covered with 25 mm (1 in.) of 2 lb/ft^3-density urethane insulation. This, in turn, is overlaid with 3mm (1/8 in.) of FRP fire-retardant resin.

Gravity vessels and process vessels running at atmospheric pres-sure together with duct systems operating at a slight negative pres-sure (3 to 4 in. water) are admirably suited for such construction. Vessel cores cut from process units running at 95°C (200°F) and 5% sulfuric acid were in excellent shape after a year of service.

The use of sandwich laminates under high-vacuum conditions (2 to 4 in. Hg absolute) should be approached with caution. The key to superior performance under these conditions is a perfect bond of the inner laminate to the core. Calculations such as those performed by Sturm and later adapted to easy graphical solution by Shadduck indi-cate the size of debonding that can occur before trouble ensues. At or near a complete vacuum, and with 9.5-mm (3/8-in.) skins and a 38-mm (1.5-in.) balsa core laminate, the critical diameter of unbonded surface is about 254 mm (12 in.). Isolated spots of 25 to 50 mm (1 to 2 in.) at that thickness are of no concern.

Case histories exist of successful applications of balsa-cored laminates under full vacuum conditions up to five years in service, so even though extra care is necessary, this has been demonstrated to be a feasible approach.

Such construction is particularly adaptable to computer calcualtions, where a whole range of choices for any set of service conditions permits the engineer the optimum choice. Debonding in thin skins is easily seen as a white spot or area in an otherwise clear laminate.

Tap testing with a mallet or with a key to detect voids (hollow versus solid) are present methods. Sonic tesing in the hands of a skilled operator is also feasible. As the laminate becomes heavier, physical inspection is more difficult. Although most fabricators do not have the capability, a 48-h test at 50 mm (2 in. Hg absolute) would be ideal to detect loss of bond of sufficient size to cause trouble (this is for full vacuum service).

The observed mode of failure is debonding of the inner laminate from the core, which forms a bubble. Quite often a crack will appear in the laminate in the bubble or at its base. This permits the liquor to penetrate the core material. Depending on the core material, various sequences may ensue. Balsa-cored material is highly vulnerable to wicking, closed-cell urethane much less so. The repair consists of removing all damaged core and replacement of core and interior laminate. An adhesive intensifier may be used to obtain a good bond with the repair. Make sure that the core surface is thoroughly saturated with resin. Layup on the wet resin core with a chopped strand mat (at least four layers of 1.5-oz material). Make sure that a good bond exists before proceeding with the layup.

For example, in the final inspection of a large vapor head running 3 m (10 ft) in diameter by 5 m (15 ft) high it was determined that there were half a dozen spots 2.5 to 5 cm (1 to 2 in.) in diameter that were not adequately bonded and one spot 15 cm (6 in.) in diameter which was not adequately bonded. It is believed that debonded areas such as these would present no problem under vacuum. It is absolutely essential that every care be taken to obtain a perfect bond of the sandwich material to the laminate, so the sandwich material should be rolled into the lamainate in a very thorough fashion. Patting it down with the hand is not adequate. The outer limit of visually cured laminates is about 0.63-cm (1/4-in.) skins—above that some kind of sound procedure must be followed. It is paramount that the vendor guarantee the fabrication for one year against defects, and in the event that they occur, be prepared to repair them promptly.

Cored construction has great cushioning properties, which can be felt as well as heard; audible reactions in a vessel continue in a very muted atmosphere.

SELECTING VENDORS

Know the Capabilities of Your Plastics Vendor

1. Look at their mold capability.
2. Look at their plant construction and safety program.

3. Look at their research effort.
4. Look twice at quality control. Does the head of quality control report to the production department or to a vice president? Quality control and production should be two individual groups. Having quality control report to production is like having the cat watch the canary.
5. What is the labor situation? Seasoned workers make quality products. In many kinds of plastics manufacturing there is a great deal of difference in the quality of output between workers with a few months' experience and those with a few years' experience.
6. While you are in their shop, look at the equipment that they have built for others.
7. Do they have adequate technical support?
8. Correlate their bids with their labor rates.
9. Have they done anything to automate?

Investigate the Vendor's Financial Reliability—And Other Considerations

1. Ask for banking references.
2. Do they live up to their delivery schedules?
3. Are shipping costs going to be a factor in your decision? They will be if you are on the East Coast and the vendor is on the West Coast.
4. Limit your price inquiry to no more than five or six vendors.
5. Do not get stuck with having to pay for circuitious shipping. Have them quote F.O.B. your plant.
6. Plan ahead so that the vendor does not have to use overtime to meet your completion date.

REFERENCES

1. J. H. Mallinson, Plastics: Trends in CPI Materials, *Chemical Engineering* Deskbook Issue, December 4, 1972.
2. J. H. Mallinson, *Chemical Plant Design with Reinforced Plastics*, McGraw-Hill, New York, 1969.
3. *Modern Plastics Encyclopedia*, Vol. 54, No. 10A, McGraw-Hill, New York, 1977.
4. *Controlling Corrosion with Reinforced Atlac 382*, Atlas Chemical Co., Bulletin, Wilmington, Del., 1966.
5. W. A. Szymanski, Ashland Chemical Co., Columbus, Ohio, personal communication, 1979.
6. R. M. Webster, Safety in the use of fiber reinforced plastic equipment for chemical service, in *Managing Corrosion with*

Plastics, Vol. I., National Association of Corrosion Engineering, Houston, 1977.

7. W. R. Vandall, Fiber reinforced plastics duct fires [personal communication], Factory Mutual Engineering, 1974.

8. *Code for Blower and Exhaust Systems*, Natl. Fire Prot. Assoc. No. 91, 1973.

9. J. H. Mallinson, Fire retardancy—duct systems in the chemical industry, 30th Tech. Conf., Soc. Plast. Ind., 1975.

10. J. H. Mallinson, Increasing the abrasion resistance of reinforced plastics, 31st Tech. Conf., Soc. Plast. Ind., 1976.

11. *Modern Plastics Encyclopedia*, Vol. 56, McGraw-Hill, New York, 1979.

12. J. H. Mallinson, Reinforced plastic failures: causes and cures. The zero maintenance concept, in *Managing Corrosion with Plastics*, Vol. II, National Association of Corrosion Engineers, Houston, 1977.

13. *Installed Cost of Corrosion Resistant Piping*, Dow Chemical Co., 1978.

14. J. A. Rolston, Fiberglass composite materials and fabrication processes, *Chem. Eng.*, January 28, 1980.

15. Unpublished work by the author, 1979, on linings.

16. J. H. Mallinson, Introduction to Section X: Reinforced plastic pressure vessels, ASME Meet., Atlanta, 1977.

5

Development and Application of Inorganic Nonmetallic Materials

RONALD A. MCCAULEY / Rutgers, The State University, New Brunswick, New Jersey

FUNDAMENTALS OF CORROSION

Introduction

Throughout the history of the materials industries, various material types or compositions have been used because of some particular advantageous intrinsic property. High strength, low electrical conductivity, or some other property may be the primary concern for a particular application. However, excellent resistance to attack by the environment (corrosion) always plays a role and may, in some cases, be the prime reason for the selection of a particular material. This is especially true for those materials selected for furnace construction in the metals and glass industries. Almost all environments are corrosive to some extent. For practical applications it comes down to a matter of kinetics—how long will a material last in a particular environment? In some cases corrosion may be beneficial, such as in the preparation of samples by etching for microscopic evaluation. The proper selection of materials and good design

practices can greatly reduce the cost caused by corrosion. To make
the proper selection, engineers must be knowledgeable in the fields
of thermodynamics, physical chemistry, and electrochemistry.

The corrosion of nonmetallic materials has not been as well cata-
gorized as it has been for metals. Similar terms do, however,
appear in the literature. The more common types referred to in the
literature are *diffusion corrosion*, which is very similar to *concentration cell corrosion* in metals; *galvanic cell corrosion; grain
boundary corrosion*; and *stress corrosion*. A more common trend in
inorganic nonmetallic or ceramic materials is to group corrosion under a more general mechanism, such as *dissolution corrosion* (i.e.,
corrosion of a solid by a liquid). In this type of corrosion, diffusion, galvanic cell, grain boundary, and stress corrosion may all
be present.

Thermodynamics and Kinetics

In the selection of materials, an engineer wishes to select those
materials that are thermodynamically stable in the environment of
service. Since this is a very difficult task, a knowledge of thermodynamics and kinetics is required so that materials can be selected that have slow reaction rates and/or harmless reactions.
Thermodynamics provides a means for the engineer to understand
and predict the chemical reactions that take place. The reader is
referred to any of the numerous books on thermodynamics for a
more detailed discussion of the topic. Only a brief summary is
given below.

The driving force for corrosion is the reduction in free energy
of the system. In general, a reaction can occur if the free energy
of the reaction is negative. The free energy of the reaction is
given by

$$\Delta F = \Delta H - T \, \Delta S$$

where

ΔF = Gibbs free energy
ΔH = heat of formation
T = absolute temperature
ΔS = entropy of reaction

The quantities necessary to calculate ΔF are available for most pure
compounds. These published values are for standard states and
any changes from standard states require calculation using

$$\Delta F = \Delta F^\circ + RT \ln k$$

where

ΔF° = free energy change at standard state
R = gas constant
k = concentration constant

At equilibrium the change in free energy must be zero; therefore,

$$\Delta F^\circ = -RT \ln k$$

Thus the value of k can be calculated at any temperature.

The importance of the entropy term increases with temperature. The reactions of concern involving ceramic materials are predominantly those at temperatures where the entropy term may have considerable affect on the reactions. In particular, species with high entropy values have a greater effect at higher temperatures.

The real problem with predicting whether a reaction may take place or not is in selecting the proper reaction to evaluate. Care must be taken not to overlook some possible reactions.

Since the corrosion of ceramics in service may never reach an equilibrium state, thermodynamic calculations cannot be strictly applied because these calculations are for systems in equilibrium. Many reactions, however, closely approach equilibrium, and thus the condition of equilibrium should be considered only as a limitation, not as a barrier.

It is normally expected that materials will corrode, and thus it is important to know the kinetics of the reaction so that predictions of service life can be made. Reaction rates for condensed-phase processes normally involve the transport of products away from the boundary. Thus the rate of the overall process is determined by the rate of each individual step and on the reaction rate constant and concentration of reactants for that step. The reaction with the lowest rate determines the overall rate of the corrosion process. Some of the more important factors that may influence the rate of reaction are diffusion rates, viscosity, particle size, and the degree of contact or mixing.

The Arrhenius equation

$$K = A \exp\left(\frac{-Q}{RT}\right)$$

where

K = reaction rate constant
A = constant
Q = experimental activation energy
R = gas constant
T = absolute temperature

describes the effect of temperature on the rate process.

For corrosion rates to be useful to practicing engineers it is best that they be expressed in a meaningful manner. In most cases the engineer is involved with the amount of material corroded away during a specified time period, or the depth of penetration per unit time. In the literature, corrosion rates are often given in the weight of material reacted per unit area for a unit time. These can easily be converted to the depth of penetration per unit time by dividing by the density of the material. Another useful rule for the engineer to remember is that the temperature at which diffusion-controlled reactions first become observable is at about one-half the melting point in degrees kelvin.

Polycrystalline Materials

General

Most, if not all, polycrystalline materials can be looked on as being composed of several components. The minor components may make up as much as 50% of the total or as little as a fraction of 1%. It is the combination of properties of these various components and how they are bonded together that gives a material its unique properties.

If one were to group together all the secondary additives or components of a particular composition, which are used for any number of reasons (bonding agents, mineralizers, grain-growth inhibitors, etc.), a material would consist of two groups of components: the major components or phases that exhibit the desired properties for a certain application, and the minor components or phases that have been added to enhance or modify the desired properties. It is these minor components that are, in most cases, of primary concern to corrosion resistance. Porosity should be included as one of these components and, in most cases, would be a minor component. The corrosive attack of a nonmetallic inorganic polycrystalline material starts with the least corrosion-resistant component, which is normally that component used for bonding or, more generally, the minor components of the material.

The effects of these minor components have presented the industries that manufacture and consume these materials with a constant struggle to optimize material properties for a particular application. More recently the trend has been toward purer starting materials and higher firing temperatures. In this way, less modification of the desired final properties is caused by the secondary components.

The corrosion of polycrystalline materials will be discussed in a general fashion taking into account the state of the environment: vapor, liquid, or solid. In some cases the environment may not provide any real corrosion effects of its own but may function only in relation to temperature. Any corrosion that does take place will occur at a faster rate at higher temperatures. It is generally accepted that many reactions double their rate for every 10° rise in temperature. There may be a situation, however, where corrosion does not take place unless the temperature somehow alters the polycrystalline material to the point where it is susceptible to corrosion.

A good source of information on the effects of temperature are equilibrium phase diagrams. Thousands of these diagrams have been published by the American Ceramic Society [1]. The most obvious effect of temperature is that of melting. A more subtle effect in real materials is the partial melting of the secondary phases or bonding phases. One should, however, have an understanding of how the properties are affected by temperature. Polymorphic transformations play an important role in the design of high quality materials. Thermal expansion mismatch of various components is also important. High thermal expansion anisotropy of individual crystals can cause severe cracking, allowing a corrosive medium to enter into a material.

To understand how corrosion takes place, one must have a thorough understanding of how the material is constructed. The distribution of phases is very important in describing bonding mechanisms and corrosion. In a system with a secondary liquid phase, the degree of solid-solid contact is extremely important and desirable for high temperature strength. A small amount of liquid, if it completely wets the solid phase, can produce a weak material, as opposed to one that may contain a larger amount of liquid but also contains some degree of solid-solid contact. This fact was brought out by Van Vlack [2], who reported that a higher-liquid-content silica refractory did not deform during heating, whereas a magnesiowustite refractory slumped during heating, although it contained a lower liquid content. Van Vlack stated: "Less energy is required to form two interfaces between periclase and liquid silicates than to form one grain boundary between two periclase grains." This energy requirement, however, is removed in the silica case. Van Vlack reported that even though a silicate liquid will penetrate between two periclase or magnesiowustite grains, it will not penetrate to as great extent between a magnesiowustite and a spinel grain.

The distribution of phases, which is obviously very important, is determined by a balance of surface tension forces between the interfaces or boundaries of the various phases. According to White [3], for a phase β to surround completely a second phase α, we need $\gamma_{\alpha\alpha} \geq 2\gamma_{\alpha\beta}$, where $\gamma_{\alpha\alpha}$ is the interfacial surface tension between two α grains and $\gamma_{\alpha\beta}$ is the interfacial surface tension between an α and a β grain. If $\gamma_{\alpha\alpha} < 2\gamma_{\alpha\beta}$, complete penetration will not occur. When $\gamma_{\alpha\alpha} = 2\gamma_{\alpha\beta} \cos(\theta/2)$, a balance of forces is obtained. The angle θ is the equilibrium dihedral angle between intersecting α and β interfaces. As θ increases from 0 to 60°, penetration of β phase between the α grains will decrease. When $\theta > 60°$, however, the β phase will occur as discrete inclusions. Thus the obvious step to take to increase solid-solid contact would be to change those variables that yield $\theta > 60°$. It has been found by White in MgO systems that by lowering the density and the mean grain size, θ will increase. This in turn will decrease the grain surface curvature. White also found that introduction of chromic oxide will increase θ, whereas iron oxide will decrease θ and that at a given composition θ will decrease with increasing temperature.

It may not always be desirable to keep the secondary β phase as discrete inclusions, as pointed out by Rigby [4]. He described a *thin-film theory* where a silicate phase becomes molten at high temperature and coats the individual grains and thus, upon cooling, acts as a strong-bonding thin film. As long as service temperatures are below the point where this thin film remains solid, materials of this type will retain their strength. The strength of the material will then depend on the ability of the silicate to coat the individual grains.

Attack By Vapor

The corrosion of a polycrystalline material by vapor attack can be very serious, much more so than attack by either liquids or solids. One of the most important material properties related to vapor attack is that of porosity or permeability. If the vapor can penetrate the material, the surface area exposed to attack is greatly increased and corrosion proceeds rapidly. It is the total surface area exposed to attack that is important. Thus not only is the volume of porosity important, but the pore size distribution is also important.

Vapor attack can proceed by producing a reaction product that may be either solid, liquid, or gas, as in the equation

$$A_s + B_g \longrightarrow C_{s,l,g}$$

As an example, the attack of SiO_2 by Na_2O vapors can produce a liquid sodium silicate.

In another type of vapor attack, the vapor may penetrate a material under thermal gradient to a lower temperature, condense, and then dissolve material by liquid solution. The liquid solution can then penetrate further along temperature gradients until it freezes. If the thermal gradient of the material is changed, it is possible for the solid reaction products to melt, causing excessive corrosion and spalling at the point of melting.

A possible rate-controlling step in vapor attack is the rate of arrival of a gaseous reactant and also possibly the rate of removal of a gaseous product. It is obvious that a reaction cannot proceed any faster than the rate at which reactants are added, but it may proceed much more slowly. The maximum rate of arrival of a gas can be calculated from the Hertz-Langmuir equation:

$$Z = \frac{P}{\sqrt{2\pi MRT}}$$

where
  Z = moles of gas that arrive at surface in unit time and over unit area
  P = partial pressure of reactant gas
  M = molecular weight of gas

R = gas constant

T = absolute temperature

Using P and M of the product gas, the rate of removal of gas product can be calculated using the same equation. To determine if service life is acceptable, these rates may be all that is needed. Actual observed rates of removal may not agree with those calculated if some surface reaction must take place to produce the species that vaporizes. The actual difference between observed and calculated rates depends on the activation energy of the surface reaction. If the gaseous reactant is at a lower temperature than the solid material, an additional factor of heat transfer to the gas must also be considered and may limit the overall reaction.

Pilling and Bedworth [5] have reported the importance of knowing the relative volumes occupied by the reaction products and reactants. Knowing these volumes can aid in determining the mechanism of the reaction. When the corrosion of a solid by a gas produces another solid, the reaction proceeds only by diffusion of a reactant through the boundary layer when the volume of the solid reactant is less than the volume of the solid reaction product. In such a case the reaction rate decreases with time. If the volume of the reactant is greater than the product, the reaction rate is usually linear with time. These rates are only guidelines, since other factors can keep a tight layer from forming (e.g., thermal expansion mismatch).

Attack by liquid

The basic theory of corrosion by liquids in ceramic systems has been described in detail by Cooper and Kingery [6]. The basic equation describing the rate of solution under free convection with density being the driving force is

$$j = 0.505 \left(\frac{g \, \Delta p}{v_i x} \right)^{1/4} D_i^{3/4} C^* \, \exp \left(\frac{\delta^*}{R + \delta^*/4} \right)$$

where

g   = acceleration due to gravity

Δp = $\dfrac{p_i - p_\infty}{p_\infty}$ (p_i = saturated liquid density and p_∞ = original liquid density)

v_i = kinematic viscosity

x   = distance from surface of liquid

D_i = interface diffusion coefficient

C^* = a concentration parameter

δ^* = effective boundary layer thickness

R   = solute radius

The exponential term was introduced as a correction for cylindrical surfaces. Since experimental tests often involve cylindrical specimens, these equations have been developed for that geometry. In practical applications the condition relating to the corrosion of slabs is most predominant. However, if the sample diameter is large compared to the boundary layer thickness, the two geometries give almost identical results.

After a short induction period (which is of no consequence in practical applications) in which molecular diffusion predominates, the rate of corrosion becomes nearly independent of time. Many corrosive environments involving ceramic materials involve diffusion in the corroding medium, and thus increased velocity of the medium increases corrosion. Thus if transport in the liquid is important, the corrosion rate must be evaluated under forced convection conditions. In such cases the rate will depend on the velocity of forced convection.

Experiments using rotating disks conducted by Gregory and Riddiford [7] have verified the theory of Levich [8]. These studies lead Cooper and Kingery [6] to derive an equation for the rate of corrosion of the disk face under forced convection:

$$j = 0.61 D^{*2/3} v^{*-1/6} w^{1/2} c^{*}$$

The terms D^{*} and v^{*} were introduced since diffusivity and viscosity may be composition dependent. The important point of this equation is that the rate of corrosion depends on the square root of the angular velocity w.

Hrma [9] has used the work of Cooper and Kingery to discuss further the rates of corrosion of refractories in contact with glass. The following equation given by Hrma describes the corrosion under the condition of free convection due to density difference:

$$j_c = k \, \Delta c \left(D^3 \frac{\Delta p g}{v L} \right)^{1/4}$$

where

j_c = rate of corrosion
Δc = solubility of material in liquid
D = coefficient of binary diffusion
g = acceleration due to gravity
v = kinematic viscosity
L = distance from surface of liquid
Δp = relative variation of density
k = constant

This is essentially the same equation as that of Cooper and Kingery, without the exponential term.

In the majority of practical cases the solubility of the material in the liquid and the density of the liquid change much slower than the viscosity of the liquid. Under isothermal conditions the viscosity change is due to compositional changes. Thus the predominant factor in the corrosion of a material by a liquid is the viscosity of the liquid.

This relationship holds quite well for the corrosion of a solid below the liquid surface. At the surface where three states of matter are present, the corrosion mechanism is different and much more severe.

At the liquid surface a sharp cut normally develops in the vertical face of the solid material being corroded. This region has been called *flux line, metal line,* or *glass line* corrosion. Pons and Parent [10] reported that the flux line corrosion rate was a nonlinear function of the oxygen potential difference between the surface and the interior of a molten sodium silicate. Cooper and Kingery [6] reported that flux line corrosion was the result of natural convection in the liquid caused by surface forces. Hrma [9] reported that the additional corrosion at the flux line depended only on the variation in surface tension and density, with surface tension being the more important factor.

Downward-facing horizontal surfaces also exhibit greater corrosion than does a vertical or upward-facing horizontal surface. In general, this is due to convection caused by density differences. As a surface corrodes the interface, if denser than the corroding medium, will be eroded away due to free convection caused by density variation. A downward-facing surface can exhibit excessive corrosion if bubbles are trapped beneath the horizontal surface. This is known as *upward drilling* since it results in vertically corroded shafts. Surface tension changes around the bubble cause circulating currents in the liquid which cause excessive corrosion very similar to flux line corrosion.

The solubility of materials in liquids can be obtained from phase diagrams which give the saturation composition at a given temperature. Unfortunately, for many practical systems phase diagrams are either very complex or nonexistent. Many data are available, however, for two- and three-component systems, and these should be consulted before attempting to evaluate the corrosion of a specific material. A good discussion of the use of phase diagrams in dissolution studies is that by Copper [11].

The corrosion of multicomponent materials proceeds through the path of least resistance. Thus those components with the lowest resistance are corroded first. The corrosion of a fusion cast alumina-zirconia-silica refractory will be used as an example. This particular material is manufactured by fusing the oxides, casting into a mold, and then allowing crystallization to occur under controlled conditions. The final microstructure is composed of primary zirconia, alumina, alumina with included zirconia, and a glassy phase that surrounds all the other phases. The glassy phase (about 15% by volume) is necessary for this material to provide a cushion for the polymorphic transformation of zirconia during cooling and subsequent use. This material is widely used as a basin-wall material in the soda-lime silica glass furnace. The corrosion proceeds by the diffusion of Na+ ions from the bulk glass into the glassy phase of the refractory. As Na+ is added to the glass, its viscosity is lowered and it becomes corrosive toward the refractory. The corrosion next proceeds by solution of the alumina and finally by partial solution of the zirconia. Under stagnant conditions an interface of zirconia embedded in a high-viscosity glass is formed. In actual service conditions, however, the convective flow of the bulk glass erodes this interface, allowing continuous corrosion to take place until the refractory is consumed. This type of corrosion can take place in any multicomponent material where the corroding liquid diffuses into a material that contains several phases of varying corrosion resistance.

Much work has been done over the past 30 years on the galvanic corrosion of refractories by glasses. Godrin [12] has published a review of the literautre on electrochemical corrosion of refractories by glasses. It has been shown that a potential difference does exist in such systems; however, the application of a bias potential has been unsuccessful in eliminating corrosion. Even though not totally foolproof, Godrin concluded that refractories that have an electrical potential with respect to glass that is positive 0.4 to 0.7 V are fairly resistant to corrosion, that refractories with a potential greater than 1.0 V have rather poor resistance, and that refractories that have a negative potential with respect to glass should not be used.

Attack by solids

Many applications of materials involve two dissimilar solid materials in contact. Corrosion can occur if these materials react with one another. Common types of reactions involve the formation of a third phase at the boundary, which can be a solid, a liquid, or a gas. In some cases the boundary phase may be a solid solution of the original two phases. Again, phase diagrams will give an indication of the type of reaction and the temperature where it occurs.

When the reaction that takes place is one of diffusion as a movement of atoms within a chemically uniform material, it is called *self-diffusion*. When a permanent displacement of chemical species occurs, causing local composition change, it is called *interdiffusion* or *chemical diffusion*.

The driving force for chemical diffusion is a chemical potential gradient (concentration gradient). When two dissimilar materials are in contact, chemical diffusion of the two materials in opposite direction forms an interface reaction layer. Once this layer has been formed, additional reaction can take place only by the diffusion of chemical species through this layer.

Solid-solid reactions are predominantly reactions involving diffusion. Diffusion reactions are really a special case of the general theory of kinetics discussed previously. Thus diffusion can be represented by an equation of the Arrhenius form:

$$D = D_0 \exp\left(\frac{-Q}{RT}\right)$$

where

  D = diffusion coefficient
  D_0 = constant
  Q = activation energy
  R = gas constant
  T = absolute temperature

The larger the value of Q, the activation energy, the more strongly the diffusion coefficient depends on temperature.

The diffusion in polycrystalline materials can be divided into bulk diffusion, grain boundary diffusion, and surface diffusion. Diffusion along grain boundaries is greater than bulk diffusion because of the greater degree of dissorder along grain boundaries. Similarly, surface diffusion is greater than bulk diffusion. When grain boundary diffusion predominates, the log concentration decreases linearly with the distance from the surface. When bulk diffusion predominates, however, the log concentration of the diffusion species decreases with the square of the distance from the surface. Thus by determining the concentration gradient from the surface (at constant surface concentration) one can determine which type of diffusion predominates.

Since grain boundary diffusion is greater than bulk diffusion, it would be expected that the activation energy for boundary diffusion would be lower than that for bulk diffusion. The boundary diffusion is more important at lower temperatures, and bulk diffusion is more important at high temperatures.

Chemical reactions wholly within the solid state are less abundant than those which involve a gas or liquid, owing predominantly to the limitation of reaction rates imposed by slower material transport. The solid-solid contact of two different bulk materials also imposes a limitation on the intimacy of contact—much less than that between a solid and a liquid or gas.

Applications of nonmetallic materials commonly involve thermal gradients. Under such conditions it is possible for one component of a multicomponent material to diffuse selectively along the thermal gradient. This phenomenon is called *thermal diffusion* or the *Soret effect*. This diffusion along thermal gradients is not well understood, especially for inorganic nonmetallic materials.

The stability of various materials to graphite is a good example of a solid-solid reaction. In this case, however, at least one of the products is a gas. The stability of a few selected refractory oxides in contact with graphite increases in the order TiO_2, Al_2O_3, ThO_2, MgO, $MgAl_2O_4$, SiO_2, and BeO, as reported by Klinger et al [13].

Miller et al [14] have shown that carbon reacts with SiO_2 to form the intermediate phase SiC, which then reacts with silica to form the gaseous phase SiO. The following equations were given to represent the reaction:

$$SiO_2 + 3C \longrightarrow SiC + 2CO$$

$$2SiO_2 + SiC \longrightarrow 3SiO + CO$$

They stated that these reactions were sufficiently rapid at 1000°C (1830°F) and in the presence of iron, which acts as a catalyst for the reduction of silica by SiC, to cause failure of silicate refractories in coal gasification atmospheres.

Because of the prevalence of platinum metal in various research and manufacturing operations, the reactions of various refractory oxides with platinum is of considerable importance. Ott and Raub [15] reported that platinum acts as a catalyst for the reduction of refractory oxides by hydrogen, carbon, CO, and organic vapors. These reactions can occur as low as 600°C (1110°F) and result due to the affinity of platinum for the metal of the oxide by forming intermetallic compounds and crystalline solutions.

Glassy Materials

General

This discussion will be concerned only with those materials that fall under the conventional definition of a glass: specifically, amorphous materials produced by the fusion of inorganic materials and then cooled at a rate that prevents crystallization. A complete discussion of all noncrystalline, nonmetallic, inorganic materials could very easily occupy an entire book; thus those materials that do not fall under the definition of a glass given above are excluded here.

Clark et al [16] have recently published a book on the corrosion of glass in which they not only describe the mechanisms of corrosion but also discuss various techniques to analyze the type and degree of corrosion. However, their discussion is limited to silicate glasses.

Attack by vapor

The corrosion of glasses by atmospheric conditions, referred to as *weathering*, is essentially attack by water vapor. Weathering occurs by one of two mechanisms. In both types, condensation occurs on the glass surface; however, in one type it evaporates, whereas in the other it collects to the point where it flows from the surface, carrying any reaction products with it. The latter type is very similar to corrosion by aqueous solutions. The former type is characterized by the formation of soda-rich films, according to Tichane and Carrier [17]. This soda-rich film has been shown to react with atmospheric gases such as CO_2 to form Na_2CO_3, according to the work of Simpson [18] and Tichane [19].

Metcalfe and Schmitz [20] studied the stress corrosion of E-glass fibers in moist ambient atmospheres and proposed that ion exchange of alkali by hydrogen ions led to the development of surface tensile stresses that could be sufficient to cause failure. In soda-lime-silica glasses, atmospheric corrosion by water vapor has been reported to halve the breaking strength [21].

Attack by liquid

Probably the most abundant examples of glass corrosion are those caused by a liquid. Release of toxic species (such as PbO) from various glass compositions has received worldwide interest during the past 10 to 15 years. Although glass is assumed by many to be inert to most liquids, it does slowly dissolve. In many cases, however, the species released are not harmful.

The corrosion resistance of glasses is predominantly a function of composition. Glasses can be soluble under a wide range of pH values from acids to bases, including water. Water-soluble sodium silicates form the basis of the soluble silicate industry, which supplies products for the manufacture of cements, adhesives, cleansers, and defloccu-lants. At the other extreme are glasses designed for maximum resistance to corrosion. In general, very high silica ($>96\%$ SiO_2), alumino-silicate, and borosilicate compositions have excellent corrosion resistance to a variety of environments. Silicate glasses, in general, are less resistant to alkali solution than they are to acid solution. A list of about 30 glass compositions with their resistance to weathering, water, and acid has been published by Hutchins and Harrington [22].

The attack of silica or silicate glasses by aqueous hydrofluoric acid provides a good example of how the mechanism of corrosion proceeds. Ernsberger [23] has described this attack in detail and related it to the structure of silica glasses. The silicon-oxygen tetrahedra are exposed at the surface in a random arrangement of four possible orientations. Protons from the water solution will bond with the exposed oxygens, forming a surface layer of hydroxyl groups. The hydroxyl groups can be replaced by fluoride ions in aqueous hydrofluoric solutions. Thus the silicon atoms may be bonded to an OH^- or an F^- ion. The replacement of exposed oxygens of the tetrahedron by $2F^-$ causes a deficiency in the silicon atom coordination, which is 6 with respect to fluorine. This causes the additional bonding of fluorine ions, with a particular preference for bifluoride. Thus the four fluoride ions near the surface provide an additional four-coordinated site for the silicon. A shift of the silicon to form SiF_4 can take place by a small amount of thermal energy. The ready availability of additional fluoride ions will then cause the $(SiF_6)^{2-}$ ion to form. This mechanism is supported by data that show a maximum in corrosion rate with bifluoride ion concentration.

The rate of corrosion of a glass that contains a cation that can increase its coordination will be greater than one that contains a cation that cannot increase its coordination. This coordination increase was discussed above for SiO_2 attacked by aqueous hydrofluoric acid.

According to Charles [21], the corrosion of an alkali-silicate glass by water proceeds through three steps:

1. H^+ from the water penetrates the glass structure. This H^+ replaces an alkali ion, which goes into solution. A nonbridging oxygen is attached to the H^+.
2. The OH^- produced in the water destroys the Si-O-Si bonds, forming nonbridging oxygens.
3. The nonbridging oxygens react with an H_2O molecule, forming another nonbridging oxygen-H^+ bond and another OH^- ion. This OH^- repeats step 2. The silicic acid thus formed is soluble in water under the correct conditions of pH, temperature, ion concentration, and time.

The development of films on the glass surface has been described by Sanders and Hench [24]. The nonbridging oxygen-H^+ groups may form surface films or go into solution. The thickness of this film and its adherence greatly affect the corrosion rate. In Na_2O-SiO_2 glasses, Schmidt [25] found that films formed only on glasses containing more than 80 mole percent SiO_2 at 100°C (212°F) for 1 h.

Budd [26] has described the corrosion of glass by either an electrophilic or a nucleophilic mechanism, or both. The surface of the glass has electron-rich and electron-deficient regions exposed. Various agents attack these regions at different rates. Exposed negatively charged nonbridging oxygens are attacked by H^+ (or H_3O^+), whereas exposed network silicon atoms are attacked by O^{2-}, OH^-, and F^-.

Budd and Frackiewicz [27] found that by crushing glass under various solutions, an equilibrium pH value was reached after sufficient surface area was exposed. The value of this equilibrium pH was a function of the glass composition, and it was suggested that it was related to the oxygen ion activity of the glass. When foreign ions were present, the amount of surface required to reach an equilibrium pH was greater.

The rate of hydrolysis of a glass surface is one of the major factors that delineates the field of commercial glasses. The rate of hydrolysis is of great importance because it determines the service life of a glass with respect to weathering or corrosion and also because it influences the mechanical properties. Glass fracture is aided by hydrolysis. The rate of hydrolysis of alkali-silicate glasses of the same molar ratios proceeds in the order Cs > K > Na > Li.

Sanders and Hench [24] showed that a 33 mole percent Li_2O glass corroded more slowly than a 31 mole percent Na_2O glass by two orders of magnitude. This difference was caused by the formation of a film on the Li_2O glass with a high silica content. Scratching the glass surface produced an unusually high release of silica.

In borosilicate glasses that require a heat treatment step after initial melting and cooling to produce phase separation, a surface layer is formed by selective evaporation of Na_2O and B_2O_3. These surface layers have been observed by several workers. This silica-rich surface layer can influence the subsequent leaching process that would be needed to produce Vycor-type glass [28]. If the hydrated surface layer is removed before heat treatment, the silica-rich layer is almost entirely eliminated.

Yoon [29] found that lead release was a linear function of pH when testing lead-containing glasses in contact with various beverages. Low pH beverages such as orange juice or colas leached lead more slowly than did neutral pH beverages such as milk. This dependence on pH was also reported by Das and Douglas [30] and by Pohlman [31]. Later, Yoon [32] reported that if the ratio of moles of lead plus moles of alkali per moles of silica were kept below 0.7, release in 1 h was minimized. If this ratio is exceeded, lead release increased linearly with increasing PbO content. Lehman et al [33] reported a slightly higher threshold for more complex compositions containing cations of Ca^{2+} and Al^{3+} or B^{3+} in addition to the base Na_2O-PbO-SiO_2 composition. The lead release in these complex compositions was not linear but increased upward with increased moles of modifiers. Lehman et al related the mechanism of release or corrosion to the concentration of nonbridging oxygens. A threshold concentration was necessary for easy diffusion of the modifier cations. This threshold was reported to be where the number of nonbridging oxygens per mole of glass-forming cations equaled 1.4.

In general, it has been determined that mixed alkalies lower the release of lead by attack from acetic acid below that of a single alkali-PbO-silicate glass; lead release increased with increasing ionic radius of the alkaline earths; however, combinations of two or more alkaline earths exhibited lower lead release; Al_2O_3 and ZrO_2 both lowered the lead release; and B_2O_3 increased the lead release. Thinner glaze coatings on clay-based ceramic bodies decreased lead release due to interaction of the glaze and the body, providing higher concentration of Al_2O_3 and SiO_2 at the glaze surface [34].

The mechanism of release or corrosion for these glasses containing lead is similar to those proposed by Charles [21] for alkali-silicate glasses. The rate of this reaction depends on the concentration gradient between the bulk glass and the acid solution and the diffusion coefficient through the reacted layer. In general, maximum durability can be related to compact strongly bonded glass structures, which in turn exhibit low thermal expansion coefficients and high softening points [35].

Several workers have investigated the concentration profiles of glass surfaces after leaching by water. Boksay et al. [36] postulated a theory to explain the profiles in K_2O-SiO_2 glass, Doremus [37] developed a theory to explain the profiles in Li_2O-SiO_2 glass, and Das [38] attempted to explain the profiles in Na_2O-SiO_2 glasses.

Douglas and co-workers [39-42] found that alkali removal was a linear function of the square root of time in alkali-silicate glass attacked by water. At longer times the alkali removal was linear with time. Wood and Blachere [43] investigated a $65SiO_2$-$10K_2O$-$25PbO$ (mole percent) glass and did not find a square root of time dependence for removal of K or Pb but found a dependence that was linear with time. This behavior has also been reported by Eppler and Schweikert [44] and by Douglas and co-workers. Wood and Blachere [43] proposed that an initial square root of time dependence occurred but that the corrosion rate was so great that it was missed experimentally.

During the study of the corrosion of soda-lime-silica glasses containing P_2O_5 by water, Clark, et al [45] found that a double reaction layer was formed consisting of a silica-rich region next to the glass and a Ca-P-rich reaction next to the water solution. This Ca-P film eventually crystallizes into an apatite structure and provides a good mechanism to bond the glass to bone in implant applications.

Katayama et al [46] determined that the corrosion of a barium borosilicate glass decreased in the order acetic, citric, nitric, tartaric, and oxalic acid, all at a pH of 4 at 50°C (122°F).

A discussion of glass would not be complete if some mention of glass fibers were not made. The corrosion of fibers is inherently greater than bulk glass simply because of the larger surface-to-volume ratio. Since one of the major applications of fibers is as a reinforcement to some other material, the main property of interest is that of strength. Thus any corrosion reactions that would lower the strength are of interest. This effect is important both when the fiber is being manufactured and after it has been embedded in another material. The strength of E-glass fibers in dry and humid environments was studied by Thomas [47], with the observation that humid environments lower strength.

The study of phosphate glass corrosion has shown that the glass structure plays a very important role in the rate of dissolution. Phosphate glasses are characterized by chains of PO_4 tetrahedra. As the modifier (alkalies or alkaline earths) content of these glasses is increased, there is increased cross linking between the chains. When very little cross linkage exists, corrosion is high. When the amount of cross linkage is high, corrosion is low. Simiar phenomena should exist for other glass-forming cations that form chain structures (B^{3+} and V^{5+}).

Carbon and Carbon- and Nitrogen-Containing Materials

General

The corrosion of carbon and carbon- and nitrogen-containing materials is synonymous with oxidation. In some cases (e.g., SiC) initial oxidation forms a protective layer that prevents further oxidation. This is called *passive corrosion*. The rate of oxidation depends on the stability and texture of the oxide layer formed and is low when a coherent film forms and high when a porous film forms or one that is highly volatile or melts. In nonoxidizing environments the corrosion is one of dissociation. The mechanisms of oxidation vary among these materials and are still not fully understood for many of them.

Carbon materials

The predominant corrosion mechanism of carbon-type materials is one of oxidation. The oxidation of pyrolytic graphite has been separated into two regions with increasing temperature [48]. Oxidation is reaction-rate-controlled at low temperatures and diffusion-controlled at high temperatures, with the changeover slightly above 840°C (1545°F). Above 2500°C (4530°F) sublimation rates exceed oxidation rates.

The oxidation of pitch-coke graphite is much greater than that of pyrolytic graphitic in the low temperature region because of the greater ease of oxidation of the binders used in pitch-coke graphite and the greater inherent porosity.

Diefendorf [49] studied the effect of atmosphere on the strength of graphite and concluded that it was a stress corrosion phenomenon. Tests conducted in vacuum and/or argon gave higher strengths than in air.

Some aspects of the corrosion of carbon brushes have been discussed by Millet [50]. He concluded that humid air produced oxidation of the surface between the carbon brush and the metal ring and that a surface layer of water molecules formed. His data showed that wear rates were also greater for humid air than for dry air at 80 to 120°C (175 to 250°F) but lower for temperatures between 40 and 70°C (105 and 160°F). Moberly and Johnson [51] reported a critical humidity level of 0.12 gH_2O/ft^3 to avoid excessive electrographitic brush wear rates on copper rings. The critical level was 0.8 g/ft^3 for graphite collector brushes on steel rings. The ideal level was stated to be 2 g/ft^3.

Robinson et al [52] studied the alkali attack of carbon refractories and found that the primary mechanism of attack by alkali was as a catalyst for oxidation. They also found that CO_2 vapors oxidize carbon at 1000°C (1830°F) with a rate that increases as the CO_2/CO ratio increases. It was indicated that a 100% CO environment caused zero attack. Alkali impurities (as carbonates, chlorides, or hydroxide) within the carbon refractory acted to lower the initiation temperature of reaction. Robinson et al also investigated the attack of molten alkali hydroxides and carbonates on carbon. The molten hydroxide penetrated the porosity, reacted with the carbon to form a carbonate, and caused cracking due to the expansion on conversion from hydroxide to carbonate. Several authors [53-55] have studied the corrosion of carbon by various alkali compounds, but Robinson et al [52] apparently were the most definitive, especially for reactions that occur in a blast furnace.

Carbides and nitrides

None of the carbides and nitrides are stable in oxygen-containing environments. Silicon carbide exhibits the greatest resistance by forming a protective layer of silica on the surface. Some other carbides and nitrides also form a protective metal oxide layer which allows them to exhibit moderately good oxygen resistance (e.g., $Si_3Al_3O_3N_5$ forms a protective layer of mullite).

The oxidation of Si_3N_4 has been described as occurring by either an active or a passive mechanism [56]. The active mechanism is one where the fugitive SiO forms in environments with low partial pressures of oxygen by the reaction

$$2Si_3N_4(s) + 3O_2(g) \longrightarrow 6SiO(g) + 4N_2(g)$$

Passive corrosion occurs in environments with high partial pressures of oxygen by the reaction

$$Si_3N_4(s) + 3O_2(g) \longrightarrow 3SiO_2(s) + 2N_2(g)$$

The SiO_2 that is produced forms a protective coating and further oxidation is limited. The pressure of N_2 formed at the interface can be large enough to form cracks or pores in the protective coating, which subsequently allow additional oxidation.

The rate-determining step in Si_3N_4 containing MgO impurities was reported by Cubicciotti and Lau [57] to be the diffusion of MgO from the bulk material into the oxide surface layer. This surface layer was composed of SiO_2, $MgSiO_3$, or glass phase and some unoxidized Si_3N_4 and was porous due to released N_2.

The oxidation of SiC has been described by Ervin [58] and is similar to that of Si_3N_4. McKee and Chatterji [59] described several different modes of behavior of SiC in contact with gas-molten salt environments relating to the formation of various interfacial reaction layers. An SiO_2 protective layer will corrode in a basic salt solution but not in an acid salt solution. With low oxygen pressures, active corrosion takes place by formation of SiO.

DEVELOPMENT AND APPLICATION OF CORROSION-RESISTANT MATERIALS

Introduction

The most obvious method of providing better corrosion resistance is to change materials; however, this can be done only to a certain extent. There will ultimately be only one material that does the job best. Once this material has been found, additional corrosion resistance can be obtained only by property improvement or in some cases by altering the environment.

Polycrystalline materials

Property Optimization

Since exposed surface area is a prime concern in corrosion, an obvious property to improve is the porosity. Much work has been done in finding ways to make polycrystalline materials less porous or more dense. The most obvious is to fire the material during manufacture to a higher temperature. Other methods of densification have also been used. These involve various sintering or densification aids: liquid-phase sintering, hot pressing, and others. If additives are used to cause liquid phase sintering, care must be exercised that not too much secondary phase forms, which might lower corrosion resistance, even though porosity may be reduced.

Alterations in major component chemistry may aid in increasing corrosion resistance, but this is actually a form of finding a new or different material, especially if major changes are made.

The history of glass-contact refractories is a good example of corrosion resistance imporvement in a polycrystalline material. Porous clay refractories were used originally. Changes in chemistry by adding more alumina were made first to provide a material less soluble in the glass. The first major improvement was the use of fusion-cast aluminosilicate refractories. These provided a material of essentially zero porosity. The next step was the incoporation of zirconia into the chemistry. Zirconia is less soluble than alumina or silica in most glasses. Because of the distructive polymorphic transformation of zirconia, a glassy phase had to be incorporated into these refractories. This glassy phase added a less corrosion resistant secondary phase to the refractory. Thus the higher resistance of the zirconia was somewhat compromised by the lower resistance of the glassy phase. The final product, however, still had a corrosion resistance greater than the old product without any zirconia. Today several grades of ZrO_2-Al_2O_3-SiO_2 fusion-cast refractories are available. Those with the highest amount of zirconia and the lowest amount of glassy phase have the greatest corrosion resistance.

The development of regenerator refractories was through the optimization of materials made of fireclay by using higher purity raw materials and then higher firing temperatures. Changes in chemistry were then made by switching from the fireclay products to magnesia-based products. Again improvements were made by using hgher purity raw materials and then higher firing temperatures. Minor changes in chemistry were also made during the process of property improvement. The evolution of regenerator refractories for the flat glass industry has been described by McCauley [60].

A part of the concept of improvement through chemistry changes is that of improving resistance to corrosion of the bonding phases. Bonding phases normally have a lower melting point and lower corrosion resistance than does the bulk of the material. The development of high alumina refractories is a good example of improvement based on the bonding phase. The best conventional high alumina refractories are bonded by mullite or by alumina itself. To change this bond to a more corrosion resistant material compatible with alumina, knowledge of phase equilibria played an important role. Alumina forms a complete series of crystalline solutions with chromia, with the intermediate compositions having melting points between the two end members. Thus a bonding phase formed by adding chromia to alumina would be a solution of chromia in alumina with a higher melting point than the bulk alumina and thus a higher corrosion resistance. In addition to the more resistant bonding phase, these materials exhibit a much higher hot modulus of rupture (more than twice mullite or alumina-bonded alumina). Nothing is ever gained, however, without the expense of some other property. In this case the crystalline-solution-bonded alumina has a slightly lower thermal shock resistance than does the mullite bonded alumina. Owing to the excellent resistance of these materials to iron oxide and acid slags, they have found applications in the steel industry.

The development of tar-bonded and tar-impregnated basic refractories to withstand the environment of the basic oxygen process of making steel is yet another example of a way to improve the corrosion resistance of a material. Tar-bonded products are manufactured by adding tar to the refractory grain before pressing into shape. In this way each and every grain is coated with tar. When the material is heated during service, the volatiles burn off, leaving carbon behind to fill the pores. An impregnated product is manufactured by impregnating a finished brick with hot tar. This product, once in service, will similarly end up with carbon in the pores. Impregnated products do not have as uniform a carbon distribution as do the bonded types. The carbon that remains within the refractory increases the corrosion resistance to molten iron and slags by physically filling the pores, by providing a nonwetting surface, and by aiding in keeping iron in the reduced state, which then does not react with the oxides of the refractory. A thin layer on the hot face (1 to 2 mm) does loose its carbon to oxidation and various slag components penetrate and react within this layer. This corrosion, however, is much slower than with a product that contains no carbon.

External methods of improvement

In the section on fundamentals, the importance of temperature was stressed several times. Various techniques have been used to lower the temperature of the interface or hot face of the material (lower hot face temperatures mean less corrosion). Many applications of a ceramic material subject the material to a thermal gradient. By altering the material or providing a means to increase the heat flow through the material, the hot-face temperature can be lowered significantly. One means of doing this is by forcibly cooling the cold face. This provides faster heat removal and thus lowers the hot-face temperature. Most industrial furnaces use some means of forced cooling on the cold face by cooling-air systems or water-cooled piping. If the thermal gradient through the material becomes too steep, failure may occur through thermally induced stresses. In general, if the gradient is above 595°C (1100°F)/in. problems of cracking may occur (this depends on the thermal expansion characteristics of the material).

Another method that has been used to lower the hot-face temperature is to place metal plates either within individual bricks or between them. A large portion of the heat is thus conducted through the metal plate. A similar technique has been used by manufacturing a product containing oriented graphite particles.

Another way to take advantage of increased cooling is to use a thinner material in the beginning. This will automatically cause a thinner reaction layer to form on the surface. In general, furnace linings should not be greater than 10 to 12 in. thick. Anything greater than about 12 in. does not normally increase overall life but adds an economic penalty in refractory cost per campaign. In fact, most linings could probably be less than 10 in.; however, the thermal environment will determine the ultimate thickness that should be used.

If a refractory lining is insulated, a greater portion of the refractory will be at a higher temperature and corrosion will proceed at a faster rate. In these cases a balance must be obtained between service life and energy conservation. Because of the potential for increased corrosion of insulated linings, the properties of the lining material must be carefully evaluated before insulation is installed. In many cases the engineer may want to upgrade the lining material if it is to be insulated.

Glassy materials

Property optimization

The development of more resistant glasses has been predominantly through optimization of compositions. In general, lowering the alkali content increases the durability. This, however, has practical limits based on melting temperatures, viscosities, softening points, and working ranges. Borosilicate glasses are, in general, more resistant than soda-lime silica glasses. In general, silicate glasses are less re-sistant to alkali solutions than they are to acid solutions. Reference 22 lists the corrosion resistance of many glasses of varying composi-tions.

External methods of improvement

The development of coating technology has provided a means to im-prove corrosion resistance, abrasion resistance, and strength. Com-binations of coatings applied while the glass is hot and after it has cooled have been developed that form a permanent bond to the glass. These coatings are not removed by cooking or washing.

The most commonly used metallic hot end coatings are tin and titanium. As the piece goes through the annealing lehr, the metal oxidizes, forming a highly protective ceramic coating. Tin is easier to work with since a thicker coating can be applied before problems of iridescence occur. These hot end metallic coatings give the glass a high glass-to-glass sliding friction and thus a cold end coating must be applied over these metallic coatings. The cold end coatings usually have a polyethylene or fatty acid base.

Another type of coating is one that reacts with the surface of the glass to form a surface layer that is more corrosion resistant that the bulk composition. Chemically inert containers are needed to contain various beverages and pharmaceuticals. To provide increased corro-sion resistance, these containers are coated internally to tie up the leachable components. Internal treatment with a fluoride gas provides a new surface that is more corrosion resistant than the original and is more economical than the older sulfur treatment.

Carbon and carbon- and nitrogen-containing materials

Property optimization

Not much can be done with graphite to improve its high temperature properties, especially its resistance to oxidation.

Most of the items discussed earlier can also be applied to these materials. The one property improvement that should be discussed a little further is that of porosity. For example, Si_3N_4 is predominantly covalent and does not densify on sintering as do conventional ionic ceramics. In applications such as turbine blades, a theoretically dense material is desired. Only through hot pressing with small amounts of additives at very high temperatures and pressures can theoretically dense material be obtained. The additives in this process cause a small amount of liquid phase to form at high temperatures, and therefore densification can proceed through liquid-phase sintering. The material that eventually crystallizes from the liquid phase is of great importance and much work has been done investigating these phases.

The development of Si_3N_4-based materials today has progressed to the point of studying materials in $Si_aM_bO_cN_d$ systems, where M has been confined mostly to trivalent cations. Most work has been in systems where M = Al, Y, and/or Be. These materials form secondary grain boundary phases which are highly oxidation resistant and thus provide a better material than materials made from Si_3N_4 densified with MgO.

External methods of improvement

One approach to the oxidation problem of graphite has been the development of protective coatings. If the coating is applied to the outside surface only (as in vapor deposition), failure may result due to coating failure from mechanical or chemical means. A way to circumvent this problem is to provide an additive of refractory metals to the graphite so that upon oxidation of the metal exposed to the surface, a liquid will form and seal the graphite surface. If the surface coating is destroyed by mechanical or chemical means, a new surface can be regenerated in situ.

The first successful composition of this type contained B_2O_3; however, its service temperature was limited to 1400°C (2550°F). Above this temperature B_2O_3 volatilizes and the viscosity of the liquid is sufficiently low to allow drip from the surface, and above 1595°C (2900°F) the reaction pressure of B_2O_3 with respect to graphite is greater than 1 atm.

To improve this composition even further, addition of oxides of zirconium, hafnium, and thorium have been found to be very helpful giving oxidation protection to 1900°C (3450°F). These oxides, when heated in contact with graphite, form highly refractory carbides; however, CO and CO_2 also form in the reaction. The evolution of CO and CO_2 may cause spalling of the coating if the reaction pressure exceeds ambient pressure.

REFERENCES

1. *Phase Diagrams for Ceramists*, American Ceramic Society, Columbus, Ohio, 1964; 1969 Supplement; and 1975 Supplement.

2. L. H. Van Vlack, Microstructures of refractories, *Ceram. Age*
    12: 29-32, 1962.
3. J. White, Magnesia-based refractories, in *Refractory Materials*,
    Vol. 5-1 (A. M. Alper, ed.), Academic Press, New York, 1970,
    pp. 110-112.
4. G. R. Rigby, Mechanical properties of basic bricks: III. The
    role of the silicate phases, *Trans. J. Br. Ceram. Soc. 70*(5):
    151-162, 1971.
5. N. B. Pilling and R. E. Bedworth, The oxidation of metals at
    high temperature, *J. Inst. Met. 29*: 529-591, 1923.
6. A. R. Cooper and W. D. Kingery, Dissolution in ceramic systems,
    J. Am. Ceram. Soc. 47(1): 37-43, 1964.
7. D. P. Gregory and A. C. Riddiford, Transport to surface of a
    rotating disk, *J. Chem. Soc. 1956*, 3756-3764, 1956.
8. B. G. Levich, Theory of concentration polarization, *Discuss.
    Faraday Soc., 1*: 37-43, 1947.
9. P. Hrma, Contribution to the study of the function between the
    rate of isothermal corrosion and glass composition, *Verres Re-
    fract. 24*(4-5): 166-168, 1970.
10. A. Pons and A. Parent, Oxygen ion activity in glass and its in-
    fluence on refractory corrosion, *Verres Refract. 23*(3): 324-333,
    1969.
11. A. R. Cooper, The use of phase diagrams in dissolution studies,
    in *Refractory Materials*, Vol. 6-III (A.M. Alper, ed.), Academic
    Press, New York, 1970, p. 237-250.
12. Y. Grodrin, *Review of the Literature on Electrochemical Pheno-
    mena*, International Commission on Glass, Paris, 1975.
13. N. Klinger, E. L. Strauss, and K. L. Komarek, Reaction between
    silica and graphite, *J. Am. Ceram. Soc. 49*(7): 369-375, 1966.
14. P. D. Miller, J. G. Lee, and I. B. Culter, The reduction of
    silica with carbon and silicon carbide, *J. Am. Ceram. Soc.
    62*(3-4): 147-149, 1979.
15. D. Ott and C. J. Raub, The affinity of the platinum metals for
    refractory oxides, *Platinum Met. Rev. 20*(3): 79-85, 1976.
16. D. E. Clark, C. G. Pantano, and L. L. Hench, *Corrosion of
    Glass*, Books for Industry & the Glass Industry, New York,
    1979.
17. R. M. Tichane and G. B. Carrier, The microstructure of a soda-
    lime glass surface, *J. Am. Ceram. Soc. 44*(12): 606-610, 1961.
18. H. E. Simpson, Study of surface structure of glass as related to
    its durability, *Ceram. Bull. 41*(2): 43-49, 1958.
19. R. M. Tichane, Initial stages of the weathering process on a
    soda-lime glass surface, *Glass Technol. 7*(1): 26-29, 1966.
20. A. G. Metcalfe and G. K. Schmitz, Mechanism of stress corro-
    sion in E glass filaments, *Glass Technol. 13*(1): 5-16, 1972.
21. R. J. Charles, Static fatigue of glass: I, *J. Appl. Phys.*

29(11): 1549-1553, 1958.

22. J. R. Hutchins, III and R. V. Harrington, Glass, in *Encyclopedia of Chemical Technology*, 2nd ed., Vol 10, Wiley, New York, 1966, p. 572.

23. F. M. Ernsberger, Structural effects in the chemical reactivity of silica and silicates, *J. Phys. Chem. Solids* *13*(3-4): 347-351, 1960.

24. D. M. Sanders and L. L. Hench, Mechanisms of glass corrosion, *J. Am. Ceram. Soc.* *56*(7): 373-377, 1973.

25. Yu. A. Schmidt, *Structure of Glass*, Vol. 1, trans. from the Russian, Consultants Bureau, New York, 1958.

26. S. M. Budd, The mechanism of chemical reaction between silicate glass and attacking agents; Part 1. Electrophilic and Nucleophilic Mechanism of Attack, *Phys. Chem. Glasses* *2*(4): 111-114, 1961.

27. S. M. Budd and J. Frackiewicz, The mechanism of chemical reaction between silicate glass and attacking agents; Part 2. Chemical Equilibria at Glass-Solution Interfaces, *Phys. Chem. Glasses* *2*(4): 115-118, 1961.

28. H. P. Hood and M. E. Nordberg, Method of Treating Borosilicate Glasses, U.S. Patent 2,215,039, September 17, 1940.

29. S. C. Yoon, Lead release from glasses in contact with beverages, M.S. thesis, Rutgers University, New Brunswick, N.J., 1971.

30. C. R. Das and R. W. Douglas, Reaction between water and glass: III, *Phys. Chem. Glasses* *8*(5): 178-184, 1967.

31. H. J. Pohlman, Corrosion of lead-containing glazes by water and aqueous solutions, *Glastech. Ber.* *47*(12): 271-276, 1974.

32. S. C. Yoon, Mechanism for lead release from simple glasses, Univ. Microfilms Int. (Ann Arbor, Mich.) Order No. 73-27, 997; *Diss. Abstr. Int. B34*(6): 2599, 1973.

33. R. L. Lehman, S. C. Yoon, M. G. McLaren, and H. T. Smyth, Mechanism of modifier release from lead-containing glasses in acid solution, *Ceram. Bull.* *57*(9): 802-805, 1978.

34. *Lead Glazes for Dinnerware*, International Lead Zinc Research Organization Manual, Ceramics I, International Lead Zinc Research Organization and Lead Industries Association, New York, 1974.

35. H. Moore, The structure of glazes, *Trans. Br. Ceram. Soc. 55*, 589-600, 1956.

36. Z. Boksay, G. Bouquet, and S. Dobos, Diffusion processes in the surface layer of glass, *Phys. Chem. Glasses* *8*(4): 140-144, 1967.

37. R. H. Doremus, Interdiffusion of hydrogen and alkali ions in a glass surface, *J. Non-Cryst. Solids* *19*: 137-144, 1975.

38. C. R. Das, Reaction of dehydrated surface of partially leached glass with water, *J. Am. Ceram. Soc.* *62*(7-8): 398-402, 1979.

39. M. A. Rana and R. W. Douglas, Reaction between glass and water: I, *Phys. Chem. Glasses, 2*(6): 179-195, 1961.

40. M. A. Rana and R. W. Douglas, Reaction between glass and water: II, *Phys. Chem. Glasses 2*(6): 196-205, 1961.

41. C. R. Das and R. W. Douglas, Reaction between water and glass: III, *Phys. Chem. Glasses, 8*(5): 178-184, 1967.

42. R. W. Douglas and T. M. M. El-Shamy, Reaction of glass with aqueous solutions, *J. Am. Ceram. Soc. 50*(1):1-8, 1967.

43. S. Wood and J. R. Blachere, Corrosion of lead glasses in acid media: I, leaching kinetics, *J. Am. Ceram. Soc. 61*(7-8): 287-292, 1978.

44. R. A. Eppler and W. F. Schweikert, Interaction of dilute acetic acid with lead-containing vitreous surfaces, *Ceram. Bull. 55*(3): 277-280, 1976.

45. A. E. Clark, C. G. Pantano, and L. L. Hench, Spectroscopic analysis of bioglass corrosion films, *J. Am. Ceram. Soc. 59*(1-2): 37-39, 1976.

46. J. Katayama, M. Fukuzuka, and Y. Kawamoto, Corrosion of heavy crown glass by organic acid solutions, *Yogyo Kyokai Shi 86*(5): 230-237, 1978.

47. W. F. Thomas, An investigation of the factors likely to affect the strength and properties of glass fibers, *Phys. Chem. Glasses 1*(1): 4-18, 1960.

48. E. J. Nolan and S. M. Scala, The aerothermodynamic behavior of pyrolytic graphite during sustained hypersonic flight, *ARS J. 32*(1): 26, 1962.

49. R. J. Diefendorf, The effect of atmosphere on the strength of graphite, *Proceedings of the 4th Conference on Carbon*, Pergamon Press, Elmsford, N.Y., 1960, p. 489-496.

50. J. Millet, Behavior of carbon brushes in dry and wet atmospheres, *Proceedings of the 4th Conference on Carbon*, Pergamon Press, Elmsford, N.Y., 1960, p. 719-725.

51. L. E. Moberly and J. L. Johnson, Wear and friction characteristics of carbon brushes as a function of air humidity, *Proceedings of the 4th Conference on Carbon*, Pergamon Press, Elmsford, N.Y., 1960.

52. G. C. Robinson, P. Schroth, and W. D. Brown, Alkali attack of carbon refractories, *Ceram. Bull. 58*(7): 668-675, 1979.

53. A. Harold, M. Colin, N. Dumas, R. Diebold, and D. Saehr, The chemical properties of heavy alkali metals inserted in graphite, in *The Alkali Metals—An International Symposium*, The Chemical Society, Oxford, 1967, p. 309-316.

54. R. J. Hawkins, L. Monte, and J. J. Waters, Potassium attack of blast furnace refractory carbon, *Ironmaking Steelmaking 1*(3): 151-160, 1974.

55. S. Wilkening, Resistance to chemical attack of carbon bricks

used in blast furnaces, *Tonind. Ztg. Keram. Rundsch.* *96*(7): 198-205, 1972.

56. S. C. Singhal, Oxidation of silicon nitride and related materials, in *Nitrogen Ceramics* (F. L. Riley, ed.), NATO Adv. Study Inst. Ser.: Ser. E, Appl. Sci., No. 23, Noordhoff, Leyden, 1977.

57. D. Cubicciotti and K. H. Lau, Kinetics of oxidation of hot-pressed silicon nitride containing magnesia, *J. Am. Ceram. Soc.* *61*(11-12): 512-517, 1978.

58. G. Ervin, Oxidation behavior of silicon carbide, *J. Am. Ceram. Soc.* *41*(9): 347-352, 1958.

59. D. W. McKee and D. Chatterji, Corrosion of silicon carbide in gases and alkaline melts, *J. Am. Ceram. Soc.* *59*(9-10): 441-444, 1976.

60. R. A. McCauley, Evolution of flat glass furnace regenerators, *Glass Ind.* *59*(10): 26-28,34, 1978.

6

Development and Application of Elastomers

PHILIP A. SCHWEITZER / Chem-Pro Corp., Fairfield, New Jersey

INTRODUCTION

The term *elastomer* has come into use, particularly in scientific and technical literature, as a name for both natural and synthetic materials which are elastic or resilient and in general resemble natural rubber in feeling and appearance. The American Society for Testing and Materials (ASTM) definition of *elastomer* is "a polymeric material which at room temperature can be stretched to at least twice its original length and upon immediate release of the stress will return quickly to approximately its original length." Compounding increases the utility of rubber and synthetic elastomers. In the raw state elastomers are soft and sticky when hot and hard and brittle when cold. Vulcanization extends the temperature range within which they are flexible and elastic. In addition to vulcanizing agents, ingredients are added to make elastomers stronger, tougher, and harder, to make them age better, to color them, and in general to modify them to meet the needs of service conditions. Today, few rubber products are made from rubber or elastomers alone.

SPECIFICATIONS

Specifications for rubber goods may cover the chemical, physical, and mechanical properties such as elongation, tensile strength, permanent set, and oven tests, minimum rubber content, exclusion of reclaimed rubber, maximum free and combined sulfur contents, maximum acetone and chloroform extracts, ash content, and many construction requirements.

It is preferable, however, to specify properties required, such as resilience, hysteresis, static or dynamic shear and compression modulus, flex fatigue and cracking, creep, electrical properties, stiffening, heat generation, compression set, resistance to oils and chemicals, permeability, and brittle point, in the temperature range prevailing in service and to leave the selection of the elastomers to a competent manufacturer.

NATURAL RUBBER

Natural rubber (NR) of the best quality is prepared by coagulating the latex of the *Hevea brasiliensis* tree, cultivated primarily in the Far East. Purified raw rubber becomes sticky in hot weather and brittle in cold weather. Its valuable properties become apparent after vulcanization.

Unloaded vulcanized rubber can be stretched to approximately 10 times its length and at this point will bear a load of 10 tons/in.^{2}. It can be compressed to one-third of its thickness thousands of times

without injury. When most types of vulcanized rubber are stretched,
their resistance increases in greater proportion than the extension.
Even when stretched almost to the point of rupture, they recover
almost their original dimension on being released, and then gradually
recover a part of the residual distortion.

Depending on the degree of curing natural rubber is classified as
soft, semihard, or hard. Only soft rubber meets the ASTM definition
of an elastomer, and therefore the information that follows pertains
only to soft rubber. The properties of semihard and hard rubber are
somewhat different, particularly in the area of chemical resistance.

Dry heat up to 50°C (120°F) has little deteriorating effect; at
temperatures of 180 to 200°C (360 to 400°F) rubber begins to melt and
becomes sticky. At higher temperatures it becomes entirely carbon-
ized.

The electrical insulation properties of natural rubber are good.

The flame resistance of natural rubber is poor.

Chemical Resistance

Mineral and vegetable oils, gasoline, benzene, toluene, and chlorinated hydrocarbons affect rubber. Strong oxidizing materials such
as nitric acid, concentrated sulfuric acid, permanganates, dichromates,
chlorine dioxide, and sodium hypochlorites severely attack rubber,
but rubber offers excellent resistance to most inorganic salt solutions,
alkalies, and nonoxidizing acids. Hydrochloric acid reacts with soft
rubber to form rubber hydrochloride and therefore is not recommended.

Resistance to Sun, Weather, and Ozone

Cold water preserves rubber, but if exposed to the air, particularly
to the sun, rubber tends to become hard and brittle. In general, it
has poor weathering and aging properties. Its resistance to ozone is
only fair. Unlike the synthetic elastomers, natural rubber softens
and reverts with aging in sunlight or ozone.

Recovery from Deformation

Resilience is the primary superior property of natural rubber over
the synthetics. It has excellent rebound properties either hot or
cold.

Applications

The major uses for natural rubber are as pneumatic tires and tubes,
power transmission belts and conveyor belts, gaskets, mountings,

hose, chemical tank linings, printing press platens, sound and/or shock absorption, and seals against air, moisture, sound, and dirt.

ISOPRENE

Natural rubber chemically is based on natural cis-polyisoprene. Isoprene (IR) is synthetic cis-polyisoprene.

The properties of isoprene are similar with those of natural rubber, with the advantage that it has no odor. This permits the use of isoprene in certain food-handling applications.

BUTYL RUBBER

Butyl rubber (11R) is a general purpose synthetic rubber whose outstanding physical properties are low air permeability (approximately one-fifth that of natural rubber) and high-energy-absorbing qualities. The former property has resulted in butyl rubber replacing natural rubber for inner tubes in tires. Its resistance to abrasion and tear is about equal that of natural rubber.

Butyl rubber can be used to temperatures of 150°C (300°F) without having any effect. Its resistance to temperature and heat aging is good.

The electrical properties of butyl rubber are generally good but not outstanding in any one area.

It has poor flame resistance.

Chemical Resistance

Butyl rubber exhibits excellent resistance to dilute mineral acids, phosphate ester oils, acetone, ethylene, ethylene glycol, and water. Unlike natural rubber, it is very resistant to swelling by vegetable and animal oils.

Its resistance to petroleum oils and gasoline is very poor. Butyl rubber is also very nonpolar.

Resistance to Sun, Weather, and Ozone

The resistance of butyl rubber to sun, weathering, and ozone is excellent.

Recovery from Deformation

At room temperature the resiliency of butyl rubber is low but it increases with increasing temperature.

Applications

Butyl rubber finds use as truck and automobile tire inner tubes, curing bags for tire vulcanization and molding, steam hoses and diaphragms, flexible electrical insulation, shock, and vibration absorption.

SBR STYRENE

SBR styrene (Buna S) is a butadiene-styrene copolymer which is manufactured by many companies. It was developed by the government during World War II because of the natural rubber shortage and became known as Government Rubber Styrene type (GR-S). It is also known as Buna S.

Compared to natural rubber, it is lacking in tensile strength, elongation, resilience, hot tear, and hysteresis, but these are offset by its low cost, cleanliness, slightly better heat aging, slightly better wear than natural rubber for passenger tires, and availability at a stable price.

SBR has relatively poor resistance to temperature, with a maximum operating temperature of 80°C (180°F), which is somewhat lower than that of natural rubber. At lower temperatures [-90°C (-120°F)] SBR products are more flexible than those of natural rubber.

The electrical properties of SBR are generally good but not outstanding in any one area.

SBR has poor flame resistance.

Chemical Resistance

In general, the chemical resistance of SBR is similar to that of natural rubber, being unaffected by water, alcohol, dilute acids, or alkalies but being attacked by oils and solvents.

Resistance to Sun, Weather, and Ozone

The resistance of SBR to sun, weather, and ozone is generally poor.

Recovery from Deformation

SBR is not as resilient as natural rubber.

Applications

SBR is used for the same applications as natural rubber.

NEOPRENE

Neoprene (CR) was actually the first commercial synthetic rubber and is still the workhorse of industry because of its outstanding properties.

It has excellent abrasion resistance, and resists damage from impact, flexing, and twisting. Because of its low heat buildup when subjected to flexing, maximum fatigue from dynamic operations results.

Neoprene has a practical high temperature range of 80 to 90°C (176 to 203°F) for continuous service. For intermittent use, products made of neoprene can operate at a temperature of 120°C (248°F). Heat exposure above these limits does not soften or melt neoprene; rather, it causes a neoprene part to harden and lose resilience.

Neoprene products show little change in performance characteristics down to -20 to -25°C (-4 to -13°F). Below that point, they stiffen until their brittle temperature is reached at about -40°C (-40°F). Specially formulated compositions, however, permit service as low as -55°C (-67°F).

Neoprene can be compounded to satisfy a number of needs in electrical applications. Its dielectric characteristics limit its use as an insulator to low-voltage (600 V) and low-frequency (60-Hz) applications. As a protective outer jacket for insulated cable, however, neoprene is used at all voltages. It is relatively unharmed by corona attack.

In gum polymer form, neoprene, by virtue of its chlorine content, is inherently more resistant to burning than are exclusively hydrocarbon polymers.

In laboratory tests, normally compounded solid neoprene products can be ignited by an open flame but will stop burning if the flame is removed. Under the same conditions, natural rubber and many other synthetic rubbers will continue to burn. However, despite its superiority over these other materials in such laboratory tests, neoprene will burn in an actual fire situation.

Chemical Resistance

Resistance to deterioration from waxes, fats, oils, greases, and many other petroleum products is one of the best known properties of neoprene. Neoprene was originally developed as an oil-resistant substitute for natural rubber and is still widely used for this purpose. Its use is limited, however, to nonaromatic hydrocarbons and it will not withstand chlorinated solvents.

Neoprene is resistant to alkalies, dilute mineral acids, and inorganic salt solutions. Acid and salt solutions of a highly oxidizing nature will cause surface deterioration and loss of strength. Unlike natural rubber and other general purpose elastomers, neoprene gives excellent service in contact with aliphatic hydrocarbons, aliphatic hydroxy compounds, and most Freon refrigerants. In contact with these fluids, neoprene displays minimum swelling, relatively little loss of strength, and virtually complete recovery of initial properties after removal of the liquid by drainage or evaporation.

Resistance to Sun, Weather, and Ozone

The resistance of properly compounded neoprene to ozone, sun, and weather accounts for its excellent aging characteristics.

Recovery from Deformation

Neoprene exhibits a relatively low degree of permanent deformation from compression (compression set). Permanent deformation due to elongation (permanent set) will be approximately 5% for most neoprene products.

Applications

Products made from neoprene include wire and cable jackets, industrial coated fabrics, gaskets, hose and belting, glazing and paving seals, structural bearings, latex foam products, a wide variety of adhesives and coatings, molded and extruded goods for industrial, architectural, and automotive use, and a series of consumer products in the shoe, paper, and tire fields.

Neoprene automotive parts have been in common usage for over 40 years. Both engine and body components benefit from its heat, oil, and weather resistance. Typical uses include hose, tubing, sponge door and truck seals, dust boots, window channels, wire jacketing, adhesives, grommets, and seals. Neoprene-jacketed power cable is used extensively in underground and underwater service.

NBR-NITRILE

NBR-nitrile (Buna N) is one of the four most widely used elastomers. The big advantage of NBR is its improved resistance over neoprene to oils, fuels, and solvents. Its physical properties closely match those of SBR except that it has much better heat resistance and considering cost (slightly above that of neoprene) it has the broadest resistance to chemicals with a balance of properties than any other elastomer.

The heat resistance of NBR is high, having an operating range of -15 to +150°C (-60 to +300°F).

The electrical properties of NBR are only fair and consequently it does not find wide usage in the electrical field.

The flame resistance of NBR is poor.

Chemical Resistance

The nitrile rubbers show good resistance to alkalies, aqueous salt solutions, oils, and solvents. A very slight swelling occurs in the presence of aliphatic hydrocarbons, fatty acids, alcohols, and glycols. The reduction of physical properties as a result of swelling is small, making NBR suitable for gasoline- and oil-resistant applications.

NBR is attacked by strong oxidizing agents, ketones, ether, and esters.

NBR has excellent resistance to water.

Resistant to Sun, Weather, and Ozone

NBR has poor resistance to sunlight and ozone and weathering in general.

Recovery from Deformation

The resiliency of NBR, either hot or cold, is good.

Applications

NBR finds use as carburetor diaphragms, aircraft hose, gaskets, self-sealing fuel tanks, gasoline and oil hose, cables, printing rolls, and machinery mountings.

CHLOROSULFONATED POLYETHYLENE (HYPALON)

Hypalon (CSM) is very similar to neoprene but does possess some added advantages in terms of extra properties. It has improved heat and ozone resistance, improved electrical properties, improved color stability, and improved chemical resistance.

Because of its chlorine content, hypalon is inherently more resistant to burning than exclusively hydrocarbon polymers. It is classified as self-extinguishing.

General purpose compounds can operate continuously at temperatures of 120 to 135°C (248 to 275°F). Special compounds of Hypalon can be used in intermittent service up to 150°C (302°F).

Conventional compounds of Hypalon are serviceable down to -18 to -23°C (to -10°F). Special compounds retain their flexibility down to -40°C (-40°F).

The electrical properties of Hypalon are good. It has proven to be a valuable insulation, in service less than 600 V, for control cable and secondary wire. For higher voltages, its use as a protective covering over other insulating materials has been successful because of its outstanding resistance to weathering, ozone, and corona.

Chemical Resistance

Compositions of Hypalon are very resistant to attack by oxidizing chemicals such as concentrated sulfuric acid and hypochlorite solutions, properly compounded, it is especially useful in contact with oils at elevated temperatures. It is also resistant to aqueous salt solutions, alcohols, and weak and concentrated alkalies. It has poor resistance to aliphatic, aromatic and chlorinated hydrocarbons, as well as aldehydes, ketones, and fuel oils.

Resistant to Sun, Weather, and Ozone

Hypalon synthetic rubber can resist the long-term effects of weathering.

Hypalon is so extraordinarily resistant to ozone that although failures can be induced with certain formulations in laboratory tests, no known product failure due to ozone attack has ever been reported.

Recovery from Deformation

Hypalon shows good resiliency.

Applications

Hypalon is used as a jacketing and insulating material for wire and cable; in fluid coating and laminated roofing systems; as sheeting for pond, pit, and reservoir liners and covers; for industrial products such as hose, tarpaulins, equipment linings, gaskets, seals, and diaphragms, maintenance coatings, and industrial rolls; and for a variety of consumer items such as whitewall tires, waterproof garments, appliance parts, coated fabrics, and shoe heels and soles.

POLYSULFIDE (THIOKOL ST)

This elastomer is commonly known as Thiokol ST and has the ASTM designation T. Compared with nitrile rubber it has poor tensile strength, pungent odor, poor rebound, high creep under strain, and poor abrasion resistance.

Polysulfides can be used in temperature range -15 to +120°C (-60 to +250°F). They are also capable of short-term use at 150°C (300°F) without any effect on its properties.

The electrical insulating properties of polysulfides are fair.

Polysulfides have poor flame resistance.

Chemical Resistance

Common alcohols, ketones, and esters used in paints, varnishes, and inks have little effect on the polysulfides. It is especially resistant to aliphatic liquids, hydrocarbon solvents, or blends of aliphatics and aromatics. Some chlorinated hydrocarbons can be handled with little effect, but tests should be conducted before using in this service.

Resistance to Sun, Weather, and Ozone

The weathering properties and resistance to sun and ozone of the polysulfides are excellent; consequently, it has wide applications for caulking purposes.

Recovery from Deformation

Polysulfide has good resiliency either hot or cold.

Applications

Caulking uses, seals, gaskets, diaphragms, valve seat disks, flexible mountings, solvent-carrying hose, printers rolls, balloons, life vests, and rafts.

SILICONE RUBBERS

The silicone rubbers (FSI, PSI, VSI, PVSI, and SI) are a series of compounds whose polymer structure consists of silicone and oxygen atoms rather than the carbon structure of all other elastomers. Silicon is in the same chemical group as carbon, but is a more stable element and therefore more stable compounds are produced. The silicones are the most heat resistant of the elastomers presently available and the most flexible at low temperatures. The basic structure can be modified with such groups as vinyl or fluoride, which improves such properties as tear resistance and oil resistance. This results in a family of silicones that covers a wide range of physical and environmental requirements.

Silicone rubbers have the best electrical properties of any of the elastomers. The decomposition product of organic compounds is conductive carbon black, which can sublime and thus leave nothing for insulation, whereas the decomposition product of the silicone rubbers is an insulating silicon dioxide. This property is taken advantage of in the insulation of electric motors.

Silicone rubbers exhibit good flame resistance.

Chemical Resistance

Silicone rubbers can be used in contact with dilute acids and alkalies, alcohols, animal and vegetable oils, and lubricating oils. Aromatic solvents such as benzene, toluene, gasoline, and chlorinated solvents will cause excessive swelling. They are not resistant to steam at elevated temperatures.

Resistance to Sun, Weather, and Ozone

The series of silicone rubbers exhibit excellent resistance to sun, weathering, and ozone.

Recovery from Deformation

Silicone rubbers have good rebound properties.

Applications

Because of their unique thermal stability properties and/or their insulating values, the silicone rubbers are used primarily in electrical applications such as appliances, heaters, furnaces, aerospace devices, and automotive parts.

POLYBUTADIENE

Like isoprene, polybutadiene (BR) is a sterospecific controlled structure having outstanding properties of resilience and hysteresis, almost equivalent to that of natural rubber. It is most similar to SBR, and because it is so difficult to process it finds wide use as an admixture with SBR and other elastomers. It is rarely used in a larger amount than 75% of the total polymer in a compound.

Its range of operating temperatures is only slightly better than that of natural rubber, ranging from -65 to +90°C (-150° to +200°F).

Polybutadiene possesses good electrical insulating properties.

The flame resistance of polybutadiene is poor.

Chemical Resistance

The chemical resistance of polybutadiene is similar to that of natural rubber. It shows poor resistance to aliphatics and aromatic hydrocarbons and a fair resistance to animal and vegetable oils.

Resistance to Sun, Weather, and Ozone

Polybutadiene has poor sunlight aging characteristics, but has good oxidation resistance.

Recovery from Deformation

The resiliency of polybutadiene is excellent either hot or cold.

Applications

Polybutadiene is usually used as a blend with other elastomers to impart better resiliency, abrasion resistance, and low-temperature properties, particularly in the manufacture of automobile tire trends, shoe heels and soles, gaskets, seals, and belting.

URETHANE RUBBERS

Very few materials offer the combination of hardness and resilience possessed by urethane rubber (U). Compositions with a durometer hardness of 95A (harder than a typewriter platen) are elastic enough to withstand stretching to more than four times their normal lengths.

Standard compounds maintain their resilience with changing temperature far better than most other rubbers through the range 10 to 100°C (50 to 212°F). This stability is valuable in certain shock-mounting applications.

The load-bearing capacity of products made with urethane rubber is far above that of conventional elastomers and only slightly below that of structural plastics, yet the material possesses the resilience of a true elastomer.

The outstanding abrasion resistance of urethane rubber has led to many important applications where severe wear is a problem.

Urethane rubber has a low unlubricated coefficient of friction that decreases sharply as hardness increases. This characteristic, coupled with its superior abrasion resistance and load-carrying ability, is an important reason why urethane rubber is used in bearings and bushings.

At room temperature, a number of the raw polyurethane polymers are liquid, simplifying the production of many large and intricately shaped molded products. When cured, these elastomeric parts are hard enough to be machined on standard metalworking equipment. Cured urethane rubber does not require fillers or reinforcing agents.

Products of urethane rubber perform well at moderately elevated temperatures; 85°C (185°F) is the usual temperature limit for continuous operation and 100°C (212°F) for intermittent exposure.

Urethane rubber remains flexible at very low temperatures and possesses outstanding resistance to thermal shock. Standard compositions do not become brittle at temperatures below -62°C (-80°F), although stiffening gradually increases as the temperature is reduced below -18°C (0°F).

Urethane rubber has been used successfully at cryogenic temperatures in handling nonoxidizing liquefied gases.

Urethane rubber has been used in potting and insulating compositions at frequencies up to 100 kHz and temperatures not exceeding 100°C (212°F). The basic polymer may be modifed with epoxy resins for improved electrical properties.

The urethane rubbers are classified as slow burning to self-extinguishing but may be compounded with flame retardents to produce nonburning products.

Chemical Resistance

The urethane rubbers are resistant to most mineral and vegetable oils, greases and fuels, and aliphatic, aromatic, and chlorinated hydrocarbons. Esters, ethers, and ketones attack urethane rubber; alcohols soften and swell them.

They have limited service in weak acid solutions, can not be used in concentrated acids, and are not resistant to steam or caustics.

Resistance to Sun, Weather, and Ozone

Weather extremes do not impair the serviceability of urethane rubber in outdoor applications. Prolonged exposure to ultraviolet light will darken and somewhat reduce the physical properties of products made of urethane rubber, but no significant surface deterioration occurs. They can be protected against the effects of severe weathering by pigmentation or by using ultraviolet screening agents.

Oxygen and ozone in atmospheric concentrations have no observable effect on urethane rubber.

Recovery from Deformation

Products made from urethane rubber have good resistance to compression set. This is shown by their successful use in die pads for metal forming, where tens of thousands of indentations are routine.

Applications

The versatility of urethane rubber has led to a wide variety of applications. Typical products are industrial tires and rolls, mining industry wear parts, sleeve bearings and bushings, die pads for metal forming, equipment linings, transmission and conveyor belts, marine products, a variety of sports and consumer items, coatings, gaskets, seals, and shoe heels.

POLYESTER ELASTOMER

Polyester (PE) is a relatively new elastomer manufactured by Dupont under the trade name Hytrel. It combines characteristics of thermoplastics and elastomers, is structurally strong, resilient, and resistant to impact and flexural fatigue.

Polyester elastomers possess mechanical properties that make them practical for a number of demanding applications. They combine such useful features as excellent flex-fatigue resistance, high resistance to deformation under moderate strain conditions, retention of flexibility at low temperatures, and good abrasion resistance.

Resistance of PE to fatigue in cyclic load-bearing applications is outstanding. This is particularly true in deformation in low-strain ranges, where it may be considered to act like a perfect spring with no, or very low, hysteresis. This means that a part made of PE, engineered to operate in low-strain levels, can usually be expected to exhibit complete recovery from deformation and to continue to do so under repeated cycling for extremely long periods without heat buildup or distortion.

PEs show high moduli in tension, compression, and flex. At comparable hardnesses, they are superior to commercial polyurethane rubbers in load-bearing capacity. They can be used in smaller, thinner sections to give equivalent performance.

PEs possess exceptional strength at elevated temperatures, particularly the harder polymers. Above 120°C (248°F) their tensile strengths far exceed those of other polymers.

All the polymers of PE have solenoid brittle points below -70°C (-94°F). As would be expected, the softer members exhibit the best low-temperature flexibility.

Chemical Resistance

In general, the fluid resistance of polyester elastomers increases with increasing hardness of the polymer. Since they contain no plasticizers, products of PE are not susceptible to solvent extraction or heat volatilization of such additives. PE withstands a wide variety of oils, fuels, solvents, and chemicals but is attacked by concentrated mineral acids and bases. It is soluble in phenols, cresols, and certain chlorinated hydrocarbons.

Resistance to Sun, Weather, and Ozone

For resistance to ultraviolet radiation, screening agents are required with PE. Properly stabilized, these elastomers will provide satisfactory service under extreme climatic conditions.

Applications

Today, products being made of PE include hydraulic hose, air and heater hose, vacuum and industrial tubing, seals, gaskets, specialty belting, noise damping devices, low-pressure tires, industrial solid tires, wire and cable jacketing, pump parts, electrical connectors, flexible shafts, sports equipment, piping clamps and cushions, gears, flexible couplings, and fasteners.

PERFLUOROELASTOMERS

Because the monomers used in the production of the base polymer are expensive and the fabrication of finished parts of perfluoroelastomer (FPM) requires difficult and complex techniques, these rubber parts command very high prices. However, initial cost usually is not a factor with FPM parts becuase they normally are specified only where no other material can do the job or where overall cost-life considerations are especially important.

The halogen groups present in these polymers provide stability. The fluorinated rubbers are especially good for high-temperature service, but do not have as high an operating temperature as the silicone rubbers. They may be used from -50 to +200°C (-90 to +400°F), have low permeability rates with air, and extremely low water absorption. The latter property makes them highly desirable for use in electrical and electronic equipment, which requires excellent insulation.

Aging characteristics are excellent. Tensile strength and tear resistance is good, but if these properties are important in the application, fillers such as finely divided silica or carbon black are used in the compounding.

FPM has excellent resistance to flame, but will burn if involved in an actual fire situation.

Typical products in this category are Viton (Dupont), Kel-F (3M), Fluorel (3M), and Kalrez (Dupont).

Chemical Resistance

A major advantage of FPM is that its nearly universal chemical resistance makes it suitable for virtually all chemical process streams.

In general, FPM is suitable for service at temperatures 40 to 60°C (100 to 150°C) higher than any other commercial elastomer. Based on retention of elastic properties, FPM perfluoroelastomer parts can often be used continuously at temperatures up to 316°C (600°F) in many fluid environments.

The fluoroelastomers are extremely resistant to aliphatic hydrocarbons, chlorinated solvents, animal, mineral and vegetable oils, gasoline, jet fuels, dilute acids alkaline media, and aqueous inorganic salt solutions. In the presence of oxygenated solvents, alcohols, aldehydes, ketones, esters, and ethers, their resistance is fair to poor.

Resistant to Sun, Weather, and Ozone

Perfluoroelastomer is extremely resistant to the adverse effects of the environment, including sunlight, ozone, weather, and moisture. With a brittle point of -37°C (-35°F), it exhibits only moderately useful low temperature characteristics. Electrical properties are not too different from those of other synthetic rubbers.

Recovery from Deformation

The resiliency of FPM depends on the particular elastomer and ranges from poor to good.

Applications

Perfluoroelastomer parts are a practical solution wherever the sealing performance of rubber is desirable but not feasible becuase of severe chemical or thermal conditions. Alternatively FPM is the O-ring used for seals on equipment in the petrochemical industry. O-rings of FPM are used in mechanical seals, pump housings, compressor casings, valves, rotameters, and other instruments. Custom-molded parts also are used as valve seats, packings, diaphragms, gaskets, and miscellaneous sealing elements, including U cups and V rings.

Other industries where FPM contributes importantly are aerospace (versus jet fuels, hydrazine, N_2O_4 and other oxidizers, Freon 21 fluorocarbon, etc.), nuclear power (versus radiation, high temperatures), oil, gas, and geothermal drilling (versus sour gas, acidic fluids, amine-containing hydraulic fluids, extreme temperatures and pressures), and analytical and process instruments (versus high vacuum, liquid and gas chromotography exposures, high-purity reagents, high-temperature conditions).

ETHYLENE-PROPYLENE RUBBER

Ethylene-propylene (EPM) rubbers have good tear resistance, even at elevated temperatures, and are very durable when exposed to abrasion and other forms of mechanical abuse. Their resilience is between that of products made from SBR and those of natural rubber.

EPM rubber can be compounded in any color, including white and delicate pastels. These colored products are highly resistant to fading and discoloration when exposed out of doors.

General purpose compounds of EPM hydrocarbon rubber are serviceable at temperatures of 125 to 145°C (257 to 293°F). Special compounds can be formulated for continuous use up to 155°C (311°F) and in intermittent service up to 170°C (338°F).

Conventional compounds of EPM rubber are flexible and serviceable at temperatures as low as -50°C (-58°F). Special compounds can function down to -68°C (-90°F).

EPM rubber has excellent electrical properties and is suitable for high-voltage cable insulation. It also withstands heavy corona discharge without sustaining damage.

EPM rubber is not flame resistant.

Chemical Resistance

Products made from EPM rubber resist attack by many acids and alkalies, detergents, phosphate esters, ketones, alcohols, and glycols. They give particularly outstanding service with hot water and high-pressure steam. However, EPM rubber should not be used in contact with hydrocarbon solvents and soils, chlorinated hydrocarbons, or turpentine.

Resistance to Sun, Weather, and Ozone

The resistance of EPM rubbers to sun, weather, and ozone is excellent.

Recovery from Deformation

After being held under compression for an extended period of time, EPM rubber shows a low degree of permanent deformation.

Applications

EPM rubber is widely used in automotive and appliance components, garden and industrial hose, belts, wire and cable insulation, electrical accessories, bicycle tires, and a variety of other molded and extruded parts for marine, agricultural, industrial, and consumer applications, especially those requiring outstanding ozone resistance.

Hose, belts, diaphragms, gaskets, and other industrial rubber goods made of EPM rubber give excellent service with high-pressure steam, hot water, and chemicals.

Good electrical and physical properties of EPM rubber suit it for use in high-voltage accessories for cable and power equipment.

ETHYLENE-ACRYLIC ELASTOMER

Ethylene-acrylic (E-A) elastomer has good tear and tensile strengths and high elongation at break, which is expected of a tough rubber. In addition, exceptionally low compression set values are an added advantage, making the product suitable for many hose, sealing, and cut gasket applications.

A unique feature is its practically constant damping characteristics over broad ranges of temperature, frequency, and amplitude. Very little change in damping value takes place between -20 and 160°C (-4 and 320°F). This property, which shows up as a poor rebound in resilience tests, is actually a design advantage. Combined with E-A's heat and chemical resistance, it allows the use of E-A in vibration damping applications.

This elastomer provides heat resistance surpassed by only the more expensive specialty polymers, such as the fluorocarbon and fluorosilicone elastomers. Parts retain elasticity and remain functional after continuous air-oven exposures from 18 months at 121°C (250°F) to 7 days at 204°C (400°F).

Parts of E-A will perform at least as long as parts of Hypalon or general purpose NBR, but at exposure temperatures 10 to 40°C (50 to 100°F) higher.

E-A is inherently superior to most other heat- and oil-resistant polymers, including standard fluoroelastomers, chlorosulfonated polyethylene, polyacrylates, and polyepichlorohydrin, in performance at low temperatures. Typical unplasticized compounds are flexible to -20°C (-20°F) and have brittle points as low as -60°C (-75°F). Compounding E-A with ester plasticizers will extend its low-temperature flexibility limits to -46°C (-50°F).

Chemical Resistance

Products of E-A have very good resistance to hot oils and hydrocarbon- or glycol-based proprietary lubricants, transmission, and power steering fluids. E-A's swelling characteristics permit it to retain its original physical properties better than silicone after oil immersion.

Ethylene-acrylic elastomer also has outstanding resistance to hot water. E-A is not recommended for immersion in esters, ketones, or highly aromatic fluids. Neither should it be used in applications calling for long-term exposure to high-pressure steam.

Resistance to Sun, Weather, and Ozone

E-A provides excellent resistance to sun, weathering, and ozone.

Applications

Ethylene-acrylic elastomer should be considered for any application calling for a tough, set-resistant rubber with good low-temperature properties and excellent resistance to the combined deteriorating influences of heat, oil, and weather.

7
Coatings

KENNETH B. TATOR / KTA-Tator Inc.

INTRODUCTION

Paints (or coatings if used primarily for corrosion protection) are something that we are all familiar with to some degree. Virtually everywhere one looks, paint can be seen: furniture, houses, automobiles, trucks, ships, airplanes, bridges, chemical plants, nuclear power

plants—literally everywhere. Additionally, paint is used for a variety
of purposes—for example, in the house alone, paints can be used to
provide an aesthetically appealing interior, a protective durable exterior, to provide mildew and rot resistance to wood, to seal masonry
from entering water, and to seal the substrate and improve sanitation
and cleanup in such areas as the kitchen and bathroom.

The major advantage of paints and coatings accounting for this widespread use is that they relatively inexpensively provide good gloss,
color, and decorative effects, while also protecting from the effects of
the environment. Most important, however, is that paints are readily
available and can be applied by a variety of means, ranging from simple brush and roller (commonly employed by the do-it-yourself house-
painter) to the rather sophisticated automated finishing lines employing
electrostatic spray, fluidized bed (for powder coatings), coil coating
and baking lines, and other technically complex application methods.
Accordingly, the ability of a diverse variety of users to apply satisfactorily an almost endless variety of paint formulations by a plethora
of different means has led to the development of the painting industry
(and allied industries) as major factors in the economy. Of all methods
of corrosion prevention described in this book, painting and protection
by coatings is certainly the most widely used.

Coating for corrosion protection should be considered an engineering function consisting of design considerations; selection of a suitable
coating system; surface preparation requirements; coating application
considerations, including control of ambient conditions during application; certain special considerations (such as thickness, variances for
application condition, substrate condition, etc.); and finally scheduled
inspection and maintenance. Each of these subjects is considered in
order.

COATING SYSTEM SELECTION

Factors important in the selection of a coating system for corrosion protection are enumerated and discussed below. The importance, and
therefore weight, of any single factor will vary from application to
application but, for most uses, the factors are listed in decreasing
order.

Service Environment

It is almost self-evident that different environments will require different paints to protect from that environment. Not so self-evident is that
many environments are too severe for protection by any kind of paint.
As a general rule, corrosion rates exceeding 50 mils per year (mpy)
are too severe for protection by coatings—and other protection methods,

such as lining, choice of a more resistant substrate, other protection
methods, or a combination of these in conjunction with coating should
be used. As a general rule, the more severe environments (chemical
immersion, splash, or spillage; salt- and freshwater immersion) require more durable coating systems, specially pigmented using synthetic resins as binders. Such coatings often require more extensive surface preparation and more exacting application parameters and techniques. In less severe environments, appropriate compromises can be
made in both the coating type and surface preparation and application requirements. Table 1 illustrates this principle.

Expected Longevity

The duration of corrosion protection afforded by the coating system is
of major importance. Most commonly, once a decision has been made
to coat, it is desired to have the coating last as long as is reasonably
possible. On the other hand, if protection longevity is not of utmost
concern (which happens more often than one might think), less expensive systems may be chosen. The automotive industry has long been
accused of planned obsolescence—and auto body rust-through and
corrosion deterioration are said to be factors in this. Some automobile
manufacturers are now advertising their use of more corrosion resistant
paints and materials—but for the most part, automotive finishes do not
last more than a few years. Similarly, porch and lawn furniture, and
original equipment manufactured (OEM) items such as motor housing,
pipe, conduit, and electrical boxes are painted to look good at the time
of sale. However, if long-term corrosion protection is needed, a
special painting order must be placed far in advance, or additional
painting for protection must be done by the purchaser. Many industrial organizations and nuclear power facilities purchase unpainted or
OEM—painted items—and routinely repaint with a more protective coating
system prior to placing the item in service.

Cost

This is an obvious, but often unplanned consideration. Most painting
operations can be best and most economically done in a fabricating
shop or commercial coating facility. Surface preparation can be done
by chemical cleaning in large vats or by mechanical blast cleaning in
shop facilities. Both chemical and shop blast cleaning can be done in
controlled environments using automated facilities. Correspondingly,
the application of protective coating can also be done under relatively
controlled conditions during both application and cure. The net effect
to the consumer is a superior coating job at a reduced cost. Too often,
however, a specifier disregards or is not knowledgeable of the appropriate coating economics, and unnecessarily specifies surface

Table 1. Painting Systems for Environmental Zones

Zone conditions	Painting system suggestions
0 Dry interiors where structural steel is embedded in concrete, encased in masonry, or protected by membrane or contact-type fireproofing	Leave unpainted
1A Interior, normally dry (or temporary protection); very mild (oil-base paints now last 10 years or more)	Latex or one-coat
1B Exteriors, normally dry (includes most areas where oil-base paints now last 6 years or more)	Oil base
2A Frequently wet by fresh water; involves condensation, splash, spray, or frequent immersion (oil-base paints now last 5 years or less)	Vinyl, coal tar epoxy, epoxy, chlorinated rubber
2B Frequently wet by salt water; involves condensation, splash, spray, or frequent immersion (oil-base paints now last 3 years or less)	Zinc-rich, vinyl, coal tar epoxy, epoxy
3A Chemical exposure, acidic (pH 5 or lower)	Vinyl
3B Chemical exposure, neutral (pH 5 to 10)	Zinc-rich
3D Chemical exposure, presence of mild solvents; intermittent contact with aliphatic hydrocarbons (mineral spirits, lower alcohols, glycols, etc.)	Epoxy
3E Chemical exposure, severe; includes oxidizing chemicals, strong solvents, or combinations of these with high temperatures	Use specific exposure data

preparation and coating in the field at the time of or after, erection or installation. In almost every case, painting in a shop or commercial facility is less expensive than painting at a job site. When painting structural steel, painting on the ground is almost always cheaper than painting in the air. In corrosion environments, surface preparation, often as a rule of thumb, costs as much as 50% or more of the total painting cost. Too often, specifiers are "penny wise and pound foolish" when they specify a good commercial or near-white metal blast

Table 2. Cost of Painting: Cheap Versus Expensive Paint

	Cheap	Expensive
Prime		
Cost per gallon	—	$20.00
Coverage per gallon	—	250 ft^2
Thickness per coat		1 mil
Body		
Cost per gallon	$10.50	$16.00
Coverage per gallon	200 ft^2	175 ft^2
Thickness per coat	1 mil	2 mils
Material cost/coat for prime	—	5.6 cents
Material coast/coat for body	5.25 cents	9.0 cents
Number of coats for 5 mils	5	3
Paint thickness obtained	5 mils	5 mils
Cost per square foot		
Material	26.25 cents	23.6 cents
Surface preparation	40.0 cents	40.0 cents
Application labor	35.5 cents	20.5 cents
Scaffolding, misc.	6.0 cents	6.0 cents
Total direct applied cost	107.75	90.1

cleaning followed by the application of an oil-base paint. Such paints oxidize and age in the atmosphere and by virtue of their oil have good wetting and penetrating properties (thus enabling their application over poorly prepared surfaces). Perhaps a more suitable choice over a thoroughly blast cleaned surface would be the choice of a synthetic resin coating, or zinc-rich primer, which can often be applied at approximately the same cost but give far superior corrosion protection.

Table 2 illustrates a cost comparison using expensive and cheap paint. The initial savings using the cheap paint usually cannot be justified for corrosion protection. The cost of paint is only a minor cost in the cost of a total coating job.

Suitability (or Availability) of Necessary Surface Preparation and Application Means

In some environments, certain surface preparation and coating application techniques are not permissible. For example, many companies do not permit open blast cleaning where there is a prevalence of electrical motors or hydraulic equipment. Refineries, as a general rule, do not permit open blast cleaning, or for that matter, any method of surface preparation that might result in the possibility of a spark, static electricity buildup, or an explosion hazard. During the course of construction or erection, many areas requiring protection are enclosed or covered, or so positioned that access is difficult or impossible. Consideration must be given to painting such structures prior to installation.

By virtue of the equipment used, some methods of coating cannot be done anywhere but at a specialized facility. Such facilities are not readily transportable to field sites, and accordingly, these methods are precluded from field use. Good examples might be most chemical cleaning, including pickling and acid etching, automatic rotary wheel blast cleaning, and automatic spraying, electrostatic, or high-speed roller coating application.

Safety

Safety is, of course, an overriding concern in all aspects of commerce. Occasionally, however, beside the normal requirements for ventilation, removal of solvents from a coating application area, suitable and safe access to the work being painted, and so on, other safety considerations must be anticipated. For example, most steelworkers on high steel such as tall buildings or bridges dislike to walk on painted steel because of its slickness, and depending on the paint color, concealment of puddled water or surface ice. Such a concern might eliminate most "barrier"-type coatings, but would allow most zinc-rich coatings. Similarly, some coatings (notably zinc-rich coatings) are formulated and applied as preconstruction primers—to allow flame cutting without detrimental fumes or weld quality deficiencies.

Local and federal jurisdictions have outlawed certain type of paints primarily those containing lead and asbestos) and are in the process of restricting others with potentially harmful pigments or other constituents.

Ease of Maintenance/Repair

Many coatings that offer good long-term protection after application (particularly the thermosetting and zinc-rich coatings) are more difficult to touch up or repair in the event of physical damage or localized failure. Adhesion of a subsequently applied paint coat to an older aged epoxy, urethane, or other catalyzed coating often results in diminished adhesion leading to peeling. Thermoplastic coatings, as a rule, do not suffer this disadvantage as solvents in the freshly applied coating soften and allow for intermolecular mixing of the new and old coatings—with good intercoat adhesion resulting. Heavily pigmented coatings (such as zinc-riches) require agitated pots to keep the pigment in suspension during application. Accordingly, touchup and repair of large areas, unless done by spray using an agitated pot, is not recommended when using zinc-rich coatings.

Generally, oil-base coatings (alkyds, epoxy esters, and modifications thereof) have a greater tolerance for poor surface preparation and an ability to wet, penetrate, and adhere to poorly prepared surfaces or old coatings. Often these coatings are specified for these purposes, even though it is known that they offer lesser long-term environmental protection.

Decoration/Aesthetics

Probably of least importance from a corrosion protection viewpoint are color, gloss, and overall appearance. However, corrosion protection coatings are available that not only offer good chemical and environmental resistance, but are aesthetically pleasing as well. A good example of this is the aliphatic urethane, which at present is quite expensive—but can often be justified solely on the basis of superior appearance. The corrosion protection capability of such coatings is not much different from some members of the epoxy family of coatings which are about half the cost. However, the superior tinting ability, color, and gloss retention of the aliphatic urethane has led to its widespread use on railroad cars, water and fuel oil storage tanks close to public thoroughfares, aircraft, and many structures where public visibility is high and appearance is important.

Certain other coatings can be modified with silicones, acrylics, and other resins at increased cost to enhance their aesthetic appeal.

WHAT IS A PAINT/COATING?

A paint is generally thought of as a liquid coating applied to a surface that dries or cures to form an aesthetically pleasing protective film. The film thickness may range from a few tenths of a mil to a quarter of an inch or more, although thicknesses between 2 and 40 mils are by far the most common (1 mil = 1/1000 in. = 25.4 μm).

The "paint" really consists of various pigments, a resin system, special additives of various types, and in most cases, water or a solvent that evaporates shortly after application.

Resin Component

Commonly, paints are designated by their resin component (such as vinyls, chlorinated rubbers, epoxies, etc). The resin is the film-forming agent of the paint, and generally the chemical and atmos-

Table 3. Kinds of Industrial Paint and Their Characteristics

Coating type (formulation)	Resistance
Vinyls: polyvinyls dissolved in aromatics, ketones, or esters	Insoluble in oils, greases, aliphatic hydrocarbons, and alcohols; resistant to water and aqueous salt solutions; not attacked at room temperature by inorganic acids and alkalies; fire resistant
Chlorinated rubbers: resins dissolved in hydrocarbon solvents	Chemically resistant to acids and alkalies; low permeability to water vapor; abrasion resistant; fire resistant
Epoxies: polyamine plus epoxy resin (amine epoxy)	Chemically resistant to acids, acid salts, alkalies, and organic solvents
Polyamide plus epoxy resin (polyamide epoxy)	Resistant to moisture; partially resistant to acids, acid salts, alkalies, and organic solvents
Aliphatic polyamine plus partially prepolymerized epoxy (amine adduct epoxy)	Partially resistant to acids, acid salts, alkalies, and organic solvents
Esters of epoxies and fatty acids modified (epoxy ester)	Resistant to weathering
Coal tar plus epoxy resin (amine or polyamide cured)	Excellent resistance to fresh water, salt water, inorganic acids, and various chemicals
Oil-base: coating formulations with vehicles (alkyd, epoxy, urethane) combined with drying oils	Resistant to weather

Limitations	Features--Uses
Adhesion may be poor until all solvents have vaporized from coating; aromatics, ketones, and ester solvents redissolve the coating; thermoplastic; relatively low thickness per coat; temperature resistance: 180°F dry, 140°F wet	Low toxicity, tasteless, and odorless; used on surfaces exposed to potable water and on sanitary equipment; widely used as industrial coatings
Degraded by ultraviolet light; hydrocarbon solvents attack the coating; thermoplastic; temperature resistance: 200°F dry, 120°F wet	Excellent adherence to metallic concrete and masonry; nontoxic; used on structures exposed to water and marine atmospheres (i.e., swimming pools)
Harder and less flexible than other epoxies; less tolerant of moisture during application; temperature resistance: 225°F dry, 190°F wet	Greatest chemical and solvent resistance of epoxies
Chemical resistance is inferior to that of the polyamine epoxies; temperature resistance: 225°F dry, 150°F wet	Resistant to moisture; used on wet surfaces or under water, as in tidal-zone areas of pilings, oil rigs, etc.
The adduct film has greater permeability than the uniformly cross-linked film of the amine epoxies; temperature resistance: 225°F dry, 150°F wet	Flexible film; used on surfaces requiring latitude of application
Resistance to chemicals and solvents is limited; attacked by alkalies; temperature resistance: 225°F dry; application time interval between coats is critical	On surfaces requiring the properties of a high-quality oil-base paint
Temperature resistance: 225°F dry, 150°F wet	Thick films to 10 mils per coat; used on clean blast-cleaned steel without a primer for immersion or below grade service
Chemical resistance is inferior to that of above coatings; attacked by alkalies; temperature resistance: 225°F dry	Lower cost than most coatings and the best penetrating power; used on exterior wood surfaces, for primers requiring penetrability and in less severe chemical environments

Table 3. (Continued)

Coating type (formulation)	Resistance

Urethanes
  Moisture cured
    Isocyanate prepolymer reacts
    with atmospheric moisture
	Resistant to abrasion; resistant to chemicals and solvents when cross linked

Catalyzed
    Aliphatic or aromatic isocya-
    nate reacted with polyester,
    epoxy, or acrylic polyhydroxyls
	Very good chemical resistance (similar to polyamide epoxy)

Silicones: two types with siloxane bond: (1) high temperature; (2) water repellent in water or solvent	Resistant (in aluminum formulation) up to 1200°F, and to weathering at lower temperatures; resistant to water
Water-base: aqueous emulsions of polyvinyl acetate, acrylic, or styrene-butadiene latex	Generally resistant to weather
Polyesters: organic acids condensed with polybasic alcohols. Styrene is a reactive diluent	Excellent resistance to acids and aliphatic solvents; good resistance to weathering
Coal tar: distilled coking by-product in aromatic solvent	Excellent resistance to moisture; good resistance to weak acids and alkalies, petroleum oils, and salts
Asphalt: solids from crude oil refining in aliphatic solvents	Some moisture resistance; good resistance to weak acids, alkalies, salts
Zinc-rich: metallic zinc in a vehicle of organic or inorganic type	Best resistance to galvanic and pitting-type corrosion

Limitations	Features--Uses
Moisture-cured types require humidity during application; may yellow under ultraviolet light; temperature resistance: 250°F dry, 150°F wet	Ease of application, fast drying, toughness, clear or colorful formulations, high gloss, and ease of cleaning; used on furniture and floors
Not recommended for immersion or exposure to strong acids/alkalies; temperature reistance: 225°F dry, 150°F wet. Quite expensive	Excellent gloss and color retention; used as a decorative coating of tank cars and steel in highly corrosive environments
High temperature type requires baking for good cure; water repellant; water-solvent formulations should be used on limestone, cement, and nonsilaceous materials; solvent formulations on bricks and noncalcareous masonry	Stable up to 1200°F; used on surfaces exposed to high temperature as water repellant; ease of application, low cost, porosity (for breathing), durability; used on masonry surfaces
Limited penetrating power because of water surface tension; may flash rust as a primer over bare steel; not suitable for immersion service	Ease of application, minimal odor, low cost, easy cleanup, compatible with other coatings; used in general applications
Alkalies and most aromatic solvents soften and swell these coatings; also, they have a short pot life and must be applied with special equipment	Inertness, tilelike appearance, good adhesive and cohesive strength; used as lining materials for tanks and chemical process equipment
Exposure to ultraviolet light and weathering will degrade these coatings, and temperatures of 100°F will soften them	Used on submerged or buried steel and concrete
Heavy dark color hides corrosion under the coating; softens at 230°F	Better weathering and softening characteristics than coal tar; used in aboveground weathering environments and chemical fume atmospheres
Must have top coat in severe environments or when pH is below 6 or above 10.5; requires clean steel surfaces, relatively difficult to apply and topcoat	Organic type more tolerant of surface preparation and topcoat; inorganic type more resistant to temperature (to 700°F) and abrasion; best primer for severely corrosive nonchemical environments

pheric resistances of the paint are dependent on the specific properties
of the film former. Types of film formers and their characteristics are
presented in Table 3, and some of the most common are further
described later in the chapter.

Pigments

Pigments function to reinforce the film structurally and to provide
color and opacity as well as to serve special purposes for metal protection.

Color or hiding pigments are selected to provide aesthetic value,
retention of gloss and color, as well as helping with film structure and
impermeability. Examples are iron oxides, titanium dioxide, carbon or
lampblack, and others.

Extenders lower the cost but are also beneficial in adding sag resistance to the liquid paint so that edges remain covered. In the dried
paint, they reduce the permeability to water and oxygen and provide
reinforcing structure within the film. Two extenders extensively
used are talc and mica, the latter limited to about 10% of the total pigment. Both, but mica in particular, reduce the permeability through
the film as platelike particles block permeation, forcing water and oxygen to seek a longer path through the binder around the particle.

Pigments, of course, must be compatible with the resin, and also
should be somewhat resistant to the environment (e.g., calcium carbonate, which is attacked by acid, should not be used in an acidic
environment). Water soluble salts are corrosion promotors, so that
special low-salt-containing pigments are used in primers for steel.

For special protective properties, primers contain one of three
kinds of pigments as follows:

1. *Inert or chemically resistant*. These are for use in barrier coatings in severe environments such as conditions below an acidity
   of pH 5 or above an alkalinity of pH 10; or as nonreactive extender
   hiding or color pigments in neutral environments.
2. *Active*. Leads, chromates, or other inhibitive pigments are used
   in linseed oil/alkyd primers.
3. *Galvanically sacrificial*. Zinc is employed at high concentrations
   to obtain electrical contact for galvanic protection in environments
   between pH 5 and 10.

Types and characteristics of these pigments are presented in Table 4.

Solvents

Organic solvents (water is considered either as a solvent or an emul-
sifier) usually are only required in order to apply the coating and,

after application, are designed to evaporate from the wet paint film. The rate at which the solvents evaporate strongly influences the application characteristics of the coating, and if the solvents are partially retained and do not completely evaporate, quite often the coating will fail prematurely by blistering or pinholing. As a general rule, the synthetic resins (vinyls, chlorinated rubbers, epoxies, etc.) are more polar in nature and therefore are more readily dissolved by polar solvents. However, polar solvents are most apt to be retained by a polar resin system, and therefore it is imperative when using such resins, particularly in immersion service, to allow sufficient time for the coating to cure or dry. Because such resins are more dependent on solvents for penetration and flow, they generally require a greater degree of surface preparation than do oleoresinous or oil-modified coatings.

Most coatings are formulated to be most successfully applied at ambient conditions of approximately 24°C (75°F) and 50% relative humidity. At ambient conditions considerably higher or lower than these optimum ranges, it is quite conceivable that the "solvent balance" should be changed for better coating application and solvent release. Generally, in colder weather, faster-evaporating solvents should be used; conversely, in hot weather, slower solvents are required. Classes and characteristics of some common solvents are listed in Table 5.

Additives

Most additives are formulated into a paint in relatively minor, often even trace amounts, to provide a specific function. For example, cobalt and manganese naphthanates are used as dryers for alkyds and other oil-base coatings to facilitate surface and thorough drying. These drying additives are added to the paint in amounts usually less than 0.1%.

Other additives are added for different purposes; for example, zinc oxide can be added to retard deterioration of the resin by heat and actinic rays of the sun. Mildew inhibitors (phenylmercury zinc and cuprous compounds) are commonly added to oil-base and latex paints. Latex paints (water emulsion) invariably have a number of additives acting as surfactants, coalescing aids, emulsion stabiliziers, freeze-thaw stabilizers, and so on. Vinyl paints often have a 1% carboxylic acid (generally maleic acid) modification to the vinyl resin to promote adhesion to metals. Conversely, a hydroxyl modification (generally an alcohol) aids in adhesion of vinyls to organic primers. The use of a particular additive may be crucial to the performance of the paint, and because additives are usually added in trace quantities, they may be most difficult to detect upon analysis of the paint.

Table 4. Characteristics of Pigments for Metal Protective Paints

Pigment	Density	Color	Opacity	Special contribution to corrosion resistance
			Active pigments	
Red lead	8.8	Orange	Fair	Neutralizes film acids; insolubilizes sulfates and chlorides; renders water noncorrosive
Basic silicon lead chromate	3.9	Orange	Poor	Neutralizes film acids; insolubilizes sulfates and chlorides; renders water noncorrosive
Zinc yellow (chromate)	3.3	Yellow	Fair	Neutralizes film acids; anodic passivator; renders water noncorrosive
Zinc oxide (French process)	5.5	White		Neutralizes film acids; renders water noncorrosive
Zinc dust at low concentration in coatings for steel	7.1	Gray	Good	Neutralizes film acids

Pigment	Sp. gr.	Color	Hiding	Function
Galvanically protective pigments				
Zinc dust sacrificial at high concentration	7.1	Gray	Good	Makes electrical contact; galvanically sacrificial
Barrier pigments				
Quartz	2.6	Nil	Translucent	Inert; compatible with vinyl ester additives
Extenders				
Mica	2.8	Nil	Translucent	Impermeability and inertness
Talc	2.8	Nil	Translucent	Impermeability and inertness
Asbestine	2.8	Nil	Translucent	Impermeability and inertness
Barytes	4.1	Nil	Translucent	Impermeability and inertness
Silica	2.3	Nil	Translucent	Impermeability and inertness
Colorants				
Iron oxide	4.1	Red		Impermeability and inertness
Iron oxide	4.1	Ochre		Impermeability and inertness
Iron oxide	4.1	Black		Impermeability and inertness
Titanium dioxide	4.1	White	Excellent	Impermeability and inertness
Carbon black	1.8	Black	Good	Impermeability and inertness

Note: Titanium dioxide has better "hiding" than any other pigment, by far.

Table 5. Characteristics of Solvent Classes

Class	Solvent name	Strength/solvency	Polarity	Density	Boiling range (°F)	Flash point of TCC	Evaporation rate[a]
Aliphatic	VM&P naptha	Low (32 KB)[b]	Nonpolar	0.74	246-278	52	24.5
	Mineral spirits	Low (28 KB)	Nonpolar	0.76	351-395	128	9.0
Aromatic	Toluene	High (105 KB)[c]	Intermediate polarity	0.87	230-233	45	4.5
	Xylene	High (98 KB)	Intermediate polarity	0.87	280-288	80	9.5
	High solvency	High (90 KB)	Intermediate polarity	0.87	360-400	140	11.6
Ketone	Methyl ethyl ketone (MEK)	Strong	High polarity	0.81	172-176	24	2.7
	Methyl isobutyl ketone (MIBK)	Strong	High polarity	0.80	252-266	67	9.4
	Cyclohexanone	Strong	High polarity	0.95	313-316	112	4.1
Ester	Ethyl acetate	Intermediate	Intermediate polarity	0.90	168-172	26	2.7
Alcohol	Ethanol	Weak	Intermediate polarity	0.79	167-178	50	6.8
Unsaturated aromatic	Styrene	Strong	Intermediate polarity	0.90			
Glycol ethers	Cellosolve	Strong	High polarity	0.93	273-277	110	0.3
	Butyl Cellosolve	Strong	High polarity	0.90	336-343	137	0.06

[a]Butyl acetate equals 1.

[b]KB, Kauri-Butanol; a measure of solvent power of petroleum thinners (ml of thinner required to produce cloudiness when added to 20 g of a solution of karigum in butyl alcohol).

[c]TCC-TAE closed cup.

Table 6. Typical Zone Defense System

	Environment	Recommendation
I.	Dry interior: building interior or outside exposure in an arid climate, etc.	Minimal hand or power tool cleaning; painting for color, aesthetics only
II.	Normally dry but exposed to weather: some bridges, tanks, building steel, etc.	Hand or power tool cleaning; alkyd- or oil-base coating system
III.	Frequently wet or exposure to high humidity: bridges, tanks, topside of ships, steel in paper mills, sewage treatment plants, etc.	Blast clean or pickle; alkyd or oil base with inhibitive pigments, zinc dust, zinc oxide paints
IV.	Continuously wet or immersion: tank interiors, cold steel, cooling towers, etc.	Blast clean: near white or white metal; phenolic, epoxy, urethane, vinyl, or zinc-rich paint
V.	Special: exposure to specific chemicals (acids, alkalies), heat, abrasion, etc.	Generally blast clean; specific coating for environment

PAINT SYSTEMS

A paint system consists of the surface preparation, pretreatment, if any, primer, and subsequent coats. In specifying a paint system, it is good practice to require a color difference for each successive coat of paint, and to specify the thickness of each coat. In milder environments (such as building interiors) a less thorough surface preparation and less resistant coating system may be specified compared to the surface preparation and coating types required for more severe environments. This philosophy is called the zone defense and a typical example is illustrated in Table 6.

SELECTION OF A SUITABLE COATING SYSTEM

Perhaps the first thing that should be done in the selection process is to define the environment around the structure or item to be protected. Is it predominantly wet (salt or fresh water?) or exposed to a chemical contamination (acid or alkaline?)? Is the atmosphere or environment uniform over the entire structure? Is the environment predominantly a

weathering environment subject to heat, cold, daily or seasonal
temperature changes, precipitation, wind (flexing), exposure to sunlight, or detrimental solar rays? If an exterior environment, are
there nearby chemical plants, pulp and paper mills, or heavy industries
that might provide airborne pollutants? Is color, gloss, and overall
pleasing effect more important than ultimate corrosion protection—or
are the normal grays, whites, and pastels of most corrosion—resistant
coatings acceptable?

All of these questions do not have to be answered in explicity detail, but the specifier should be generally aware of the environment the
protective coating is expected to serve in—even before paints are
chosen as the method of corrosion protection.

After the corrosive environment is categorized, a suitably resistant
coating can be selected. As mentioned previously, the resistance of
most coatings is determined by their organic constituents (resin or
binder). Characteristics of some of the more commonly used binders
are outlined in Table 3. This information, however, will not be in-
clusive enough to select a coating system, and the reader will require
further information. Such information can be obtained readily from a
number of sources, notably books and technical papers. Perhaps the
best information source will be the local sales representative of a paint
manufacturing company specializing in corrosion protective coatings.

While Table 3 provides a summary of the resistances, limitations,
and uses of most of the more common industrial/maintenance coating
resins, the following is a somewhat more detailed discussion of some of
the more common generic types.

Thermoplastic Solvent-Deposited Resins

Protective coatings formulated from these resins do not undergo any
chemical change from the time of application until the attainment of
final properties as a dried protective film. Film formation is achieved
by solvent evaporation, concentrating the binder/pigment solution and
causing it ultimately to precipitate as a continuous protective film over
the surface to which it had been applied. Shellacs and lacquers used
in wood finishing fit this description, together with other resin systems described below.

Vinyls

Vinyls are perhaps the best known and widely used resin for industrial
coatings in this class. Vinyl is a general term denoting any compound
containing the vinyl linkage ($-CH = CH_2$ group). This group, however, is contained by many compounds not commonly thought of as
vinyl coatings (such as styrene, diallylphthalate, vinyl toluene,
propylene, and many others in the ethylene family of olefins).

For the most part, vinyl coatings are considered to be copolymers of vinyl chloride and vinyl acetate copolymerized in approximately 86% vinyl chloride to 14% vinyl acetate. The chemical structure of a vinyl chloride/vinyl acetate copolymer is as follows:

Often approximately 1% of a carboxylic acid or anhydride (usually maleic acid or anhydride) is added to improve adhesion to metal surfaces. To improve adhesion to other previously applied coatings, 1% of hydroxyl modification is often added, usually in the form of an alcohol.

Because a relatively high amount of solvent (ketones and esters) is required to dissolve a vinyl copolymer high in vinyl chloride content, the volume of solids in solution is relatively low. As a result, most vinyl coatings must be applied thinly (1 to 1.5 mils per coat). Accordingly, a vinyl coating system may require five coats or more, and although protection generally is excellent in the proper environment, the system is considered highly labor intensive. High-build vinyls have been formulated, allowing application of the coating to 2 to 2.5 mils or more per coat. However, this advantage comes at the expense of lesser performance, as the thixotropes, fillers, and additives used to provide greater thickness are also more susceptible to environmental and moisture permeation.

Vinyls are widely specified in water immersion because of their extreme toughness and impermeability. However, it is important to ensure that essentially all solvents have evaporated prior to placing a vinyl-coated object in immersion service. The high polarity of the resin tends to retain the solvents used to dissolve the resin. Solvent evaporation is retarded, with the resultant effect of solvent voids within the vinyl coating, pinholes penetrating through a coat or more than one coat, and blistering caused by volatilization of retained solvents upon heating—or water-filled blisters due to hydrogen-bonding attraction of water by the retained solvents in the coating.

Chlorinated Rubber

Chlorinated rubber resins resulted originally from the chlorination of natural rubber. Today, however, the term also includes the chlorination of synthetic rubbers. The addition of chlorine to unsaturated double bonds occurs until the final product contains approximately 65%

chlorine. The chemical structure of a segment of the chlorinated
rubber resin is as follows:

The result is a hard, brittle material with poor adhesion and elasticity.
For use in surface coatings, a plasticizer must be used. Although a
number of materials can be used as plasticizers for chemical resistance,
a nonsaponifiable plasticizer usually of the chlorinated paraffin or
chlorinated diphenyl types are used. The type and amount of plasticizer in the chlorinated rubber plays a determining role in the final
resistances and properties of the chlorinated rubber paint. Chlorinated rubber resins are generally soluble in most organic solvents—
generally all but aliphatic hydrocarbons and alcohols. Chlorinated
rubber resins have good compatability with a variety of other resins,
including alkyds, phenolics (and medium to short oil resin modifications of each), acrylics, melamine and urea formaldehyde resins, and
many other natural or synthetic resins. The addition of these materials may enhance the ease of application, but in some cases may dimi-
nish chemical resistance. Unmodified chlorinated rubber resins are
generally formulated at a high molecular weight, and accordingly are
somewhat difficult to spray apply. "Cobwebbing" often occurs when
spraying, and brushing or roller application results in a noticeable
"drag." Volume solids of the coating is somewhat higher than a vinyl,
and a suitably protective chlorinated rubber system often consists of
only three coats.

Chlorinated rubber coatings are widely used on masonry surfaces
and as swimming pool paints. They, like vinyls, exhibit very good
adhesion after intial through-drying (usually 3 or 4 days to 2 weeks
after application); and the same care must be taken with respect to
solvent evaporation as described above for the vinyls. The high chlorine content in the resin accounts for its inertness, its good adhesion,
and very importantly, its fire-retardant nature.

Acrylics

Acrylics can be formulated as thermoplastic resins, thermosetting
resins, and as a water emulsion latex. The resins are formed from
polymers of acrylate esters—predominantly polymethyl methacrylate
and polyethyl acrylate. The acrylate resins do not contain tertiary
hydrogens attached directly to the polymer backbone chain—and as a
result are exceptionally stable to oxygen and ultraviolet light deterio-

ration. The repeating units of the acrylic backbone are joined to
make long polymer chains. The repeating units for the methacrylate
and the acrylate are as follows:

$$
\begin{array}{cc}
\underset{\text{Polymethylmethacrylate}}{
-CH_2-\overset{\displaystyle CH_3}{\underset{\displaystyle O=C-OCH_3}{C}}-
}
&
\underset{\text{Polyethylacrylate}}{
-CH_2-\underset{\displaystyle O=C-OCH_2-CH_3}{CH}-
}
\end{array}
$$

Acrylic resins, particularly the methacrylates, are somewhat resistant to acids, bases, weak and moderately strong oxidizing agents,
and many corrosive industrial gases and fumes. This resistance is
derived in major part from the fact that the polymer backbone has
only carbon atoms comprising it. However, pendant side-chain ester
groups, although quite resistant to hydrolysis, generally preclude the
use of these resins in immersion service, or strong chemical environments. The resins are generally soluble in moderately hydrogen
bonded solvents such as ketones, esters, aliphatic chlorinated hydrocarbons, and aromatic hydrocarbons. Acrylic resins are generally
quite compatible with most other resins (depending on the type of
acrylic), and the properties of many other resinous materials, such as
alkyds, chlorinated rubbers, epoxies, and amino resins are often
"modified" with an acrylic to improve application, lightfastness, gloss,
or color retention.

Thermosetting Resins

These resins differ from thermoplastic resins in that a chemical change
occurs after application and solvent evaporation. The coating is said
to "cure" as the chemical reaction takes place. The reaction can take
place at room temperature, or in the case of baking coatings, at elevated temperatures. The reaction is irreversible, and unlike a thermoplastic coating, high temperatures or exposure to solvents does not
cause the coating to soften or melt. Thermosetting resins used as
binders for industrial coatings have become increasingly important
for two predominant reasons:

1. They can be formulated and applied at low molecular weight with
    the adherent advantage of low solvent demand, relatively good
    penetration and wettability, and in-can stability and long shelf
    life (all but moisture-curing thermosetting resins are packaged as
    two or more components, and mixing of these components is neces-
    sary before the chemical reaction can take place).
2. Once applied and allowed to cure, the resultant film is generally a
    large macromolecule with a defined state of cross-linking density,

resulting in a flexibility and chemical resistance that can be
tailored for a given environment.

   The more tightly cross-linked thermoset coatings, as a general rule,
are more chemical and solvent resistant, but also are less flexible,
more brittle, and generally have poorer adhesion. Care must be taken
when topcoating an aged thermoset coating because solvents in the
freshly applied topcoat will not soften or swell the aged undercoat,
with the effect that poor intercoat adhesion may result. Some of the
more common thermoset resins are described below.

Epoxies

The epoxy resin itself is a common condensation product of epichlorhydrin and bisphenyl acetone.

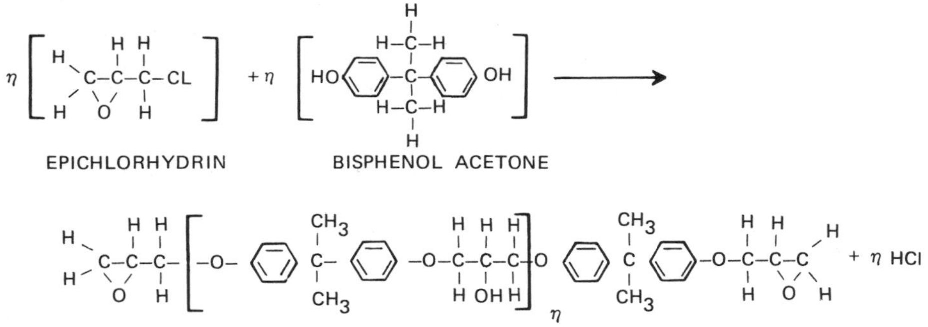

EPICHLORHYDRIN   BISPHENOL ACETONE

Epoxy resins themselves are not suitable for protective coating because
when pigmented and applied, they dry to a hard brittle film with very
poor chemical resistance. However, when suitably copolymerized with
other resins (notably those of the amine or polyamide family), or esterified with fatty acids, epoxy resins will form a durable protective coating. Epoxy resins copolymerized with amines and polyamides are discussed in this section. Esterified epoxy resins are discussed in the
next section under autooxidative cross-linking resins.
   The epoxy resin can react through pendant hydroxyl groups or the
terminal oxirane ring. The characteristics of the final film depend on
the molecular weight of the epoxy resins used, the coreacting resin,
and modifiers such as phenolic resins or coal tars.
   Amine resins, usually diethylene triamine or triethylene tetramine
or similar aliphatic polyamines, react to give a relatively highly cross-
linked chemically resistant but hard-curing and relatively inflexible
film. The following structure is that of an epoxy cross linked with
amine (e.g., diethylene triamine).

Active hydrogens from the amine nitrogen react to open epoxy rings forming hydroxyl groups, thereby cross-linking the nitrogen atom with the epoxy carbon. The hydroxyl groups may also open the epoxy ring to further cross-link and eliminate H_2O. Note that there are no ester links. Increased flexibility with only a slight loss of chemical resistance can be achieved by prereacting some of the epoxy resin with an aliphatic polyamine during paint manufacture. At the time of application, the prereacted resin is mixed with additional amine, and applied. Paints formulated in this way are called amine adducts.

Polyamide resins can also be reacted with epoxies to form durable protective coatings. These resins are somewhat bulkier than the amine by virtue of their fatty acid modification. Accordingly, they impart more flexibility to the cross-linked resin. Polyamide-cured epoxies are also more resistant to chalking, and are more receptive to topcoating after extended intervals than are amine-cured epoxies. Polyamide adducts can be made in a manner similar to that described for amine adducts. Cross linking takes place by opening the epoxy ring with active hydrogens from the polyamide nitrogen in the manner illustrated for amine epoxies. The reaction of polyamide with epoxy resin is as follows:

TYPICAL POLYAMID RESIN MOLECULE

R AND R' ARE ALKYL OR ARYL GROUPS

Ketimine-cured epoxies enable the application of a solventless or 95 to 100% high solids epoxy coating with standard spray application equipment. A ketimine under dry conditions reacts very slowly with epoxy resins, but in the presence of water or humidity, the ketimine reacts to decompose into a polyamine and a ketone.

Ketimine + Moisture ——→ Polyamine + Ketone

R, R1, and R2 are Alkyl Groups

The ketone evaporates, and the polyamine then reacts with the epoxy resin in a normal fashion. Ketimine-cured epoxies should not be applied to thicknesses greater than 8 mils or so in one coat in order to allow moisture access and complete curing.

Epoxy coatings, generally cross-linked with amines or polyamines, are widely used as heavy-duty moisture-and chemical-resistant coatings and linings in immersion and atmospheric fume environments. Amine- and polyamide-cured epoxies, when combined with approximately 50% or so of refined coal tar, comprise one of the least water permeable coatings available. Coal tar epoxies, however, because of the ultraviolet sensitivity of coal tar pitch, are normally not used in atmospheric exposures. However, for below-grade protection (e.g., buried pipelines) and in immersion service, they are considered excellent.

Other curing agents, such as imidazolines or any other resin with an active hydrogen, can be used to cross-link an epoxy resin. Such resins are of lesser commercial importance and will not be discussed.

Urethanes

Urethanes are reaction products of isocyanates with materials possessing hydroxyl groups, and simply contain a substantial number of urethane groups, regardless of what the rest of the molecule may be. Some isocyanate reactions are:

1. With hydroxyl-bearing polyesters, polyethers, epoxies, and so on:

$$R-N=C=O \;+\; R'-OH \longrightarrow R-\underset{H}{N}-\underset{O}{\overset{}{C}}-O-R'$$

ISOCYANATE   URETHANE
                                   LINKAGE

2. With an amine: $R-N=C=O \;+\; R'-NH_2 \longrightarrow R-\underset{H}{N}-\underset{O}{C}-\underset{H}{N}-R'$

                                                            A UREA

3. With an amide:

$$R-N=C=O \ + \ R'-\overset{\overset{\displaystyle H}{|}}{N}-\overset{\overset{\displaystyle O}{\|}}{C}-R'' \ \longrightarrow \ R-\overset{\overset{\displaystyle H}{|}}{N}-\overset{\overset{\displaystyle O}{\|}}{C}-\underset{\underset{\displaystyle R'}{|}}{N}-\overset{\overset{\displaystyle O}{\|}}{C}-R''$$

AN ACYL UREA

4. With moisture:

$$R-N=C=O \ + \ HOH \ \longrightarrow \ R-NH_2 \ + \ CO_2 \uparrow$$
$$+$$
$$R-N=C=O \ \longrightarrow \ A \ UREA$$

SEE 2 ABOVE

The polyol side (hydroxyl containing) may consist of a number of materials, including water (moisture-cured urethanes), as well as epoxies, polyesters, acrylics, and drying oils (discussed later in this chapter). Epoxy and polyester polyols are more chemically and moisture resistant (and somewhat more expensive) than acrylic polyols. The acrylic polyol, however, when suitably reacted to form a urethane coating is entirely satisfactory for most weathering environments. The isocyanate can be either aliphatic or aromatic. Up to the last six or eight years, aromatic isocyanates (often toluene diisocyanate and diphenylmethane diisocyanate) were most commonly used for urethane coatings. Generalizing, however, the aromatic urethanes have properties not significantly dissimilar from those of the epoxy family of coatings. However, the cost for urethanes is somewhat higher, and accordingly, widespread use of aromatic urethanes has not occurred.

However, this is changing rapidly as a result of the increasing availability of aliphatic isocyanates. While coatings formulated with aliphatic isocyanates are extremely expensive (approximately twice the cost of similarly formulated epoxy and aromatic urethanes), they have superior color and gloss retention and resistance to deterioration by sunlight. Whereas most other coatings will chalk or yellow somewhat upon prolonged exposure to sun, aliphatic urethanes continue to maintain a glossy, "wet" look years after application. Furthermore, they can be tinted readily, and a wide variety of deep and pastel colors are available. Care should be taken to ensure that the pigment used and the polyol-containing coreactant are light stable, or if not, used in small proportions. Aliphatic urethane protective coatings, due primarily to the high initial molecular weight of the aliphatic isocyanate, are generally formulated to be applied at somewhat lesser thicknesses (1.5-2.0 mils per coat) than similar aromatic urethane coatings.

Formulation of urethane coatings is important, as the isocyanate component will, to some degree, always react with moisture in the air. The reaction is accelerated by high humidities, and in the presence of sunlight and/or heat, results in the liberation of carbon dioxide gas. Poorly formulated coatings may foam, bubble, or gas, and the dried film may have numerous voids or pinholes.

Urethane coatings are not suitable for immersion service, or prolonged exposure to water or strong chemical environments.

Polyesters

The term *polyester* means many ester groups, and as such would include
alkyds, drying oils, and many ester-containing materials (ester group

$$-\overset{\overset{\textstyle O}{\|}}{C}-O-).$$ However, as used in the coatings industry, polyesters are
characterized by resins based on components that introduce unsatura-
tion ($-C\!=\!C-$) directly into the polymer backbone. The following
structure shows an isophthalic polyester resin:

n = 3 to 6

This unsaturation must be capable of direct addition copolymerization
with vinyl monomers (usually styrene). Most commonly, polyester
resins are polymerization products of maleic or isopthalic anhydride
or their acids. During paint manufacture, the polyester resin is
dissolved in the styrene monomer, together with pigment, and small
amounts of a reaction inhibitor. Additional styrene, and a free radical
initiator, commonly a peroxide, are packaged in another container. At
the time of application, the containers are mixed or, because of the
fast initiating reaction (short pot life), mixed in an externally mixing
or dual-headed spray gun. After mixing (and application) a relatively
rapid reaction occurs, resulting in cross linking and polymerization
of the monomeric styrene with the polyester resin. The resultant coating is highly cross linked, relatively brittle, and resistant to most
acids, mild alkalis, solvents, and so on. Pigmentation of sprayed polyester coating is important due to the high shrinkage of the resin system after application. Proper pigmentation reinforces the coating and
reduces the effect of the shrinkage. With spray-applied polyester coatings, a small amount of wax is often added to minimize surface evaporation of the volatile styrene monomer. Such a wax, if used, together
with a high degree of cross linking and solvent impermeability, may
interfere with the adhesion of subsequently applied coats.

Vinyl esters, a relatively new class of coatings said to have higher
temperature and greater acid resistance, are formulated in a similar
manner and have essentially the same application limitations. Instead
of using a polyester resin, the vinyl esters derive from a resin based
on a reactive end vinyl group which can open and polymerize. The

chemical structure of the vinyl ester is shown below (contrast with an isophthalic polyester shown previously).

Note that there are fewer ester groups in the molecular structure of a vinyl ester than a polyester. Additionally, there is less C=C unsaturation and a more symmetrical molecular structure, with less polarity. Accordingly, the vinyl ester has better moisture and chemical resistance, and is more stable than the polyester.

When applying both polyesters and vinyl esters, care should be taken to ensure that the surface is dry (particularly if applying over concrete) as moisture may inhibit the curing reaction. Additionally, excessive thicknesses should not be applied unless special reinforcing (such as a fiberglass cloth) is introduced. Two or three thin coats is better than one thick coat, and the attainment of a substantial surface roughness (a high anchor pattern of blast-cleaned steel, or a broom finished or otherwise roughened concrete surface) will be necessary for proper adhesion.

Autooxidative Cross-Linking Coatings

Coatings of this type all rely on the reaction of a drying oil with oxygen to introduce cross linking within the resin and attainment of final film properties. In every case, then, a drying oil is reacted with a resin, which is then combined with pigments and solvents to form the final coating. The formulated paint can be packaged in a single can. After opening the can, mixing the paint, and applying it, the solvents evaporate and the coating becomes hard. However, the attainment of final film properties may occur weeks or even months after application, as oxygen reacts with the coating, introducing additional cross linking.

For industrial maintenance paints, autooxidative cross-linking-type coatings as a class are very commonly used. Alkyds, epoxy esters, oil-modified urethanes, and so on, are frequently used in atmospheric service. These resin-fortified "oil-base paints" are widely used both in industry and around the house. By choice of an appropriate resin, and the use of the proper amount of a given drying oil, the properties of the coating can be varied considerably. The coatings can be formulated as air-drying or baking types, and with suitable pigments can be formulated to be resistant to a variety of moisture and chemical fume environments; and as well for application over wood, metal, and masonry substrates.

An example of an autooxidative coating is the epoxy ester, which is made as follows:

EPOXY RESIN   FATTY ACID   REACTION THRU HYDROXYL

EPOXY RESIN   FATTY ACID   REACTION THRU TERMINAL EPOXY

WHERE R IS A FATTY ACID SUCH AS LINOLENIC

OR A POLYBASIC ACID (MORE THAN ONE CARBOXYL GROUP)

The result is a large, bulky epoxy ester molecule with ester linkages in both the backbone and pendant side chains.

The major advantage of these coatings is their ease of application, their great versatility, excellent adhesion (wetting by virtue of the oil modification), relatively good environmental resistance (in all but immersion and high chemical fume environments), their widespread availability, and tolerance for poorer surface preparation than any of the coating systems based on synthetic resins (all described previously). The major disadvantage is the lessened moisture and chemical resistance compared with those coatings.

Water-Soluble/Emulsion Coatings

Water-soluble resins, except for some baking formulations, have not made inroads into the heavy-duty industrial or maintenance coating market. However, because of increasing governmental legislation restricting the amount of volatile organic solvents allowed in the coating formulation, or to be emitted by a prospective user, increasing attention will be given to these resins. Essentially, any type of resin can be made water soluble. Sufficient carboxyl groups are introduced into the polymer, giving it a high acid value. These groups are then neutralized with a volatile base such as ammonia or an amine, rendering the resin a polymeric salt, soluble in water or water/ether-alcohol mixtures. The major disadvantage to such resins (and the reason they are not yet widely used) is that polymers designed to be dissolved in water will remain permanently sensitive to it. Because of water's abundance in nature, such sensitivity is highly undesirable.

Water emulsion coatings, on the other hand, have gained wide acceptance and are now by far the most popular paint sold to homeowners and commercial painters.

Water-base latex emulsions consist of a high molecular weight resin in the form of microscopically fine particles of high molecular weight copolymers of polyvinyl chloride or polyvinyl acetate, acrylic esters, styrene-butadiene, or other resins. The water phase carries the pigments, plasticizers, and a variety of thickeners, coalescing aids, and other additives. Even a water-base epoxy has been manufactured and is commercially available based on emulsion principles. The epoxy resin and a polyamide copolymer are emulsified and packaged in separate containers. Upon mixing, coalescence, and ultimate drying, the polyamide reacts with the epoxy to form the final film.

Emulsion coatings form a film initially by water evaporation. As the water evaporates, the emulsified particles come closer and closer together until they touch each other. The latex particles, with the aid of a coalescing agent (usually a slow-evaporating solvent) ultimately merge to form a relatively continuous film. Because of irregularities in the physical packing of the emulsion particles, latex films are not noted for their impermeability. Also, initial adhesion may be relatively poor as the water continues to evaporate, and coalescing aids, solvents, and surfactants evaporate or are leached from the "curing" film. Later, from as little as a few days to as much as a few months after application, the latex coating attains its final adhesion and environmentally resistant properties.

The major advantage of water emulsion coatings is their ease of cleanup, their low atmospheric pollution (little or no solvent evaporation), the high inherent durability of the organic emulsion resins (due to their high molecular weight), and their wide compatibility with virtually all other coatings (except bituminous coatings—asphaltic and coal tar, which tend to bleed and discolor most other coatings—and wax- or grease-based coating systems). Disadvantages are the permanent water sensitivity of the coating, and accordingly their inability to be used in immersion or continually wet environments. Problems of initial adhesion exist, and water-base coatings cannot be applied to blast-cleaned steel surfaces with the same confidence as can solvent-based coatings.

Zinc-Rich Coatings

All the coatings described above owe their final film properties, corrosion protection, and environmental resistance in major part to the organic composition of their constituent resin or binder. The pigment, while playing an important role in the corrosion-inhibiting nature of these coatings, is secondary to the resistance of the organic binders.

On the other hand, the role of the pigment in a zinc-rich coating predominates, and the high amount of zinc dust metal in the dried film determines the coating's fundamental property—that of galvanic protection! Thus although many of the binder systems described above (chlorinated rubber and epoxies in particular) are formulated as zinc-rich coatings, the high pigment content dramatically changes the characteristics of the formulated coating.

Zinc-rich coatings are commonly subcategorized as organic or inorganic. The organic zinc-riches have organic binders, with polyamide epoxies and chlorinated rubber binders most common. Additionally, high molecular weight polyhydroxyl ether epoxy (phenoxy types) can also be used. Phenoxys are high molecular weight thermoplastic resins that do not have reactive terminal oxirane groups and are themselves thermally stable with a long shelf life. Upon solvent evaporation, the resin can be used by itself as a coating with no chemical conversion or reacted with a coconstituent.

Organic zinc-rich coatings can also be formulated with other binders, and formulations using alkyds and epoxy esters are widely used in the automotive industry (but are not recommended as suitable vehicles for field-applied, air-dried industrial or maintenance primers). Vinyl and styrene-butadiene resins also have been used for zinc-rich coatings, whereas some vinyl zinc-riches are still available, styrene-butadiene is no longer used.

Inorganic zinc-rich binders are based on silicate solutions, which after curing or drying, crystallize and form an inorganic matrix, holding zinc dust particles together and to the steel substrate. The ethyl silicate zinc-rich primer curing reaction is

$$2\begin{bmatrix} & OR & \\ RO{-}Si{-}OR \\ & OR & \end{bmatrix} + H_2O \longrightarrow \begin{matrix} OR & OR \\ | & | \\ RO{-}Si{-}O{-}Si{-}OR \\ | & | \\ OR & OR \end{matrix} + 2ROH \quad \text{Ethyl Alcohol}$$

Tetra ethyl ortho silicate
$(R = C_2H_2)$ Partially Hydrolyzed Teos

$$n\begin{bmatrix} OR & OR \\ | & | \\ RO{-}Si{-}O{-}Si{-}OR \\ | & | \\ OR & OR \end{bmatrix} + n\,H_2O \longrightarrow \begin{bmatrix} Si{-}O{-}Si{-}O{-}Si \\ | \quad | \quad | \\ O \quad O \quad O \\ | \quad | \quad | \\ O{-}Si{-}O{-}Si{-}O{-}Si{-}O \\ | \quad | \quad | \\ O \quad O \quad O \\ | \quad | \quad | \\ O{-}Si{-}O{-}Si{-}O{-}Si{-}O \\ | \quad | \quad | \\ O \quad O \quad O \end{bmatrix}_n + n\,ROH$$

Atmospheric Moisture   Ethyl Alcohol

Crosslinked Silicate Binder

The first zinc-rich coatings were postcured (by the application of heat or acid) water-based sodium silicate solutions; and later lithium, potassium, ammonium, and other alkali silicates were used. It is still felt by many that the postcured inorganic zinc-rich silicates provide the best binder and the longest protection of any zinc-rich primer. However, they are rather difficult to apply and somewhat labor intensive as, after application, a curing solution must be applied and—if top-coated—brushed off. Consequently, self-curing inorganic silicates have been developed, based on some of these same alkali silicates, and additionally, alkyl silicates (notably ethyl silicate). These organic silicates, upon curing, react with atmospheric moisture to form alcohol, which volatilizes. The resulting film is inorganic and essentially the same as that of the alkali silicate. Single-package inorganic resins have recently appeared on the market. Single-package materials are based on ethyl silicate or polyalcohol silicate binders. To be considered a zinc-rich, the common rule of thumb is that there must be at least 75% by weight of zinc dust in the dry film. Recently, however, this has been controversial, as conductive extenders (notably di-iron phosphide) have been sold to enhance weldability and burn-through, with supposed equivalent corrosion protection at lower zinc loadings. Resinous organic additives such as vinyl butyrate have been added to some inorganic formulations to form a hybrid zinc-rich between the organic and inorganic. These, together with the addition of conventional inhibitive pigments (such as the chromates, leads, etc.), may result in equivalent performance (particularly when topcoated) at lower zinc loading levels.

The major advantage of the use of zinc-rich coatings is that they protect galvanically. The zinc pigment in the coating preferentially sacrifices itself in the electrochemical corrosion reaction to protect the underlying steel (which is cathodic to zinc in the electromotive series). This galvanic reaction, combined with the filling and sealing effect of zinc reaction products (notably zinc carbonate, zinc hydroxide, and complex zinc salts), provide more effective corrosion protection to steel substrates than does any other type of coating.

Zinc-rich coatings, by virtue of the amphoteric nature of the zinc dust pigment, cannot be used in aggressive acid or alkaline environments (pH outside the range 6 to 10.5). Furthermore, because of their high pigment content, they are considered relatively porous, and are more difficult to topcoat (particularly the inorganics) than are conventionally pigmented coatings. Because of the galvanic protective nature, intimate contact with the underlying steel must be attained, thus resulting in the requirement for very thorough blast cleaning or pickling prior to zinc-rich application. The high pigment weight in the coating makes spray application more difficult than most organic coatings, and additionally increases the possibility of "mudcracking" at excessive thicknesses. Generally, thicknesses above 5 mils per coat are not recommended.

COATING APPLICATION, INSPECTION, AND MAINTENANCE

Once the coating has been selected, it is imperative that it be applied
properly to attain its maximum corrosion protection benefits. This is,
of course, where the real work begins! The definition or determination
of the environment and even the selection of the coating can, for the
most part, be done in the safety and comfort of one's office. However,
when it comes to coating application, the work must be done at the
fabricating shop, plant site, or piece of equipment, tank interior,
or whatever, that is to be coated. Here also, for the most part,
the comfort vanishes and the safety is diminished. Surface preparation (often blast cleaning) and coating application is, even when
properly done, dirty, dusty, demanding work. Proper access,
lighting, heating, and ventilation must be provided, and workers must
have safe work practices and equipment for the work intended. Only
when the proper equipment is on the job, and trained, skilled workers
are available with proper supervision to conduct the actual coating
work, should the coating application sequence of operations begin.

Initially, a "pre surface-preparation inspection" should be made.
This inspection is to determine if additional work need be done by
other crafts prior to the start of surface preparation for painting.
Such other work might include grinding and rounding of edges and
welds; removal of weld spatter, heavy deposits of oil, grease, cement
spatter, or other contaminants; moving equipment out of the work
area; masking or otherwise protecting equipment or items not to be
painted in the work area; and other such preliminary activities. After
this is done, the painters can then begin to work in earnest.

Coating application is in the most basic terms the preparation of the
surface in order to receive paint and the application of the paint in a
proper manner to the specified thickness. The surface preparation
specified is predicated by the coating system to be applied. It is usually a good idea to specify a "standard" surface preparation method.
The most common standard methods are those defined by the Steel
Structures Painting Council.* Table 7 summarizes the SSPC surface
preparation methods. These standards, and others prepared by the
National Association of Corrosion Engineers, Society of Naval Architects and Marine Engineers, various highway departments, and private
corporations are in virtually every case final appearance standards.
The standards give the desired end product but do not describe in
detail the means to achieve this end. It is important, therefore, that
the painter or party doing the surface preparation be knowledgeable.
It is important that the various pieces of equipment be sized properly;
that air and abrasives (if used) be cleaned, graded, and free of
moisture, oil, and other contaminants; and that ambient conditions be
controlled, or if not, at least closely monitored.

---

*Steel Structures Painting Council, 4400 Fifth Avenue, Pittsburgh,
Pennsylvania 15213.

Table 7. Summary of Surface Preparation Specifications

SSPC Specification	Description
SP1; Solvent cleaning	Removal of oil, grease, dirt, soil, salts, and contaminants by cleaning with solvent, vapor, alkali, emulsion, or steam.
SP 2; Hand tool cleaning	Removal of loose rust, loose mill scale, and loose paint to degree specified by hand chipping, scraping, sanding, and wire brushing.
SP 3; Power tool cleaning	Removal of loose rust, loose mill scale, and loose paint to degree specified, by power tool chipping, descaling, sanding, wire brushing, and grinding.
SP 5; White metal blast cleaning	Removal of all visible rust, mill scale, paint, and foreign matter by blast cleaning by wheel or nozzle (dry or wet), using sand, grit, or shot. (For very corrosive atmospheres where high cost of cleaning is warranted.)
SP 6; Commercial blast cleaning	Blast cleaning until at least two-thirds of the surface area is free of all visible residues. (For rather severe conditions of exposure.)
SP 7; Brush-off blast cleaning	Blast cleaning of all except tightly adhering residues of mill scale, rust, and coatings, exposing numerous evenly distributed flecks of underlying metal.
SP 8; Pickling	Complete removal of rust and mill scale by acid pickling, duplex pickling, or electrolytic pickling.
SP 10; Near-white blast cleaning	Blast cleaning nearly to white metal cleanliness, until at least 95% of the surface area is free of all visible residues. (For high humidity, chemical atmosphere, marine, or other corrosive environments.)

The same is true for coating application. The coating thickness should be achieved in one or more coats (usually three), and care should be taken to assure that there is no contamination between coats; that recoating is done within the proper time frame; and that the coating materials are properly mixed in clean, well-functioning equipment and are applied to controlled thicknesses. All reputable coating contractors are well versed in this knowledge and have the proper equipment and trained personnel available. Unfortunately, it is the irresponsible, "fly-by-night" painting contractor that has given many reputable contractors in the business a bad name. Often the relatively common practice of taking the low bid has, in the painting industry, very unfortunate ramifications. This is due to the ease of entry into the business (often a contractor will rent equipment and hire workers from a local hiring hall after being awarded a lowest price contract). Additionally, without adequate supervision or inspection, surface preparation can be skimped on, paint coats omitted or applied at low thicknesses, and in general, the work poorly done. The old adage "A painter covers his mistakes" is often all too true, and paint jobs that are intended to last 10 to 15 years or more with occasional touchup or repair sometimes fail within 2 to 3 years.

No matter what effort is put into defining the environment and selecting the proper paint, the most important part of the job is ensuring that the paint is applied properly. It is better to have less effective paint applied properly than the best paint poorly applied.

There are a wide variety of inspection instruments, aids, and standards that can be used to assess the quality of the coating system. These include devices for assessing the cleanliness of a prepared surface; the depth of a blast-cleaning anchor pattern or profile; and various magnetic, eddy current, and destructive thickness gages (capable of measuring the total paint thickness or the thickness of each coat in a multicoat system). Additionally, there are instruments and techniques available to monitor temperature, humidity, and dew point on a continual basis. After application, adhesion tests and holiday tests (for pinholes and other discontinuities) may also be specified.

Finally, it is important to realize that a paint job, even if properly done, does not last forever. Within the first six months or year after application, inadvertent misses, thin spots, or weak areas in the coating system can often be observed by simple visual inspection. Most reputable contractors will warrant work for a 1-year period, and a thorough inspection, and repair, if necessary, should be accomplished at the end of this time.

Later, as the coating breaks down and deteriorates due to the effect of the environment, scheduled inspections are desireable to assess the extent and rate of coating breakdown. Spot touchup repair should be accomplished at localized areas of failure prior to drastic deterioration of the entire coated surface. If the "planned maintenance approach" of periodic spot touchup, followed occasionally by a full coat over the entire coated surface, is done, the extensive costs of total surface preparation (such as complete blast-cleaning removal of all old coating) can be avoided, sometimes for periods approaching 30 years or more.

It is important that the entire coating sequential flow be maintained from beginning to end—definition of the environment, selection of the proper coating system, proper surface preparation and application, inspection, and periodic maintenance and repair. If all these operations are performed intelligently and competently, corrosion protection by coatings can be economically achieved for long periods.

8
Linings

8.1
Liquid-Applied Linings

DEAN M. BERGER / Gilbert/Commonwealth, Reading, Pennsylvania

INTRODUCTION

Liquid-applied linings are coatings which may be spray applied or troweled. They require the best surface preparation possible, Steel

Structures Painting Council (SSPC) SP5 white metal blast cleaning or
National Association of Corrosion Engineers (NACE) 1. The coatings
should be applied under ideal conditions with full inspection procedures
employed. These requirements are universal for any tank-lining application, owing to the severity of constant immersion (or partial immersion) that the tank lining must resist. All other types of coating applications are less severe, because they will alternate between corrosive
contact and periods of little or no corrosive contact, thereby providing
a recovery or cleaning period for the coating surface. In a tank lining,
there are usually four areas of contact with the stored product that
may lead to different types of corrosion. These areas are: the vapor
phase (the area above the liquid level), the interphase (the area where
the vapor phase meets the liquid phase), the liquid phase (the area
always immersed), and the bottom of the tank (where moisture and
other contaminants of greater density may settle). Each of these areas
can, at one time or another, be more severely attacked than the rest,
depending on the type of material contained, the impurities present,
and the amounts of oxygen and water present.

CRITERIA FOR TANK LININGS

All lining materials and tanks require intensive engineering, and various criteria must be carefully established. These criteria are listed in
Table 1.

The coating of interior surfaces of steel or concrete tanks, pipe,
and equipment is most demanding for coating application expertise,
engineering design, and material performance. Most lining materials
are selected to prevent iron contamination of the product during the
processing operation. Strict application procedures must be followed
to provide maximum performance.

Lined tanks are used when chemicals are transported or stored.
Barges, hopper cars, tank cars, drums, bins, and storage tanks may
all have chemically resistant liquid-applied linings. But how does the
engineer select the proper lining for a given chemical?

Experience over many years has produced a considerable quantity
of data concerning the chemical resistance of liquid-applied lining
materials. However, unless the application is exactly similar to one on
which data are available, it is wise to test the proposed lining before
specifying it.

Proposed linings should be tested as prescribed in the National
Association of Corrosion Engineers procedure, NACE TM-01-74 [1].
Another NACE publication, NACE TPC 2, *Coatings and Linings for
Immersion Service* [2], should be used as a guide for material selection
and for inspection practices.

Two types of tests are described in NACE TM-01-74: one-side testing and immersion testing. One-side testing consists of exposing one

Table 1. Criteria for Tank Linings

1.	Design of the tank
2.	Coating selection
3.	Tank construction
4.	Surface preparation
5.	Coating application
6.	Cure of the lining material
7.	Inspection
8.	Safety
9.	Performance history and failure analysis
10.	Operation instructions and temperature criteria

side of a coated flat panel to the chemical solution, which is contained in a special glass flask. The test coupons are bolted to the glass flanged cylinder, using a double-flange ring. This test cell is fitted with a heating mantle to control the temperature of the solution.

To provide product protection, the tank-lining coating must be impervious to surface attack and must not react with any of the materials to be contained in the tank. In the case of food products, the lining must not impart any taste, smell, or other harmful ingredient to the contained product. For these special applications, it is absolutely necessary that the surface coating in contact with the food product be manufactured and tested in accordance with Food and Drug Administration (FDA) Regulation 121.2514. To meet this requirement, all components of the coating have to be selected from an FDA-approved list of nontoxic coating ingredients. In addition to this, the formulated coating must resist certain extraction tests as specified by FDA or the Environmental Protection Agency. The American Water Works Association governs coatings for potable-water storage tanks [3].

Tank Design

In designing a steel tank that is to receive a lining, always specify welded construction. Riveted tanks will expand or contract and damage the lining, causing leakage. Butt welds should be ground smooth. If overlapping is used, specify a fillet weld and be sure that all sharp edges are ground smooth. All weld splatter must be removed before blast cleaning. A good way to judge a weld is to run your finger over the surface. Sharp edges can be detected easily.

All surfaces must be readily accessible for proper coating application. Where this is not possible, special caulking compounds may be required. These should be applied prior to the initial lining application.

Tanks larger than 25 ft in diameter may require three manways for working entrances. These are usually located two at the bottom and one at the top. The minimum opening diameter should be 20 in. but 30-in. openings are preferred. The two bottom openings should be side-mounted 180° apart. This will facilitate ease of ventilation.

Avoid the use of bolted joints. Butt welded joints are preferred. Lap-welded joints should be properly filled. If bolted joints are used, they should be made of corrosion-resistant materials and sealed shut. The mating surface of steel surfaces should be gasketed. The lining material should first be supplied.

Other appurtenances inside the tank must be located for accessibility of coating application. Also make sure that the processing liquor is not directed against the side of the tank but toward the center.

Heating elements should be placed with a minimum clearance of 6 in. Baffles, agitator base plates, gage devices, pipe, ladders, and other devices can either be coated in place or detached and coated prior to installation. The use of complex shapes such as angles, channels, and I beams should be avoided. Sharp edges should be ground smooth and they should be fully welded. Structural reinforcement should be installed on tank exteriors.

Tanks that receive heat-cured linings should have bottoms permanently insulated prior to erection.

Sharp edges and fillets must be ground smooth to a 1/8- or 1/4-in. radius. All welding should be of the continuous type. Spot welding or intermittent welding should not be permitted. Gages, hackles, deep scratches, slivered steel, or other surface flaws should be ground smooth.

Reinforcement pads, structural support members, and penetrations should be mounted on the exterior surface wherever possible. All protrusions should be ground smooth and small apurtures and pipe should be previously lined. Instructions for designing steel tanks are available from NACE [4].

Concrete tanks require special coating systems. Concrete tanks should be located above the water table. In designing a concrete tank, expansion joints should be avoided unless absolutely necessary. Normally, small tanks do not require expansion joints. In larger tanks, a chemical-resistant joint, such as PVC, has been used successfully. The concrete curing compound must be compatible with the coating system or be removed prior to coating. Form joints must be made as smoothly as possible to provide a flat surface for the coating. A strong, dense, concrete mix should be used, with adequate steel reinforcement, to reduce movement and cracking of the tanks. The coating manufacturer should be consulted for special instructions.

Concrete tanks generally are lined only by a licensed applicator, and special surface preparation is required.

Coating Selection

When selecting the proper coating system, it is necessary to determine all the conditions to which the lining will be subjected. An assessment of these conditions would involve consideration of percentages of all materials to be contained, operating temperature, degree and type of abrasion, impact, thermal shock, and whether or not the tank will be insulated for high-temperature service. If a detail in any one of these areas is neglected, or considered insignificant and not emphasized, an incorrect recommendation could result. Examples of some of these variables are a fluctuation in temperature or the presence of some minute, strong corrosive or penetrant, such as free chlorine in chlorinated solvents.

Field performance, case histories, and ease of application are important factors. Most corrosion engineers keep records of the performance of coating systems they have used in a particular service [1]. Such information is often available from the coating manufacturer. Included in the case history should be the name of the applicator who applied the coating, application conditions, type of equipment used, degree of application difficulty, and other special procedures required. A coating with outstanding chemical resistance will fail rapidly if it cannot be properly applied, so it is advantageous to gain from the experience of others.

Where no previous history exists prior to use, tank-lining materials should be tested in accordance with NACE Standard TM-01-74 [1]. This standard provides a procedure for immersion testing of the total panel or one side of the panel. The glass atlas test cell may contain any liquid, and partial immersion is also available. After testing under these conditions, one may determine the type of lining material suitable for the selected service.

Sometimes, case histories will be unavailable, or manufacturers will be unable to make recommendations, as when the material to be contained is of a proprietary nature or involves a solution that contains unknown chemicals. When this occurs, tests should be conducted by evaluating sample panels of several coating systems for a minimum of 90 days. A 6-month test would be preferable; however, due to normal time requirements, 90 days is a standard. It is important to test at the maximum operating temperatures to which the coating will be subjected. Such tests should simulate actual operating conditions, including washing cycles, cold wall, and effects of insulation.

Before applying a tank lining, a recommendation should be obtained from the coating manufacturer, a technically capable lining contractor, or from NACE literature for an identical problem. Table 2 shows various types of coatings available and their general area of application.

Table 2. Polymeric Linings and Their Use

Type of Coating	Use
Inorganic zinc water-base postcure	Jet fuel storage tanks, petroleum products
Inorganic zinc water-base self-cure	Jet fuel storage tanks, petroleum products
Inorganic zinc coatings (solvent-based, self-curing)	Excellent resistance to most organic solvents (i.e., aromatic, ketone, and hydrocarbons); excellent water resistance; poor acid and alkali resistance; difficult to clean; often sensitive to decomposition products of materials stored in tanks
Graphite alkyl silicates	Acid service
Acid-resistant cements	Acid service—processing tanks chemical treatment and acid pickling tanks
Furan	Mortar for acid brick—most acid-resistant organic polymer; stack linings and chemical treatment tanks
High-bake phenolic	Most widely used lining material; excellent resistance to acids, solvents, food products, beverages, and water; poor flexibility compared to other coatings
Modified air-dry phenolics (catalyst required)	May be formulated for excellent resistance to alkalies, solvents, fresh water, salt water, deionized water; mild-acid resistance; nearly equivalent to high-bake phenolics; excellent for dry products
Epoxy (amine catalyst)	Good alkali resistance; fair to good resistance to mild acids, solvents, and dry food products; widely used for covered-hopper-car linings and nuclear containment facilities
Epoxy polyamide	Poor acid resistance, fair alkali resistance; good resistance to water and brines; used in storage tanks and nuclear containment facilities
Epoxy polyester	Poor solvent resistance, good abrasion resistance; used for covered hopper-car linings

Table 2. (Continued)

Type of Coating	Use
Epoxy coal tar	Excellent resistance to salt water, fresh water, mild acids, and mild alkalies; solvent resistance is poor; used for crude oil storage tanks, sewage disposal plants, and water works
Coal tar	Excellent water resistance; used for water tanks
Asphalts	Good water and acid resistance
Polyester unsaturated	Excellent resistance to strong mineral and organic acids and oxidizing materials; very poor aromatic solvent and alkali resistance
Vinyl ester	Excellent resistance to strong acids and better resistance to elevated temperatures up to 350 to 400°F, depending on thickness
Modified polyvinyl chloride (polyvinyl chloracetates), air cured	Excellent resistance to strong mineral acids and water; Poor solvent resistance; used in water immersion service, potable and marine; most popular lining for water storage tanks (beverage processing)
Polyvinyl chloride (PVC) plastisols	Popular acid-resistant lining; must be heat-cured
Chlorinated rubber	Excellent water resistance; poor solvent resistance; used in marine applications and for swimming pools
Chlorosulfonated polyethylene (Hypalon)	Chemical salts
Polyvinylidene chloride	Excellent vapor barrier; good general chemical resistance; used in food packaging
Neoprene	Good acid and flame resistance; chemical processing
Polysulfide (Thiokol)	Good solvent resistance and water resistance; used for caulking and lining jet fuel tanks

Table 2. (Continued)

Type of Coating	Use
Butyl rubber	Good water resistance; used for caulking
Fluoropolymers	Lining for SO_2 scrubber service; high chemical resistance and fire resistance
Styrene-butadiene polymers	Food and beverage processing; concrete tanks
Acrylic polymers	Food and beverage processing
Urethanes	Excellent resistance to strong mineral acids and alkalies; fair solvent resistance; superior abrasion resistance; used to line dishwashers and washing machines; used to modify asphalts
Vinyl urethanes	Wood tanks, food processing, hopper cars
Rubber latex	Excellent alkali resistance; used to line 50 and 73% caustic tanks 180° to 250°F
Vinylidene chloride latex	Excellent fuel oil resistance; concrete fuel tanks
Alkyds, epoxy esters, oleoresinous primers	Water immersion service; primers for other top coats
Waxes and grease	Coating systems for water storage

Descriptions of the chemical compositions and application properties of some of the most popular coating materials can be found in NACE TPC 2. This NACE publication also lists the chemical resistance of some of these coatings. Recommendations are available for specific chemical use.

Once the coating has been selected, the various manufacturers should be consulted. Often recommendations are available as to the choice of coating applicator. A pre-job conference should be held between the owner, the engineer, the applicator, coating supplier, and the job inspector. All lining work should receive 100% inspection.

Tank Construction

In the design section, several features of construction were discussed. It is important that the coating applicator check out the tank prior to painting. Such items as sharp edges and rough welds could be overlooked. All of these should be ground smooth before the applicator starts.

Sometimes certain parts of the tank, such as a bottom plate for a center post, need to be dismantled and painted separately. This particular section would then be reassembled after the tank is blast cleaned and coated.

Penetrations are frequently coated prior to assembly even though welding may burn the lining.

For every steel-tank-lining application, all seams and joints must have continuous interior welds. All sharp edges and welds must be ground smooth to a rounded contour, and all weld splatter removed. It is not necessary that the welds be ground flush, as long as they are smooth and continuous with a rounded contour. Any sharp prominence may result in a spot where the film thickness will be inadequate and noncontinuous, and thus cause premature failure.

Surface Preparation

The key to all coating application work is in obtaining maximum adhesion of the lining material to the substrate. The basic requirement is to get the surface absolutely clean.

All steel surfaces to be coated must be abrasive-blasted to white metal in accordance with SSPC Specification SP5-63 or NACE Specification NACE 1. A white-metal blast is defined as removing all rust, scale, paint, and so on, to a clean white metal, which has a uniform gray-white appearance. No streaks or stains of rust or any other contaminants are allowed. Frequently, a near white blast-cleaned surface equal to SSPC SP-10 is used. This is satisfactory for most lining materials and is more economical. Written surface preparation standards are available from the SSPC [5].

A guide for the choice of abrasives is found in SSPC Volume I [6] or NACE TPC 2.

After blasting, remove dust and spent abrasive from the surfaces by brushing or vacuum cleaning. All workers and inspectors in contact with the blasted surface should wear clean, protective gloves and clothing to prevent contamination of the blasted surface. Any contamination may cause premature failure by osmotic blistering or adhesion loss. The first coat should be applied as soon as possible after the blasting preparation is finished and always before the surface starts to rust. If the blasted steel changes color, or rust bloom begins to form, it will be necessary to reblast the surface. Dehumidifiers and temperature controls are helpful.

Be certain that no moisture or oil passes through the compressor and on to the blasted surface. Use a white rage to determine the air quality. Also a black light may be used to identify oil contamination. Rotary-screw, two-stage, lubrication-oil-free compressors are available to provide 100% oil-free air.

To check visually the quality of the surface preparation, a number of standards are available for comparative purposes. [7-11].

Concrete

Concrete surfaces must be clean, dry, and properly cured before the coating application is performed. Remove all protrusions and form joints. All surfaces must be roughened by blasting. The surface preparation must remove all loose, weak, or powdery concrete to open all voids and provide the necessary profile for mechanical adhesion of the coating. Remove dust by brushing or vacuuming. Special priming and caulking methods are often required for concrete surfaces. Follow the coating manufacturer's recommended procedures in these cases.

Application

Coating application can be a problem, particularly where inexperienced applicators are applying the coating. It is important that the applicator be very familiar with the coating to be applied. Emphasis should be given to choosing a professionally competent coatings applicator.

Often, the lowest bidder is selected without adequately considering the quality of workmanship, with the net result of a tank-lining failure. Since a tank lining requires a nearly perfect application, a knowledgeable and conscientious applicator is needed. Evaluate the applicator before awarding the coating contract, to assure that the tank-lining contractor is experienced in applying the recommended lining.

In reviewing the qualifications of an applicator, ask what jobs he has done using the selected lining material and check out his references. Ask what equipment he has available and, if possible, visit his facilities and inspect his workmanship prior to placing him on the applicators bid list. These precautions will be rewarded in assuring total performance.

The primary concern is to deposit a void-free film of the proper thickness on the surface. Any area that is considerably less than the specified thickness may have a noncontinuous film. Also, pinholes in the coating may cause premature failure.

Films that exceed the maximum recommended film thickness may entrap solvents, which can lead to improper cure, excessive brittleness, bad adhesion, and subsequent poor performance. Avoid "dry spraying' of the coating, as this causes it to be porous. Never use thinners other than those recommended by the coating manufacturer; they, too, may cause poor film formation.

There are many types of equipment for determining the thickness of the coating on steel. During the application, a Nordsen or Elcometer wet-film thickness gage should be used as a guide for obtaining the desired dry-film thickness. If the wet-film thickness meets the recommendations, this is a good indication that the dry-film thickness will also meet the requirements.

To measure dry-film thicknesses, all gages must be calibrated prior to use by the manufacturer's suggested technique. The samplings should be numerous and at random, paying particular attention to the hard-to-coat areas. Particularly useful instruments are the Mikrotest and Elcometer. (Remember that the Elcometer has to be prestandard-ized on the tank structure itself prior to obtaining measurements.) Film thickness measurements should be taken in accordance with SSPC PA2-73T [12].

The best method for calculating film thickness on concrete tanks (also useful on steel tanks) is to measure the area covered and the material consumed, to see that the proper coverage as shown by the manufacturer is met. (You must take into consideration the material losses during application and mixing.) Although destructive, a Tooke gage will determine the approximate dry-film-coating thickness on concrete.

Safety Precautions During Application

All applicators are aware of the hazards within their own working environment. Tank-lining work involves more specific requirements and the use of specialized rigging and equipment. The Occupational Safety and Health Act of 1970 provides many requirements. These are found in the *Federal Register* 39(122), June 24, 1974. An analysis of these data and a recent review of the pertinent information is readily available from the Painting and Decorating Contractors of America (PDCA) [13].

Safety is of prime importance in any type of job, particularly where protective coatings containing solvents are being used. Obviously, all coatings and thinners must be kept away from any source of an open flame. This means that "no smoking" must be the rule during coating application. Welding in areas adjacent to that of coating application must be discontinued.

Vapor concentration inside tanks should be checked regularly to make sure that the maximum allowable vapor concentration is not reached. This lack of air observation caused five deaths in the United States in 1975. An explosion from solvent vapors will occur only within a small range of concentration—for most solvent vapors, there must be between 2 and 12% vapors in the air for an explosion to happen. As long as the vapor is kept below the lower level, there will be no explosion. All electrical equipment must be grounded and the sandblasting

unit equipped with a deadman control. Precautions must also be taken
on the exterior of the tank, becuase the flammable solvents are being
exhausted and will travel a considerable distance at ground level. No
flames, sparks, or ungrounded equipment can be nearby.

The Occupational Safety and Health Administration (OSHA) issues
a form called the Material Safety Data Sheet. The information contained
on this form is supplied by coating manufacturers. They are required
to list all toxicants or hazardous materials and provide a list of the solvents used. Included also would be the threshold limit value (TLV)
for each substance. TLVs are published annually for all chemicals [14].
Explosive hazards, flash points, and temperature limits will be established for safe application of each lining material.

Basically, all tank-lining work requires proper air supply and ventilation. Never take chances or try to skimp on the required equipment. Never allow one worker in a tank alone. Use safety belts and
all the necessary rigging for maintaining safe working conditions.
Recommended ventilation for various size tanks has been discussed by
Weaver [15] and Munger [16]. All the requirements of safety are
listed and discussed in NACE TPC 2 [2] as well as in SSPC Volume I
[6].

Scaffolding and ladders must be carefully checked to ensure that
they are sturdy and sound. Blasters must wear protective clothing,
and sprayers should wear fresh airline respirators in confined areas
such as tank interiors. Some coatings contain ingredients that may
be irritating to some people (particularly the strongly alkaline amine-
type catalysts used in epoxies).

Free toluene diisocyanate (TDI) or isocyanate monomer will also
cause severe skin irritation. All workers spraying these materials
should wear protective creams on exposed parts of the body such as
face and hands, or complete protective clothing. Water must always
be readily available for flushing accidental spills of such materials
from the skin. Safety rules of the plant where materials are being
applied must be known and observed. OSHA Material Safety Data
Sheets should be on file on the first-aid station and in the job super-
intendent's office.

Estimating

Prior to application, the coating applicator must estimate the cost.
Usually, this can be done rapidly by experienced people. In addition
to information available from SSPC and NACE, an estimating guide is
published annually by the PDCA [17]. Estimates are always relative
and based on the total surface area to be coated and the complexity of
the tank. Safety, surface preparation, material, and application costs
are basic requirements. Location of the job, rigging, and logistics
must be considered. Beware of low bidders. Make sure of the final

contract and try to obtain a performance guarantee. Guidelines for
Proposal-Contract-Specification Forms are available from the Georgia
Municipal Association [18].

Curing the Applied Coating

During and after application, thorough air circulation must be provided until the coating is cured. To obtain proper air circulation, no
tank should be lined unless it has at least two openings. The air circulation should be accomplished by a fresh air intake [temperature
over 10°C (50°F) and relative humidity less than 89%] at the top of the
tank, and an exhaust at the bottom, fed by forced-air fans whenever
possible. This requirement is necessary becuase the solvents used in
coatings are almost always heavier than air; therefore, proper exhaust-
ing can only be obtained in a downward direction.

The curing time and temperature of the substrate must be in accordance with the manufacturer's application instructions for each individual product, to prevent solvent entrapment between coats and to
ensure a proper final cure. Do not allow application to take place at
temperatures below those recommended by the manufacturer. For a
faster and more positive cure of all linings, a warm forced-air cure
should be used between coats and as a final cure prior to placing the
lining into service. This added heat provides a dense film and tighter
cross linking, which yields superior resistance to solvents and moisture permeability. Prior to placing the tanks in service, most linings
should be thoroughly washed down with water to remove any loose
overspray. For linings in contact with food products, a final warm-
forced-air cure and wash is essential.

It is most important that linings be allowed to attain a full cure
prior to use. This frequently requires 3 to 7 days for room-temperature-cured coatings.

Inspection of the Lining

It is necessary in many instances to check for porosities. The first
visual inspection is mandatory to detect pinholing and provide recoat
instructions. After repairs of visible defects, inspection may be done
by using low-voltage (75 V or less) detectors that ring, buzz, or light
up to show electrical contact through a porosity to the metal or concrete surfaces. It is preferable to check the lining in this manner
after the primer or second coat, so that such areas may be touched up
and be free of porosities prior to the final top coat [1]. If a wet-
sponge detector is used, any surface contamination must be removed
prior to application of a topcoat.

The inspector should participate in the following functions: (1)
prior work conference, (2) pre-job inspection, (3) surface preparation

Table 3. Inspection Report Items for Applied Coatings

Item	Item	Tank linings	Concrete surfacer	Concrete top coats	Inorganic zinc primer	Organic primer steel	Organic top coats steel
1	Pinholes	X	X	X			X
2	Blisters	X	X	X			X
3	Color and gloss uniformity			X			X
4	Bubbling	X	X	X		X	X
5	Fish eyes	X	X	X		X	X
6	Orange peel	X	X	X		X	X
7	Mud cracking	X	X		X		
8	Curing properties	X	X	X	X	X	X

No.	Item					
9	Runs and sags	X	X	X	X	X
10	Film thickness, dry	X	X	X		X
11	Film thickness, wet				X	X
12	Holidays, missed areas	X	X	X	X	X
13	Dry spray	X	X	X	X	X
14	Foreign containments	X	X	X	X	X
15	Mechanical damage	X	X	X	X	X
16	Uniformity	X	X	X	X	X
17	Adhesion	X	X	X	X	X

Source: From Ref. 21.

inspection, (4) coating application inspection, (5) daily inspection reports, and (6) final acceptance report.

Visual inspections are performed with either the unaided eye or by the use of a magnifying glass. Sometimes, visual inspections require the use of telescopic observation or the use of low-power magnification. A Pike magnifier is one of several types available for this use.

By employing the standard visual techniques, the inspector identifies areas that have been missed, damaged areas, or thin areas. White primers have been used to spot areas of low film thickness or inadequate coverage of the substrate. The use of instruments provides an inspector with an accurate appraisal of what dimensional requirements have been met by the applicator.

The inspector should examine coatings during and after coating application and report on the items indicated in Table 3, where applicable. Pictorial standards are available from the American Society for Testing and Materials (ASTM) [19] and the Federation of Societies for Coatings Technology (FSCT) [20] which can aid an inspector in reporting the degree of failure in terms readily interpreted.

Prior to the application of a paint or coating, the surface must be properly prepared to receive the material. This can be accomplished in a number of ways. The most common way is through the employment of abrasive blast cleaning. Prior to blast cleaning, the inspector should look for the removal of all visible oil and grease. This is done by solvent cleaning with an approved solvent or by some other approved means. Only slight oil traces will be removed by blast cleaning, especially when the abrasive is not recycled. The inspector should take a clean white rag and rub the surface to be coated. If the surface is clean, the rag should remain clean.

Welds must be ground smooth, although not necessarily flush. Sharp protrusions should be rounded and weld crevices should be manually opened enough so that the coating can penetrate. If this procedure is not done, the projections should be removed by grinding. Sharp edges should also be ground smooth. A good way to estalbish a smooth weld criteria is to rub your finger rapidly over the surface. If you can do this without cutting your finger, the weld is smooth enough. Naturally, the inspector will slowly check the smoothness of the weld.

Back-to-back angles, tape or stitch welding, and so on, cannot be properly cleaned and coated. Thus they should be sealed with caulking to prevent crevice corrosion. Also, electric welding flux should be neutralized or removed prior to painting.

The inspector is expected to work quickly so that the application of the coating to the surface is not delayed. But there is usually no harm in allowing the steel to stand unprotected for a few hours before beginning application of the coating.

In most work environments, fallout is ever present. Make sure that the surface is clean and free of dust before the coating is applied.

Brush off, blow off, or vacuum the dust collecting on the surface to be coated.

Surface Profile

Without an adequate "key" provided by abrasive blasting or mechanical grinding, many coating systems will not provide adequate long-term performance. Too little anchor pattern will result in too smooth a surface and thus poor adhesion. A deep profile will require additional paint. A good rule to go by would be to keep the profile depth about 25% of the total paint thickness. Therefore, if 6 mils is specified, the proper profile would be 1.5. mils.

Profile varies, of course, with the type of abrasive employed. There are many ways to determine surface profile in the field.

The Keane-Tator Surface Profile Comparator contains a metal disk with nominal surface profiles of 0.5, 1, 2, 3, and 4 mils. This flashlight magnifier is used as a handy pocket-type comparator to check on sandblast cleaned surfaces. Also available is a metal disk for comparison of anchor patterns prepared with shot or grit blast.

Clemtex offers a series of four steel coupons with profile gages ranging from 1 to 4 mils.

A Testex tape has been developed which is pressed into the profile. The tape is removed and the profile that remains in the tape is measured with a micrometer, substracting the thickness of the tape.

COATING APPLICATION

Good painters who have a "handle on spray application of coatings" keep their guns parallel to the surface of the work at all times. They will release the gun trigger at the end of each pass to avoid heavy deposits of paint. They will frequently use the crosshatch method of laying up high solids materials to prevent them from sagging.

When a painter applies paint or coatings far too heavily for the temperature of the substrate and for the viscosity of the material being applied, runs, sags, drips, and curtains will develop. These inconsistencies are the result of poor technique. When you see these problems, the work should be stopped and the runs, sags, and so on, should be brushed out before they are allowed to cure. Careful adjustment of the gun and a better application technique should help reduce these problems.

An inspector should witness the performance of the painters on the job. Remember that heavily applied paint may never cure due to solvent entrapment, or if it does cure, the film buildup can cohesively weaken the entire paint film.

If large areas exist where the coating system does not have the specified dry-film thickness, tell the painters that you want them to follow the minimums specified. Review the manufacturer's literature with the painters to determine what, if anything, must be done to prepare the painted surface for the additional coat that will be needed to meet the thickness prescribed by the specifications. Special considerations are required for a wide range of materials. Check with the coating manufacturer before allowing the work to continue.

An inorganic zinc primer may mud-crack if applied too heavily. Follow the manufacturer's instructions for the best results.

Overspray problems occur whenever winds, temperature, or humidity are rapidly changing. A partially dried coating material that does not wet and flow into the previously applied coat will appear on the surface.

Inorganic zinc primers are susceptible to overspray. They should never be topcoated unless the overspray is removed by abrading with a wire screen.

It should be rather obvious that paints and coatings must be suitably mixed prior to their application to a given substrate. Proper mixing redisperses the heavier pigments to ensure that the coating material is completely homogenized prior to painting. Follow the coatings manufacturer's mixing instructions.

One-Component Coatings

These materials are packaged to be mixed in their original containers. A paddle, mechanical agitator, or power mixer is employed to do the job. Thinners, if required, are added slowly after initial mixing of the material. The painter should be mixing the paint long enough so that no portion of the coating appears as "swill" on the surface. After mechanical mixing, a skilled journeyman will "box" the paint by pouring it from one clean can to another and back again. This will not be necessary where power agitation equipment is used.

Two-Component Coatings

Catalyzed coatings are normally packaged in separate containers, or in some cases in separate portions of a larger container. Be sure that the base is thoroughly mixed prior to the addition of the catalyst. See that the catalyst is added slowly and that the combined portions are thoroughly mixed. Some catalyzed epoxies must be allowed to stand for approximately 30 min before painting begins. This 30-min induction period will ensure a more unified curing of the applied film. Make sure that the temperature limitations correspond to the manufacturer's instructions. Induction periods are not required for polyester materials or amine adduct epoxies.

Cure

Many factors affect the cure of a coating or lining. Surface temperature and room temperature should be a minimum of 18°C (60°F) for best results. The thickness of the coating, the type of solvent used, and the paint vehicle characteristics will influence the cure.

Solvent Rub

After an epoxy paint or an inorganic coating cures, use a color-contrasting cloth dipped ind a strong solvent [methyl ethyl ketone (MEK) or methyl isobutyl ketone (MIBK)] to rub the surface of the coating. If the material was improperly mixed or cured, the material will redissolve and the color can be seen on the cloth.

Sandpaper Test

A number of paints and protective coatings will remain slightly tacky when they have not cured properly. When you abrade them with fine sandpaper, no coating material should be seen on the face of the sandpaper. It should be removed as a fine, powdery residue.

Hardness Test

By using your fingernail, you should soon be able to employ your judgment rather effectively. Other techniques, such as the Barcol impressor or the pencil hardness test, are also used to check on the hardness of a film. For inorganic zincs, a coin scratched against the surface should not remove the film.

Adhesion

The best method to determine adhesion is to use the pocket knife. Cut a "V" in the film and pick off the coating at the vertex. Good adhesion is expected. The coating should be very difficult to remove.

Dollies may be cemented onto the coated surface. An Elcometer adhesion test may be run. Epoxies usually far exceed 200 psi and frequently attain 800 psi adhesion. For nuclear-grade coatings, 200 psi is the lowest acceptable adhesion.

Crosshatch adhesion has also been used for thin-film coatings.

Film Thickness

Nondestructive test instruments for determining dry coating thickness on steel substrates fall into two main categories: magnetic and eddy

current. The most popular and commonly used instruments employ
the magnetic principle. The magnetic gage in its simplest form measures magnetic attraction, which is inversely proportional to coating
thickness. There are several pulloff gages available that employ
this simple magnetic principle, for example, the BSA-Tinsley thickness
gage manufactured by Evershed & Vignoles, London, England, and
the Elcometer 157 pulloff gage manufactured by Elcometer Instruments
Ltd., Manchester, England.

The pulloff gage is intended for use in the field as a rough guide
to determine if the protective coating is within the thickness specification. The manufacturer's state accuracy is ±15%, provided that the
gage is used within a true vertical plane. If the gage is used in a
horizontal or overhead position, more error will result. If the gage
must be used in positions other than the true vertical, it is recommended to plot a correction curve for each different position for best
results.

Care should be exercised to inspect visually the hemispherically
tipped magnet for dirt, small steel particles, tacky paint film, and tip
wear before use. Wear on the hemispherically tipped magnet will alter
the calibration of the gage. Never expose any magnetic gage to strong
ac or dc fields because the magnet will change, affecting the tracking
and calibration of the gage.

Besides accuracy limitations, the pulloff gage has other disadvantages: (1) the eye must record the coating thickness as the magnet
breaks away from the coating, and (2) erroneous readings will result
if the magnet is allowed to slide over the coating prior to break away
or lift-off.

The Type 7000 Tinsley gage, however, contains a diallike scale
with a balanced pointer which is not affected by angular positions. It
offers a direct readout from a lock-in zero reset which provides an
accuracy of 10%.

A more sophisticated version of the magnetic pulloff principle is
incorporated into the "banana-type" thickness gage. This particular
type of gage has been termed a "banana" gage because of the shape
of the case. A permanent magnet is mounted at one end of a balanced,
pivoted arm assembly, and a coil spring is attached to the pivot and to
a calibrated, rotable dial. The operator moves the rotable dial forward
until the magnet sticks to the coating. Variations in film thickness
above the steel substrate will alter the attractive force of the magnet.
This unknown force is determined by turning the rotable dial back-
ward, applying tension to the spring. When the spring tension exceeds the unknown magnetic attractive force, the magnet breaks contact with the coated surface. At this point, an audible click will be
heard and the coating thickness will be shown on the graduated, rotable dial.

There are several gages available using the guided or controlled
magnetic pulloff principle: Mikrotest thickness gage, Model 102/FIM,

and Mikrotest II FM manufactured by Elektro Physik, Cologne, West Germany, and Inspector thickness gage, Model 111/1E, manufactured by Elcometer Instruments Ltd.

This type of gage will measure coating thickness in any position without recalibrating, becuase the pivot arm is balanced. The Inspector gage has an external calibration adjustment (screwdriver slot) located under the nameplate.

Another version of the magnetic principle applies to the Elcometer thickness gage, which utilizes a magnetic reluctance technique. One could define reluctance as the characteristic of a material that resists the creation of a magnetic flux in that material (e.g., iron has less reluctance than air).

The Elcometer gage contains a permanent magnet that is located between two soft-iron poles resembling a horseshoe magnet. (The magnet is adjustable to produce an air gap.) A meter-pointer assembly with a soft iron vane is placed in the center of this horseshoe configuration, making a magnetic circuit with an indicating device that requires no power supply or battery.

When the Elcometer gage is placed onto a dry coating applied to steel, the magnetic flux will change in strength across the air gap in the magnetic circuit, thus moving the meter pointer across a calibrated scale, indicating the coating thickness in either mils or micrometers.

Care should be exercised when using the Elcometer always to hold the gage at a right angle to the surface to be measured because tilting the gage will give erroneous measurements. Always recalibrate the gage when changing from vertical to horizontal or perhaps to the overhead position [12].

External calibration is provided with a zero knob located on the side of the instrument. A small button must be depressed to take readings. Basically, the gage is very sensitive to surface roughness, residual magnetism in the substrate, edge effect, and tilt of head. Blind-hole measurements cannot be made with this gage.

In making film thickness measurements of any kind, avoid measurements close to the edge of a steel surface. The magnetic properties of the steel influence the reading, causing distorted results. This is also true of corners, angles, crevices, welds, and joints. A recommended practice is to keep at least 1 in. away from the edge. Always measure a clean surface, never an oily or dirty one.

There are electronic gages that possess greater accuracy than the mechanical gages mentioned previously. The General Electric (Schnectady, New York) Model B thickness gage uses the magnetic induction principle. A coating between the coil probe and steel substrate represents an air gap which is measured and displayed on the front panel meter. The General Electric gage requires a 115-V ac source and is not portable for field use.

The Model 158 Minitector thickness gage and Model 102/F100 Minitest thickness gage are in some respect similar to the GE gage, in that a probe senses and a meter displays coating thickness.

The Minitector uses a permanent magnet in the probe for a flux source, thus making a magnetic circuit that measures coating reluctance. The thicker the coating, the higher the reluctance; also, the thinner the coating, the lower the reluctance. When the Minitector probe is placed onto a dry coating over steel, the magnetic flux adjusts to the particular thickness of coating. The change in flux is then measured and displayed on the meter scale in either mils or micrometers of coating thickness. This gage is portable and uses standard transistor radio batteries that are readily available; it is manufactured by Elcometer Instruments Ltd. [22].

The Model 102/F100 Minitest thickness gage utilizes the eddy current principle. A coil of wire in the probe tip is energized with a high frequency alternating current, creating a magnetic field. When the probe tip is brought into close proximity to a metallic object, as is the case when the probe is placed onto a dry coating applied to steel, eddy currents are induced into the steel substrate, altering the electrical characteristics of the probe coil. This electrical characteristic change in the probe coil is then measured and displayed on the meter scale in mils of coating thickness. The Minitest is battery operated and comes equipped with an automatic battery power-off switch to extend battery life; it is manufactured by Elektro Physik.

Thickness standards are recommended as a ready reference to check the accuracy and calibration of gages. A range of thickness standards is commercially available from various sources. Chrome-plated steel panels are available from the National Bureau of Standards. Four plated thickness standards mounted in a plastic case are offered by Elektro Physik (102/TSFIM). A range of plastic, color-coded precision shims and color-coded glass on steel panels (101/7) is offered by Elcometer Instruments Ltd.

PERFORMANCE HISTORY AND FAILURE ANALYSIS

After the tank lining has been placed into service, a thorough inspection should be made at regular intervals for any visual defects, such as blistering, rusting, softening, or cracking. In addition, soon after the tank has been placed into service (depending on the corrosive nature of the contents), an inspection and touch-up is recommended to repair any defects resulting from application (such as holidays). Minor touch-up at this time could prevent catastrophic failure later. Inspection records should be kept to describe precisely the condition of the lining versus time. These records can be used to support failure recommendations based on actual field performance. If repairs are recommended, do them immediately. Do not wait for the condition to get worse.

Operating instructions should include the maximum temperature and service level expected. The outside of the tank should be labeled

DO NOT EXCEED X°C (X°F). THIS TANK HAS BEEN LINED WITH
_____. IT IS TO BE USED ONLY FOR _____
SERVICE.

A tank-lining failure should be prudently analyzed by the engineer to determine the cause. A tank lining may fail for numerous reasons.

The tank lining must not impart any impurities to the material contained. If the product is contaminated by color, taste, smell, or by any other means, even though the tank lining is intact, the application is a failure. This contamination can be caused by the extraction of impurities from the coating leading to blistering between coats and/or to metal. If the lining is unsuited for the service, complete failure may occur by softening, dissolution, and finally complete disintegration of the coating. This type of problem is prevalent between the interphase and the bottom of the tank. At the bottom of the tank and throughout the liquid phase, penetration is of great concern. Contrary to what one would think, failure of penetration is more prevelent with materials of mild corrosive qualities than it is with concentrated acids and alkalis. Water, for example, is very penetrating, especially when heated or deionized. The vapor phase of a tank is subject to corrosion from concentrated vapors mixed with the oxygen present and can cause extreme corrosion. It is often very difficult to determine on any given application which area of a tank will be subject to the greatest corrosion attack.

The most common types of failure are due to misuse of the tank lining, which may cause blistering, cracking, hardening or softening, peeling, staining, burning, and undercutting; overheating during operation is a frequent cause of failure. Never change chemicals for tank use without consultation with the lining-materials engineer. Cracking usually occurs when a very heavily pigmented surface or thick film begins to shrink, forming stresses on the surface. The cracks do not always expose the steel and may not penetrate. The best practice is to remove these areas and recoat per standard repair procedures prior to use. Do not take a chance.

An automatically controlled cathodic protection system of water storage tanks will greatly prolong a tank lining. A surveillance system for monitoring tank-to-water potentials can be easily installed and operated to provide practical information to field activities on the protection of tank interiors. Silver/silver chloride reference half-cells are used for such a purpose.

Hardening or softening is the result either of aging or poor resistance to the corrosive. As the coating ages (more characteristic of epoxy and phenolic amines), it becomes brittle and may chip from the surface. Peeling can result from a number of causes, such as poor surface preparation, a wet or dirty surface, or improperly cured undercoats. Staining can result from a reaction of the corrosive on the surface of the coating, or slight staining from impurities in the corrosive. It is necessary to determine the true cause by scraping or detergent-washing the film. If the stain is removed and softening of the film is not apparent, failure has not occurred.

Undercutting can result from any of the defects noted previously. After the corrosive penetrates to the substrate, corrosion will proceed to extend under the film areas that has not been penetrated or failed. Some coatings are more resistant to undercutting or underfilm corrosion than others. Generally, if the coating displays good adhesive properties and if the prime coat is chemically resistant to the corrosive environment, underfilm corrosion will be greatly retarded.

SUMMARY

A tank-lining application is probably the most critical of all coating applications. The coating itself must be resistant to the corrosive, and there must be no pinholes through which the corrosive can penetrate to reach the substrate. Thin-film linings should be used mainly when the corrosion rate of the steel is 25 mils per year or less for the chemical used. Care and observation may be required.

The severe attack that many corrosives have on the tank itself emphasizes the importance of using the correct procedure in lining a tank to obtain a perfect lining. First select a coating that resists the corrosive media and an applicator who has clearly demonstrated the ability to apply that particular type of lining properly. Consider the tank design and construction. If it is a new tank, be sure it is built with the coating application in mind; if it is an old tank, be sure adjustments are made to compensate for its design. Inspection should be made during application, after curing when the tank is ready for service, shortly after the tank has been put into service, and at scheduled intervals thereafter. Records should be kept and the coating supplier informed of the performance of the lining—both good and bad performance. Safety is an important part of a quality application.

ACKNOWLEDGMENTS

The author is grateful to Wallace Cathcart Tank Lining Corp., Oakdale, Pennsylvania, for his contributions. Much of the information contained herein is a result of over 25 years of NACE T6A Coating and Lining Materials for Immersion Service committee and subcommittee activities.

APPENDIX

Shown here in a series of 11 figures are the essentials for the design and preparation of steel tanks for lining.

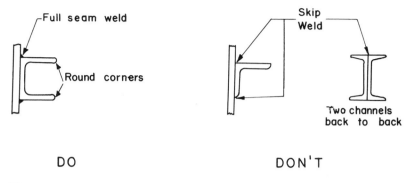

DO   **DON'T**

Figure 1. All construction involving pockets or crevices that will not drain or that cannot be properly sand blasted and lined must be avoided. (From Jack Kiewit, *The Paint Manual*, USDA Bureau of Reclamation, Denver, Colo., 1976.)

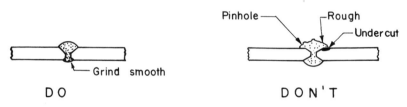

D O   **D O N'T**

Figure 2. All joints must be continuous solid welded. All welds must be smooth with no porosity, holes, high spots, lumps, or pockets. Peening is required to eliminate pososity, and grinding to remove sharp edges. (From Jack Kiewit, *The Paint Manual*, USDA Bureau of Reclamation, Denver, Colo., 1976.)

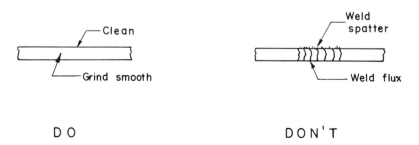

D O   **D O N'T**

Figure 3. All weld spatter must be removed. (From Jack Kiewit, *The Paint Manual*, USDA Bureau of Reclamation, Denver, Colo., 1976.)

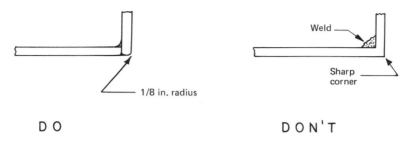

Figure 4. All sharp edges must be ground to a minimum of 1/8-in. radius. (From Jack Kiewit, *The paint Manual*, USDA Bureau of Reclamation, Denver, Colo., 1976.)

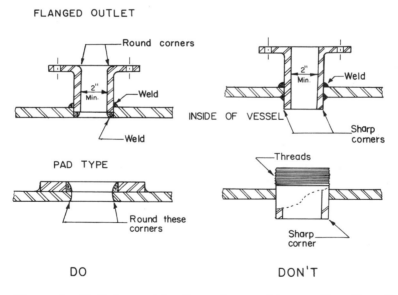

Figure 5. Outlets must be flanged or pad type rather than threaded. Within pressure limitations slip-on flanges are preferred as the inside diameter (ID) of the attaching weld is readily available for radiusing and grinding. If pressure dictates the use of weld neck flanges the ID of the attaching weld is in the throat of the nozzle. It is therefore more difficult to repair surface irregularities such as weld undercutting by grinding.

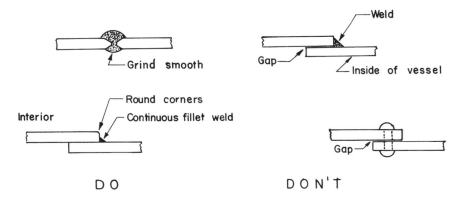

Figure 6. Butt welding should be utilized rather than lap welding or riveted construction.

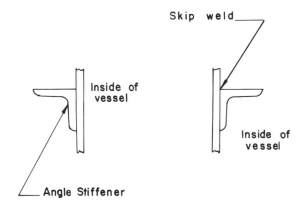

Figure 7. Stiffening members should be on the outside of the vessel or tank.

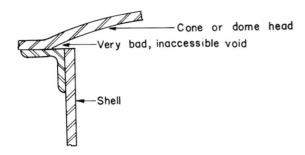

Figure 8. Standard field storage tank head. (From Jack Kiewit, *The Paint Manual*, USDA Bureau of Reclamation, Denver, Colo., 1976).

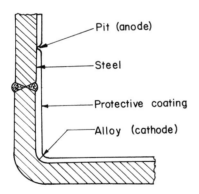

Figure 9. Dissimilar metal (galvanic) corrosion will occur where an alloy is used to replace the steel bottom of a tank. If a lining is then applied to the steel and for several inches (usually 6 to 14 in.) onto the alloy, any discontinuity in the coating will become anodic. Once corrosion starts, it progresses rapidly because of the large exposed alloy cathodic area. Without the coating, galvanic corrosion would cause the steel to corrode at the weld area, but at a much slower rate. The recommended practice is to line the alloy completely as well as the steel, thereby eliminating the possible occurrence of a large cathode-to-small anode area.

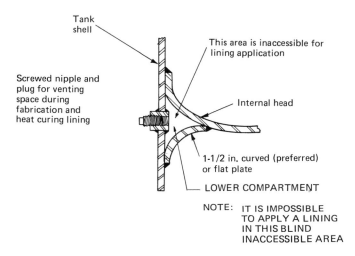

Tank
shell

This area is inaccessible for
lining application

Screwed nipple and
plug for venting
space during
fabrication and
heat curing lining

Internal head

1-1/2 in. curved (preferred)
or flat plate

LOWER COMPARTMENT

NOTE: IT IS IMPOSSIBLE
TO APPLY A LINING
IN THIS BLIND
INACCESSIBLE AREA

Figure 10. Technique (detail of fabrication) to allow for good continuity of lining application for inaccessible areas for such as multi-compartment tanks.

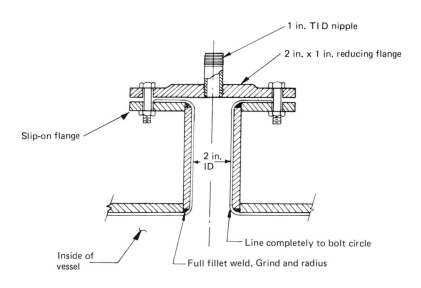

1 in. TID nipple

2 in. x 1 in. reducing flange

Slip-on flange

2 in.
ID

Line completely to bolt circle

Inside of
vessel

Full fillet weld. Grind and radius

Figure 11. Nozzle-lining detail.

REFERENCES

1. *Laboratory Methods for Evaluation of Protective Coatings Used as Lining Materials in Immersion Service Materials Performance*, National Association of Corrosion Engineers, Houston, NACE TM-01-74, January 1974.
2. *Coatings and Linings for Immersion Service*, NACE TPC 2, National Association of Corrosion Engineers, Houston, 1972.
3. *Painting Steel Water Storage Tanks*, AWWA-D102-78, American Water Works Association, Denver, Colo., 1978.
4. *Recommended Practice, Design, Fabrication, and Surface Finish of Metal Tanks and Vessels to Be Lined for Chemical Immersion Service Materials Performance*, NACE RP-01-78, National Association of Corrosion Engineers, Houston, January 1978.
5. Volume II: *Systems and Specifications*, Steel Structures Painting Council, Pittsburgh, Pa., 1982.
6. Volume I: *Good Painting Practice*, Steel Structures Painting Council, Pittsburgh, Pa., 1982.
7. *Visual Standards for Surface of New Steel Airblast Cleaned with Sand Abrasive*, NACE Standard TM-01-70, National Association of Corrosion Engineers, Houston, 1970.
8. *Pictorial Surface Preparation Standards for Painting Steel Surfaces*, SSPC Vis. I-67T, Steel Structures Painting Council, Pittsburgh, Pa., 1967.
9. *Pictorial Surface Preparation Standards for Painting Steel Surfaces*, ASTM D-2200-78, American Society for Testing and Materials, Philadelphia, 1967.
10. *Near White*, Maryland Pictorial Standard Ser. No. 001-A, and *Commercial*, Ser. No. 001B, Maryland State Highway Administration, Baltimore, 1974.
11. *Abrasive Blasting Guide for Aged or Coated Steel Surfaces*, SNAME Technical and Research Bulletin No. 4-9, Society of Naval Architects and Marine Engineers, New York, 1969.
12. *Measurement of Dry Paint Thickness with Magnetic Gages*, SSPC PA2-73T, Steel Structures Painting Council, Pittsburgh, 1973.
13. *OSHA Reference Manual*, Painting and Decorating Contractors of American, Falls Church, Va., August 1976.
14. *TLV's threshold limit values for chemical substances and physical agents in the workroom environment with intented changes for 1978*, American Conference of Governmental Industrial Hygenists, Cincinnati, Ohio.
15. Paul E. Weaver, *Industrial Maintenance Painting*, National Association of Corrosion Engineers, Houston, 1973.
16. C. G. Munger, *Corros. Control Rep. 16*(4), Ameron Corp., 1965.
17. *Estimating Guide*, 12th ed., Painting and Decorating Contractors of America, Falls Church, Va., 1979.
18. *Water Tank Maintenance and Repair Practices for Georgia*

Municipalities, Georgia Municipal Association, Atlanta, Ga., 1973.

19. *Evaluating Degree of Rusting on Painted Steel Surfaces*, ASTM
     Method D-610 (also SSPC Vis. 2-687), and *Evaluating Degree
     Blistering of Paints*, ASTM Method D-714, American Society for
     Testing and Materials, Philadelphia, 1979.
20. *Exposure Standards Manual*, Federation of Societies for Coating
     Technology, Philadelphia, 1979.
21. *Proposed Manual of Coating Work for Light Water Nuclear Power
     Plant Primary Containment and Other Safety Related Facilities*,
     American Society for Testing and Materials, Philadelphia, 1979.
22. Dean M. Berger, and Stanley E. Mroz, Instruments for inspection
     of coatings, *J. Test. Eval.* 4(1), 29-39, January 1976.
23. R. A. Mixer, and S. J. Oechsle, Jr., Materials of construction
     part I—protective lining systems, *Chem. Eng.* 181-182, November
     1956.

8.2

Cements and Mortars

GEORGE W. READ, JR. / Sauereisen Cements Company, Pittsburgh, Pennsylvania

INTRODUCTION

Corrosion-resistant cements are available for use as mortars with acid-proof brick and as monolithic linings to be poured, cast, or gunned into place. Typical installations are for lining acid tanks, waste treatment tanks, process vessels, sumps, pits, foundations, trenches, sewers, chimneys, stacks, ducts, and floors.

Several factors will influence the selection of the proper systems for a specific installation:

1. The corrosive chemicals involved, their concentrations, and the pH range.
2. The operating temperature and thermal shock conditions.
3. Mechanical abuse, abrasion, and traffic conditions.
4. Type of substrate to be protected, and its condition.
5. Nature of cleaning or washdown. There is no point in installing a lining designed to withstand acid and then using a strong caustic such as sodium hydroxide to clean it if the lining is not resistant to alkalies. In such a case the cleaning operation would do more damage to the lining than the actual service it would receive.
6. Whether the installation will be outdoors or indoors.

It is difficult to make a direct cost comparison between a brick and mortar lining and a monolithic lining because of the number of variables involved. Factors such as special brick shapes or extensive cutting which may be necessary when using brick will increase costs tremendously. Brick must be handled individually and the corrosion-resistant cement applied with a trowel, so construction is slow and the time before the installation is ready for service is extended.

The installed or in-place cost per square foot of a monolithic lining will vary with the type used and the method of installation, just as the cost of brick linings will vary with the size and quality of the brick and the type of corrosion-resistant cement used. Generally, a gunned monolithic lining will cost approximately half as much as a comparable brick and mortar installation.

CORROSION-RESISTANT MASONRY

Corrosion-resistant masonry provides maximum protection from highly corrosive chemicals, abrasion, and thermal shock. It adds years to equipment life, reduces maintenance expense, and eliminates costly shutdowns.

Brick and mortar linings provide some thermal insulation, but do transmit a great deal of heat through them to the substrate. Since the coefficient of thermal expansion of the brick and mortar lining will generally not match that of the substrate, provisions must be made to ensure that tensile stresses are not generated in the masonry due to the substrate expanding more rapidly.

High tensile stresses should be avoided in brick and mortar linings and they should be designed so that the brickwork is placed in slight compression. Cylindrical vessels and tanks employing arched or domed tops are best.

Corrosion-resistant masonry is not a liquid-tight barrier in itself and should not be designed to serve as such. For liquid-tight installations it is necessary to provide a backup impervious membrane and the masonry will protect it from mechanical shock, abrasion, and temperature.

Impervious Membranes

Impervious membranes are classified as either true membranes or semi-membranes. True membranes in probable order of use are:

1. Rubber and related synthetic elastomers
2. Polyvinyl chloride (PVC)
3. Lead
4. Various synthetic resin formulations with glass cloth reinforcing
5. Rigid or semirigid plastic sheets
6. Baked-on coatings, including resins and glass

Semimembranes are:

1. Asphalts or bituminous mastics
2. Unimpregnated asbestos felt applied with a silicate solution

Chemical-Resistant Brick

ASTM C-279 lists the two types of chemical-resistant brick used for these installations as:

Type H: brick intended for use where thermal shock is a service factor and minimum absorption is not required
Type L: brick intended for use where minimum absorption is required and thermal shock is not a factor

These brick differ from ordinary face brick in that they are manufactured of raw materials of extremely low flux content and their low porosity and nonabsorbent qualities are imparted by the method of manufacture and the high temperature at which they are fired. Both types are available as red shale or fireclay, with the latter being higher in cost and of better size tolerances. The red shale type generally contains some readily available iron which may be leached out in acid service; therefore, when halogen acids are involved, the best choice would be the fireclay equivalent for that type of brick.

Carbon brick are recommended for use in hydrofluoric acid, acid fluorides, and strong caustic solutions, and for severe thermal shock conditions. They have higher absorption than either the red shale

or fireclay types, but are more shock resistant and have lower coefficients of thermal expansion.

CHEMICAL-RESISTANT MORTARS

One of the most important steps in corrosion-proof construction is the
selection of the proper mortar for bonding the brick. Since there is
no "all-purpose" mortar, there are several different types with individual characteristics designed to meet specific service requirements.

Silicate Mortars—Air-Drying Types

Mortars based on soluble silicate comprise some of the original corrosion-
resistant cements used. The first silicate mortars were simply mixtures
of fillers such as silica, quartz, clay, or barytes, and a sodium silicate
solution. Mortars of this type harden by loss of water and require exposure to air or heat to set. Construction with such mortars is extremely slow. Although thin joints are used, the fluid mixture squeezes
out if more than three or four courses of brick are laid at one time. In
most cases brickwork heights of not more than 6 ft could be laid at one
time. Very careful drying was also necessary, and a 30-day period
was usually recommended before putting the structure in service. Because of these drawbacks and the development of improved mortars,
air-drying mortars are no longer used for brick linings.

Sodium Silicate Mortars—Chemical-Setting Types

Chemical-setting sodium silicate mortars came into use early in the
1930s. These utilize a setting agent that reacts with the soluble sodium
silicate to cause the mixture to harden. The setting agent may be
either an acid or a compound that will decompose and liberate acid to
accelerate the cure. Typical setting agents used are ethyl acetate,
zinc oxide, sodium fluosilicate, glyceryl diacetate, formamide, and
other amides and amines.

These mortars are supplied as two-component systems and consist
of the liquid sodium silicate solution and the filler powder containing
selected aggregates; or they may be one-part systems in powder form
to be mixed with water when used. Chemical-setting mortars take
initial sets in 15 to 45 min and final sets in 24 to 96 h or longer, depending on the temperature. Continuous bricklaying is possible since
the mortars harden quickly as a result of the chemical reaction and do
not require exposure to air or heat.

Large quantities of chemical-setting sodium silicate mortars have
been used in industry for the past 45 years, and mortars of this type
are still employed in many types of acid service.

Potassium Silicate Mortars—Chemical-Setting Types

When potassium silicates became commercially available early in the 1960s, new chemical-setting mortars were developed in which potassium silicates were substituted for the sodium silicates previously used. Several fundamental properties of potassium silicates combine to make them preferable to sodium silicates.

Potassium silicates have better workability due to their smoothness and lack of tackiness. They do not stick to the trowel and do not run or flow from the joints of the brickwork. They possess greater resistance to strong acid solutions as well as to sulfation, and they have greater refractoriness. Still another special value of potassium silicate mortars is that they do not effloresce or bloom and they have less tendency to form hydrated crystals in the hardened mortar.

Chemical-setting potassium silicate mortars are supplied as two-component systems consisting of the silicate solution and the filler powder. Mortars are available which utilize inorganic setting agents, organic setting agents, or a combination of the two. The properties of the mortar are determined to a great degree by the setting agent used. Such properties as absorption, porosity, strength, and water resistance are affected by the choice of setting agent. For example, organic setting agents will burn out at low temperatures, thereby increasing porosity and absorption. Organic setting agents are water soluble and can be leached out if the mortar is exposed to steam or moisture. Because of the type of crystal structure formed by organic setting agents, the mortars invariably take a longer time to gain any significant strength, as they remain in the plastic state for 96 h or more at normal temperatures.

Mortars utilizing inorganic setting agents are water and moisture resistant immediately upon final set, normally within 24 h at 21°C (70°F). They permit rapid, continuous construction without danger of brickwork slipping or sliding out of line. These mortars present no health or safety hazards when used, as they do not emit noxious fumes or gases. They have unlimited storage life and will not deteriorate under normal storage conditions.

Modified Silicate Mortars

One-part powder-form chemical-setting silicate mortars which require only mixing with water have been commerically available for several years but were not extensively used because of their higher cost. These one-part mortars are similar to the two-component mortars described previously, but have somewhat lower mechanical properties. Extensive research and field testing has recently produced a new class of modified silicate mortars which provide characteristics not available before in either the one- or two-part systems. These new modified silicate mortars utilize condense aluminum phosphates for hardening, which are obtained by subjecting acidic aluminum phosphates to thermal treatment. Potassium and sodium silicates with a ratio of SiO_2 to Na_2O or K_2O between 1.5 and 4.0 are suitable for these mortars.

One of the major advantages the new modified silicate mortars provide is a greater pH range of service than other types of chemical-setting silicate mortars. They are resistant to most acids (except hydrofluoric and acid fluoride salts), as well as most alkalies over a pH range of 0.0 to 9.0, and in some cases to a pH of 14, depending on the particular alkali. They are highly resistant to sulfation, blooming, efflorescence, and have excellent adhesion to brick, concrete, and steel.

Silica Mortars

Silica mortars contain over 95% silica in the cured state, which is provided by a silica sol instead of the potassium or sodium silicates used in other mortars. These mortars are two-component systems consisting of a powder composed of high quality crushed quartz, and a hardening agent, which are mixed with the colloidal silica solution to form the mortar. These mortars are particularly recommended for use in hot concentrated sulfuric acid, especially where iron or aluminum salts are present. They are also used for weak acid conditions at pH 5.5 to 7.0. Typical applications are in brick linings for acid concentrators, absorbers, stacks, process vessels, and storage tanks.

Sulfur Mortars

Sulfur mortars are available in powder, flake, and ingot forms. They are hot-melt compounds and must be heated to a temperature of approximately 120°C (250°F) and poured into the joints while hot. These mortars consist of inert silica, carbon fillers, and plasticizers. The plasticizers reduce brittleness and improve the mechanical properties. Sulfur mortars are particularly useful for protection against oxidizing acids. When they are all carbon filled they are suitable for protection against combinations of oxidizing acids and hydrofluoric acid. The heat resistance of sulfur mortars is relatively low, and therefore their use is limited to installations with operating temperatures below 88°C (190°F). Chemical resistance to alkaline solutions and some organic solvents is poor. The recommended pH range for use is 1.0 to 9.0. The shelf life is indefinite.

Phenolic Resin Mortars

These mortars have phenol-formaldehyde resin binders with inert powder fillers containing an acid catalyst. Fillers may be silica, carbon,

coke flour, or barytes. They have good resistance to most mineral
acids and solutions of inorganic salts and mildly oxidizing solutions,
but they are rapidly attacked by strong oxidizing agents such as
nitric, chromic, and concentrated sulfuric acids. They are satisfactory in mild alkaline solutions and in many solvents, but have poor
resistance to strong alkalies. The temperature limit is 175°C (350°F)
and they are effective from a pH of 0.7 to 9.0. They have poor storage life and must be kept under refrigeration until used.

Furan Resin Mortars

Furan resin mortars have been used with chemical resistant brick and
tile for floors and vessel linings for over 40 years. They are supplied
as two-component systems consisting of a liquid furan resin and a
powder filler containing an acid catalyst. The filler is usually carbon,
coke flour, or silica. Furan mortars have a wide range of chemical
resistance and are suitable for use in nonoxidizing acids, alkalies,
salts, gases, oils, greases, detergents, and temperatures to 175°C
(350°F). They have an effective pH range of 1.0 to 13.0. Heat
accelerates the cure or hardening and cold slows it.

Polyester Resin Mortars

Polyester resin mortars may be supplied as either one- or two-component systems. The filler powder is generally silica, but carbon can
also be used. These mortars are excellent in most acids and are
recommended for use in bleach vessels and for mild oxidizing agents.
They lack resistance to strong alkalies and many solvents. Temperature limit is 120°C (250°F) and they have an effective pH range of
0.9 to 8.0.

Epoxy Resin Mortars

Epoxy resin mortars are usually two- and three-component systems
consisting of the resin solution, a setting agent, and inert fillers.
They have excellent physical and mechanical properties and adhere
strongly to most surfaces. Resistance to nonoxidizing acids and alkalies is excellent and they have good resistance to organic solvents.
They are attacked by oxidizing acids and alkalies. Temperatures resistance will range from 95 to 120°C (200° to 250°F). The effective
pH range is 2.0 to 14.0.

Application of Mortars

In many cases, combinations of mortars offer not only the least expensive but the ideal construction as well. A combination of a sulfur cement and a furan resin cement is often used in lining acid pickling tanks in steel mills. In this two-course construction, the inner course of brick which will be in contact with the acid is bonded with the furan resin mortar; then the sulfur cement is poured in to fill the back course joints. The furan resin mortar acts as a seal to prevent leakage of the molten sulfur mortar, and the hot sulfur mortar speeds up the cure or set of the furan resin mortar.

Sulfur mortar joints must be 1/4 in. or more in thickness to allow the molten compound to flow in around the brick and completely fill the joints. With either silicate or resin mortars the joints should be kept as thin as possible to minimize shrinkage and provide the strongest possible joint at the most economical cost. These mortars are considerably higher in cost per unit volume than brick, so joints should be kept as this as possible—preferably not more than 1/8 in. thick.

It is also highly recommended that mortar be applied to five sides of the brick in floor construction by using a bed joint as well as buttering the sides of the brick. This will assure a level base on which the brick can be placed since neither the subfloor nor the brick are ever perfectly level, and without a bed joint the brick could move or wobble and eventually crack.

The same practice should be followed in wall or lining construction using two or more courses of brick. All circumferential or back vertical joints in the lining should be completely filled with mortar for maximum impermeability and strength.

MONOLITHIC LININGS

Monolithic corrosion resistant linings have been used successfully for more than 25 years for both new and existing installations. These linings may be cast poured or applied by pneumatic gunning. They have the following advantages:

1. Curved or irregular surfaces can be covered uniformly.
2. Monolithic linings bond to steel, brick, and concrete.
3. Monolithic linings can be gunited horizontally, vertically, or overhead, without complex forms, supports, or scaffolds.

Monolithic linings are composed of a single, generally massive element, with or without expansion joints, and are applied in thicknesses from 1/2 in. up to several inches, depending on the service. Linings 1 in. or more in thickness should be anchored with wire mesh, studs, or anchors, with a minimum of 1/2 in. cover over the highest point.

The anchoring system is necessary to retain the lining in intimate contact with the shell during application and curing of the lining.

Castable monolithics are mixed and cast in a manner similar to procedures used in handling concrete. They are poured or cast into forms to harden. Trowel applications are generally thin coatings of less than 3/4 in. and are troweled in place with or without reinforcement. The largest tonnage of monolithic linings for corrosion-resistant service is applied by penumatic gunning.

Monolithic linings have limitations which must be considered. For example, like all cementitious materials, they are inflexible and tend toward brittleness. The modulus of elasticity will range from 10^5 to 10^6 psi and flexural strength will range from 500 to 2000 psi. Tensile strengths will vary from 150 to 400 psi, and compressive strengths will range from 1800 to 5000 psi.

The thermal properties of monolithics vary considerably. In general, the coefficient of thermal expansion of the monolithic should be matched as closely as possible to that of the substrate over which it will be applied. If it cannot be matched, a bond breaker or membrane should be considered. The thermal conductivity of monolithic linings is lower than either steel or concrete. This is usually advantageous, as the lower temperature on the substrate reduces corrosion rates exponentially and also reduces thermal movement and stresses in the substrate.

Care must be exercised in placing monolithic linings over substrates which are still experiencing curing shrinkage in excess of 1%, they can be expected to develop shrinkage cracks if placed directly on such substrate. In such cases it is necessary to provide expansion joints and to place the monolithic over a bond breaker such as an impervious membrane.

Sodium Silicate-Base Monolithics

The original acid-resistant monolithic is an inorganic silicate base which consists of two components, a powder and a liquid. The powder is basically quartzite of selected gradation and a setting agent. The liquid is a special sodium silicate solution. When the monolithic is used, the two components are mixed together and hardening occurs due to a chemical reaction.

This monolithic may be cast or poured into construction forms, or it may be applied by guniting. It has excellent acid resistance and is suitable for use over a pH range of 0.0 to 7.0.

Modified Silicate-Base Monolithics

There are two types of modified silicate-base monolithics available. The first is supplied in powder form and must be mixed with water when used. It is designed specifically for gunite application. It

hardens within 24 h at 20°C (68°F) to produce a high-strength corrosion-resistant lining. It has extremely good adhesion to steel, concrete, and brick, and requires only a minimum of surface preparation.
It is not affected by exposure to acids (except hydrofluoric), mild
alkalies, water, solvents, and temperatures to 950°C (1740°F), and is
suitable for use in the pH range 0.0 to 9.0. This monolithic will weigh
approximately 135 lb/ft^3 in place.

The second monolithic of this type is thermally insulating and lightweight. It is supplied in powder form and must be mixed with water
when used. It weighs approximately 98 lb/ft^3 and is considerably
lighter in weight than most other corrosion-resistant linings, thereby
reducing the dead load. It will develop compressive strength over
2000 psi, has a K factor of 2.25 to 2.50, and will provide a temperature
drop of approximately 50% per inch of thickness. It has excellent acid
resistance, will withstand temperatures to 925°C (1695°F), and is
suitable for use over the pH range 0.0 to 9.0.

Calcium Aluminate–Base Monolithics

Monolithics of this type consist of a calcium aluminate-base cement and
various inert aggregates, and are supplied in powder form to be mixed
with water when used. They may be cast, poured, or applied by
guniting. Calcium aluminate-base monolithics are hydraulic in nature
and consume water in their reaction mechanism to form hydrated
phases. In this respect they are similar to portland cement compositions; however, their rates of hardening are very rapid and essentially full strength is attained within 24 h at 23°C (73°F). They have
better mild acid resistance than portland cement, but are not useful
in acids below pH 4.5 to 5.0. They are not recommended for halogen
service, nor are they recommended for alkali service above a pH of
12.0.

ADDENDUM

The American Society for Testing and Materials has a number of published standards which have been developed for chemically resistant
materials, in Part 16 of the *1979 Annual Book of Standards*:

ASTM C-279 *Chemical-Resistant Masonry Units*
ASTM C-287 *Chemical-Resistant Sulfur Mortars*
ASTM C-395 *Chemical-Resistant Resin Mortars*
ASTM C-466 *Chemically Setting Silicate and Silica Chemical-Resistant
     Mortars*

There are also "Methods of Tests" available for the following:

ASTM C-267 *Chemical Resistance of Mortars*
ASTM C-307 *Tensile Strength of Chemical-Resistant Resin Mortars*
ASTM C-308 *Working and Setting Times of Chemical-Resistant Resin*
Mortars
ASTM C-321 *Bond Strength of Chemical-Resistant Mortars*
ASTM C-396 *Compressive Strength of Chemically Setting Silicate and*
Silica Chemical-Resistant Mortars
ASTM C-413 *Absorption of Chemical-Resistant, Nonmetallic Mortars,*
Grouts, and Monolithic Surfacings
ASTM C-414 *Working and Setting Times of Chemical-Resistant Silicate*
and Silica Mortars
ASTM C-531 *Shrinkage and Coefficient of Thermal Expansion of*
Chemical-Resistant Mortars, Grouts, and Monolithic Surfacings
ASTM C-580 *Flexural Strength and Modulus of Elasticity of Chemical-*
Resistant Mortars, Grouts, and Monolithic Surfacings

There are "Recommended Practices for the Use of":

ASTM C-386 *Chemical-Resistant Sulfur Mortars*
ASTM C-397 *Chemically Setting Chemical-Resistant Silicate and Silica*
Mortars
ASTM C-398 *Hydraulic Cement Mortars in Chemical-Resistant Masonry*
ASTM C-399 *Chemical-Resistant Resin Mortars*

9

Corrosion Monitoring

GENE FREDERICK RAK / Exxon Company, U.S.A., Baton Rouge, Louisiana

INTRODUCTION

In order to monitor the condition of process equipment both on-stream and off-stream, the engineer must rely on nondestructive testing. Nondestructive testing may be defined as testing to detect internal, external, and concealed flaws in materials using techniques that do not damage or destroy the items being tested. Nondestructive tests are an essential component of production processes. They can be applied, if necessary, to all critical components and assemblies. In many applications, these indirect tests have greater sensitivity, accuracy, and reliability than do widely accepted and more costly direct tests. Nondestructive testing can be utilized as an early warning device to indicate when process equipment is approaching its retirement limit or when process conditions are such that increased corrosion rates are occurring. It is important that the engineer understand the operation methods and limitations of nondestructive testing equipment so that the data extracted from inspection reports can be analyzed properly to give meaningful results.

Often, corrosion monitoring equipment can be used to optimize process conditions to achieve maximum throughout without sacrificing the integrity of the equipment. Process changes such as increases or decreases in temperature, pressure, velocity, and concentration can accelerate corrosion rates. When these process changes occur, it is imperative that the engineer monitor the effect on corrosion rates in order to prevent failure of equipment due to corrosion and thus avoid unscheduled shutdowns of process units.

Occasionally, engineers are limited in the amount of on-stream testing they can perform and therefore must rely on off-stream testing methods. On-stream inspection tools and techniques are limited due to the accessibility of equipment, extreme temperatures, extreme pressures, physical shapes and configurations, and so on. This chapter addresses several tools and techniques available to the engineer to monitor equipment both on-stream and off-stream.

ON-STREAM MONITORING EQUIPMENT

Electrical Resistance Principle

The Rohrback Corrosometer* probe (Fig. 1) operates on the electrical resistance principle. The measuring element, consisting of a wire, tube, or strip section, is exposed to the gas or liquid stream. As the

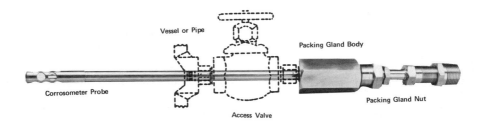

Figure 1. Rohrback Corrosometer probe assembly. (Courtesy of Rohrback Corporation, Santa Fe Springs, Calif.)

---

*Trademark of Rohrback Instrument.

Figure 2. Rohrback Corrosometer, Model CK-3 instrument. (Courtesy of Rohrback Corporation, Santa Fe Springs, Calif.)

measuring element corrodes, the cross-sectional area reduces and the electrical resistance increases. The thickness of the measuring element is directly proportional to a corrosion dial reading. A Corrosometer, Model CK-3, instrument (Fig. 2) is used to monitor the progress of corrosion in the measuring element. The difference in dial readings is plotted over a period of time as shown in Fig. 3. Slope A corresponds to a moderate corrosion rate, slope B corresponds to a slight corrosion rate, and slope C corresponds to a high corrosion rate. The actual corrosion rate, described in mils per year (mpy) can be determined as follows:

$$\text{mpy} = \frac{\Delta \text{ dial reading}}{\Delta \text{ time, in days}} \times 0.365 \times \text{probe multiplier}$$

Example 1: Corrosion rate of slope A:

$$\text{mpy} = \frac{100 - 80}{15 - 8} \times 0.365 \times 10$$

$$= 10.4 \text{ mpy}$$

Example 2: Corrosion rate of slope C:

$$\text{mpy} = \frac{158 - 98}{32 - 27} \times 0.365 \times 10$$

$$= 43.8 \text{ mpy}$$

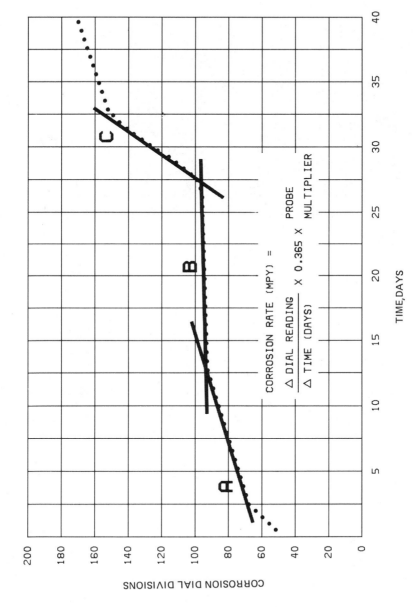

Figure 3. Plot of data from corrosion probe. (Courtesy of Rohrback Corporation, Santa Fe Springs, Calif.)

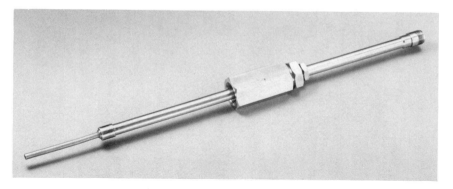

Figure 4. Rohrback Corrosometer retractable probe. (Courtesy of Rohrback Corporation, Santa Fe Springs, Calif.)

Corrosometer readings should be recorded daily until a pattern is established. At that time, reading intervals may be increased or decreased to obtain an accurate corrosion rate. The engineer should avoid determining a corrosion rate based on only two successive readings, as small deviations may distort the actual corrosion rate. Readings can be obtained using either a portable manual instrument or an automatic recorder system.

While the process unit is shut down, a 1 1/2-in. nipple and valve is usually installed in the area in which the maximum corrosion is expected. The Rohrback Corrosometer retractable probe (Fig. 4) can be inserted during a process run. Occasionally, connections are not available in areas of interest and on-stream connections can be added using hot tapping equipment. Hot tapping is a procedure for making a pipe connection into a line or vessel without removing it from service. Adequate wall thickness must be established prior to hot tapping the equipment.

The retractable probe is connected to the 1 1/2-in. valve as shown in Fig. 1, and the packing gland is then tightened to prevent leakage. The valve is opened slowly. When the valve is fully opened the packing is loosened and the probe is inserted through the valve and into the process stream. The packing is then tightened and should be checked frequently to ensure that leakage does not occur through the packing gland. Safety clamps are available to ensure that the corrosometer probe does not retreat beyond the packing gland. Although the safety clamp was designed for high pressure systems, it should be included in all installations.

Figure 5 shows a typical installation of a retractable process probe. The measuring element should be located as close as possible to the metal surface being monitored. Also, the probe should be installed in a manner to avoid impingement of high velocity streams or abrasive

Figure 5. Installation of a retractable process probe. (Courtesy of
Rohrback Corporation, Santa Fe Springs, Calif.)

particles. Electrical connections can be dropped to ground level to
minimize climbing on equipment when gathering daily readings.

The corrosometer probe can be used in most process equipment and
is available in a variety of alloys. The probe can be either fixed or
the retractable type. The fixed probe is designed to be inserted in
equipment when there is no pressure or flow in the system. As expected, the process equipment must be isolated and depressurized
prior to the removal of a fixed probe. The retractable probe can be
operated at a maximum pressure of 1000 psig and a maximum temperature of 1000°F. The fixed probe can operate at a maximum pressure
of 3000 psig and a maximum temperature of 1000°F. Both the fixed
and the retractable probe can be modfied using a Grant Oil Tool
Cosasco high-pressure access fitting to allow their use at a maximum
pressure of 6000 psig. Selection of a seal material to isolate the measuring element from the probe body is based on process conditions.
The manufacturer of the probe should be consulted in the selection of
a proper seal material.

The main disadvantage of the resistance probe is that normally it
requires several days or weeks to determine a corrosion rate. If
process conditions are such as to cause a wide swing in dial readings,
it becomes necessary to gather a significant amount of data in order

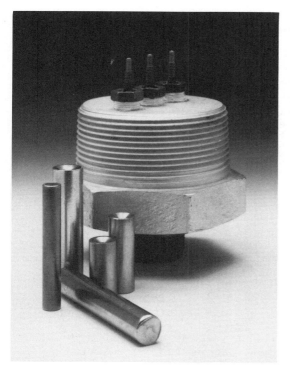

Figure 6.   Petrolite M-510 standard industrial probe.   (Courtesy of
Petrolite Corporation, Stafford, Tex.)

to establish corrosion rates.   Since the resistance wire element is
available only in 40- and 80-mil diameters, the measuring element life
is fairly short in highly corrosive environments.

Linear Polarization Principle

The Petrolite Pair* probe operates on the linear polarization principle.
Figure 6 shows the Petrolite M-510 standard industrial probe, and Fig.
7 illustrates the circuitry.   The corrosion rate is determined by measuring electrical current flow between the test and auxiliary electrodes.
That current either cathodically protects or anodically accelerates the
corrosion rate of the test electrode, depending on the flow.   The current is measured on a microammeter that has been converted to read the

---

*Trademark of Petrolite Corporation.

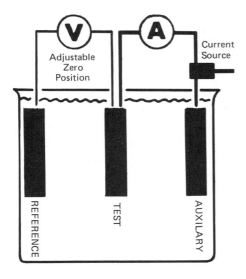

Figure 7. Principle of the three-electrode probe. (Courtesy of Petrolite Corporation, Stafford, Tex.)

corrosion rate directly (in mils per year) of the test electrode. The
Petrolite Portable Industrial Meter, Series M-212 instrument (Fig. 8)
is designed for manual, nonrecording use in the field. The Petrolite
M-510 probe is built on a 2-in. pipe plug base and is designed to be
operated at a maximum pressure of 3000 psig and a maximum temperature of 400°F. The probe can be supplied in either mild steel or 304
stainless steel.

The Rohrback Corrater* operates similarly to the Petrolite Pair
Probe. Figure 9 shows several types of Corrater probes. The two
(or three) elements are immersed in the corrosive stream and a small
millivolt potential is then supplied across the two electrodes. The
effect of this potential is to increase slightly the anodic activity that
is occurring on the surface of one of the electrodes and the cathodic
activity on the other. A small current flows through the electrolyte
and is measured by the Rohrback 1120 Portable Corrater Meter (Fig.
10).

The main advantage of the linear polarization method is that instantaneous corrosion rates can be determined. This is extremely
helpful in determining the effect of process changes. Any changes in
temperature, pressure, velocity, concentration, and pH will cause increases or decreases in corrosion rates, and a decision can be made

---

*Trademark of Rohrback Instrument.

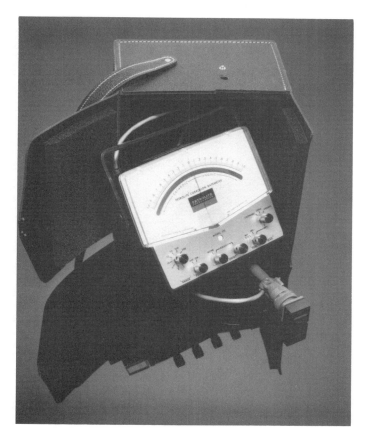

Figure 8. Petrolite M-212 portable industrial meter. (Courtesy of Petrolite Corporation, Stafford, Tex.)

instantaneously as to the effect on the integrity of the equipment. The probes are not operable in nonconductive gas or hydrocarbon process service.

The Petrolite Flush-Mounted† probe (Fig. 11), operates on the same principle as the three-electrode instantaneous polarization method. The small ring is the test electrode and the probe body is an auxiliary electrode. The planar probe surface has the reference electrode in the center. The advantage of this probe is that the monitoring device is flush with the surface in question and accurately simulates the

†Trademark of Petrolite Corporation.

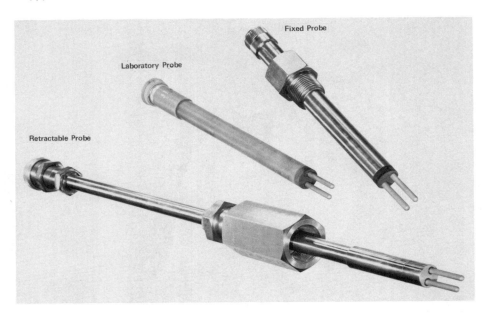

Figure 9. Rohrback Corrater probes. (Courtesy of Rohrback Corporation, Santa Fe Springs, Calif.)

corrosion rate occurring on the wall of piping or equipment. The
Petrolite Flush Mounted probes are pressure rated to 3600 psig at
21°C (70°F), with insertion pressure ratings dependent on the access
system rating. The process equipment must be isolated and depressurized prior to the removal of the probe.

Hydrogen Test Probe

The Cosasco hydrogen test probe (Fig. 12) operates on the principle
that hydrogen will diffuse through the thin wall and set up a pressure
within the tube. The rate at which the pressure increases is measured
by a pressure gauge. The rate at which hydrogen is penetrating per
unit area can then be determined, using the exposed surface area and
internal volume of the probe. If hydrogen diffusion rates exceed
0.25 cm^3/cm^2/per day, severe hydrogen damage can be anticipated.
The advantage of this instrument is that it alerts the engineer that
that hydrogen levels could lead to hydrogen blistering, hydrogen
embrittlement, or hydrogen stress corrosion cracking. A disadvantage
of the hydrogen probe is that since its measurement of corrosion is
relative, it is not directly correlatable with other monitoring techniques.

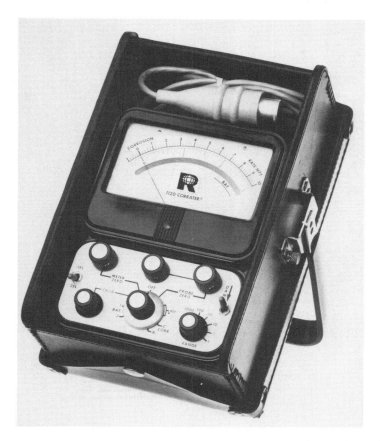

Figure 10. Rohrback 1120 portable corrater. (Courtesy of Rohrback Corporation, Santa Fe Springs, Calif.)

The Petrolite Hydrogen Patch* probe is shown in Fig. 13. A thin piece of palladium foil (0.010 in.) is placed against the pipe wall. An electrochemical patch is mechanically strapped onto the palladium foil using a thin Viton† gasket and a thick Teflon gasket which has been shaped to fit the contour of the pipewall. The cell itself contains a Hastelloy‡ B reference electrode. The palladium foil is the working electrode. The cell is filled with 20 to 30 ml of 90% sulfuric acid. The

---

*Trademark of Petrolite Corporation.
†Trademarks of E.I. duPont de Nemours.
‡Trademark of Stellite Corporation.

Figure 11. Petrolite flush-mounted probe. (Courtesy of Petrolite Corporation, Stafford, Tex.)

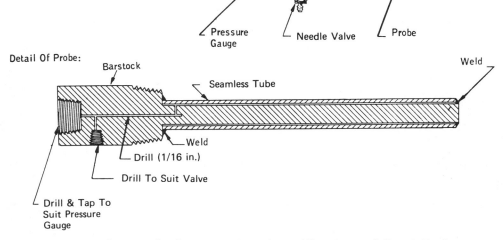

Assembly:

Pressure Gauge

Needle Valve

Probe

Detail Of Probe:

Barstock

Seamless Tube

Weld

Weld

Drill (1/16 in.)

Drill To Suit Valve

Drill & Tap To Suit Pressure Gauge

Figure 12. Cosasco hydrogen test probe. (Courtesy of Grant Tool Company, Cosasco Division.)

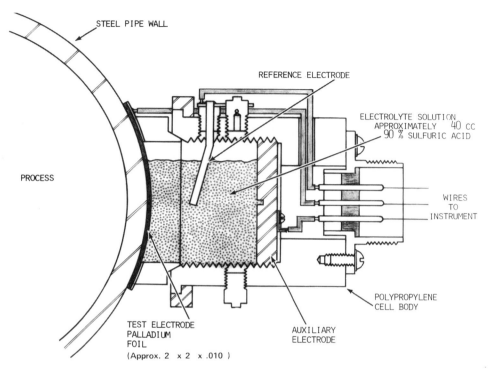

Figure 13. Petrolite Hydrogen Patch probe. (Courtesy of Petrolite Corporation, Stafford, Tex.)

instrument polarizes the palladium to a potential where hydrogen is oxidized quantitatively. Therefore, after an initial pump-down period of several hours, the current being read on the recording instrument is equivalent to the hydrogen penetration occurring at the time.

Since the hydrogen being measured by the Petrolite Hydrogen Patch Probe is being produced as a part of the corrosion reaction, the hydrogen penetration rate (which is being measured directly) is related to the internal corrosion rate. It must be noted that hydrogen measurement is only an indirect measure of corrosion rate because the amount of atomic hydrogen which dissolves in the steel, relative to that produced in the corrosion reaction, is a complex function of several variables such as stream composition, surface condition, temperature, and corrosion rate.

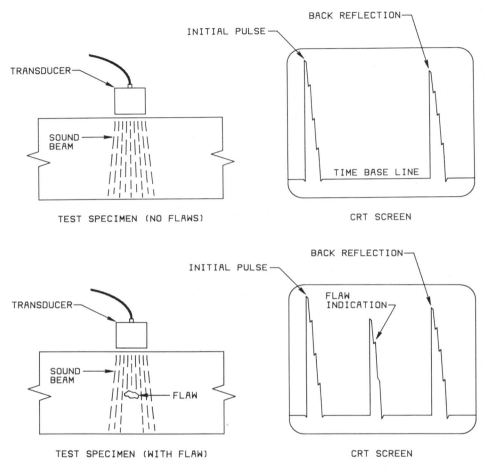

Figure 14. Principle of straight-beam ultrasonics. (Courtesy of
Krautkramer-Branson, Inc., Stratford, Conn.)

Ultrasonic Testing

Ultrasonic testing is a nondestructive method of determining wall
thickness or the location of flaws within any material capable of con-
ducting sound. In normal ultrasonic testing a beam of high frequency
(megahertz) vibrations is passed into the material under test. Reflections are obtained from the surfaces of the material and from
various physical discontinuities within it. These vibrations, by means
of a transducer, are converted into electrical signals which are amplified and fed to a cathode ray tube for interpretation. In general,

ultrasonic testing is a method of transforming electrical pulses into mechanical vibrations and transforming the mechanical vibrations back into electrical pulses. Figure 14 illustrates the principle of straight beam ultrasonic nondestructive testing. Figure 14 (top) represents the propagation of sound within a test specimen that does not contain any flaws. A typical cathode ray tube (CRT) screen presentation is illustrated to show the intial pulse, time base line, and back reflection. Figure 14 (bottom) represents the propagation of sound within a test specimen containing a known flaw. Note that the flaw indication as shown on the CRT screen display. Angle-beam or shear-wave ultra-sonic testing can be defined as testing in which the sound beam is sent into the test piece at an angle, using a type of ultrasonic sound wave known as a shear wave. Angle-beam testing is used to locate flaws or cracks that are not oriented properly in the test piece to be located by means of straight-beam tests. This method of testing is most favorable for weld inspection. Figure 15 represents ultrasonic evaluation of

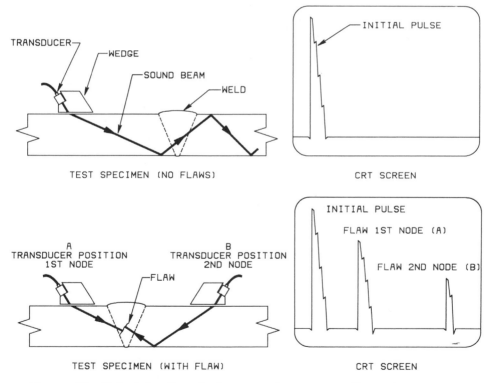

Figure 15. Principle of angle-beam or shear-wave ultrasonics. (Courtesy of Krautkramer-Branson, Inc., Stratford, Conn.)

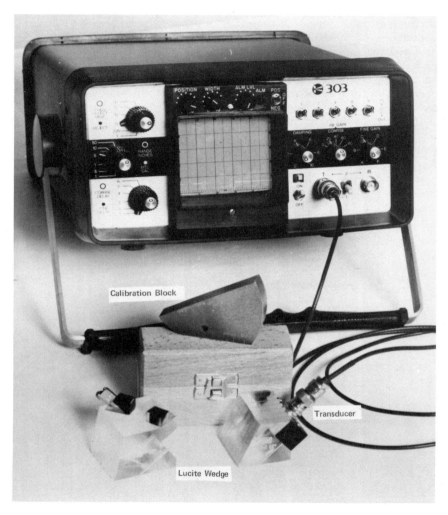

Figure 16. Krautkramer-Branson 303 ultrasonic flaw detector/thickness tester. (Courtesy of Krautkramer-Branson, Inc., Stratford, Conn.)

welded test specimens using the angle-beam (or shear-wave) method of sound propagation. Note the angular position of the transducer within the wedge. The CRT screen presentation illustrates the initial pulse of sound produced by the transducer in a welded test specimen that does not contain any flaws. The absence of the back reflection, which indicates material thickness and is usually visible in straight-beam tests, is attributed to the angle of the sound beam. A welded test specimen containing a known flaw is also illustrated. Note the transducer position and the distance between the transducer and the weld area. The flaw indication as illustrated on the CRT screen display is also shown. Figure 16 illustrates one of the Krautkramer-Branson portable ultrasonic instruments that can be utilized for both straight- and angle-beam flaw detection and wall thickness measurements.

Using an ultrasonic flaw detector such as that shown in Fig. 16, an operator can also measure wall thickness on pipes and pressure vessels as well as inspect for cracks and other material flaws. The left-to-right position of the back reflection that appears on the CRT screen during straight-beam testing indicates the thickness of the material being tested. The thinner the material, the farther to the left this signal appears; the thicker the material, the farther to the right it moves. Thus the operator can measure wall thickness by monitoring the position of this back reflection signal on the CRT.

There is, of course, a limit to how accurately an operator can visually judge the position of the back reflection, and this in turn limits the accuracy of the wall thickness measurement. Flaw detectors are now available that give digital thickness and flaw location readouts in addition to the CRT-screen presentation, and these readouts are generally more accurate than simple visual monitoring of the screen. However, even the accuracy of visual screen monitoring can be improved by use of what is known as the "multiple reflection method."

In the multiple reflection method, the flaw detector's screen is usually adjusted to show five times the wall thickness (see Fig. 17). Each multiple of the thickness is indicated by an additional back reflection known as a "multiple back reflection." The fifth back reflection will move five times more screen distance than the first reflection will move for a given change in wall thickness. Therefore, the operator can read wall thickness with five times greater accuracy by monitoring movement of the fifth rather than the first back reflection.

Ultrasonic wall thickness measurement can also be accomplished—much more easily, in fact—by the use of digital readout ultrasonic thickness gages. These lightweight, portable, battery-powered ultrasonic thickness gages are widely accepted and highly reliable. Figure 18 illustrates a typical application of the Krautkramer-Branson direct-reading (D meter) digital thickness gage.

On-stream metal thickness measurements can be obtained at temperatures above 900°F by use of cooling mediums with the transducer. The

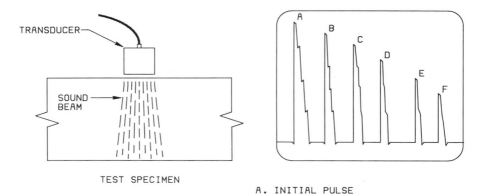

TEST SPECIMEN

A. INITIAL PULSE

B. FIRST REFLECTION FROM BACK SURFACE

C-F. MULTIPLE BACK REFLECTIONS

Figure 17. Typical CRT display of the multiple reflection method. (Courtesy of Kratkramer-Branson, Inc., Stratford, Conn.)

higher the surface temperature, the greater the potential for error due to material expansion and a lower acoustic velocity. The engineer must take this into account and adjust the readings downward to indicate actual wall thickness. Ultrasonic testing can be used to determine wall thinning, pitting, erosion, and flaws in metals, plastics, and rubbers. Several disadvantages include: numerous readings are required to describe general condition of material; pitting corrosion is not easily located; readings must be taken over a period of time to determine the corrosion rate; and high temperature measurements may have to be adjusted.

Engineers must have a good understanding of where to expect corrosion in equipment, such as towers, drums, heat exchangers, and piping, if they are to take maximum advantage of ultrasonics to monitor corrosion. In towers, areas of concern include: behind downcomers, liquid-vapor interfaces, and across from gas or liquid inlets. For drums and heat exchangers, areas of concern include: liquid-vapor interfaces, stagnant areas, and across from gas or liquid inlets. In piping systems areas of concern include: elbows, tees, downstream of valves and pumps, and small connections. Figures 19-22 illustrate typical locations where corrosion would be most likely to occur in a piping system. The number of measurements and/or locations depends on the corrosive environment and configuration, and therefore vary between piping systems. Where there are small connections and/or branch connections, size and/or configuration may not be feasible for ultrasonic measurements. In these instances, radiography should be considered for determining wall thickness.

Figure 18. Use of the Krautkramer-Branson D meter for wall thickness measurement. (Courtesy of Krautkramer-Branson, Inc., Stratford, Conn.)

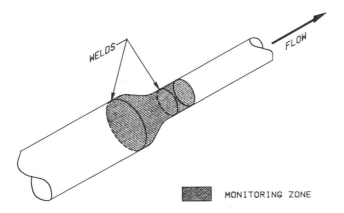

Figure 19. Typical corrosion monitoring of a reducer.

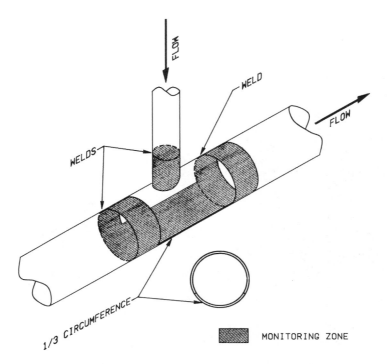

Figure 20. Typical corrosion monitoring of a tee.

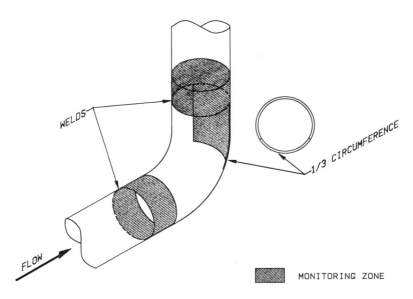

Figure 21. Typical corrosion monitoring of an elbow.

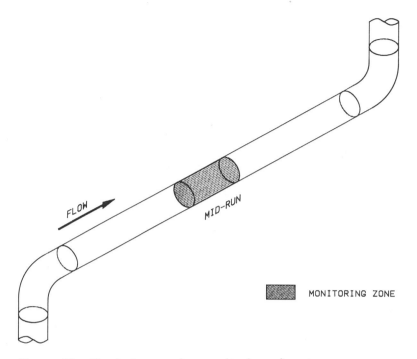

FLOW

MID-RUN

▓▓▓▓ MONITORING ZONE

Figure 22. Typical corrosion monitoring of a pipe.

Radiography

Radiography is one of the most important, and most versatile, of all the nondestructive test methods used by industry. Employing highly penetrating x-rays, gamma rays, and other forms of radiation, radiography provides a permanent visible film record of internal conditions, containing the basic information by which soundness can be determined. X-ray equipment is portable but bulky, and requires electrical connections in order to operate. X-ray equipment is normally used for the inspection of thin material, 0.125- to 0.750-in. steel.

The most widely used gamma radiographic sources are iridium 192 (^{192}Ir) and cobalt 60 (^{60}Co). Iridium 192 is used for material thickness of 0.250 in. to 3.500 in. steel and cobalt 60 for 2.500 in. to 8.000 in. steel. Gamma radiographic sources do not require electrical connections and are very portable. For this reason they are more widely used. The use of radiography for on-stream corrosion monitoring is often overlooked. Radiography is very useful in detecting corrosion or corrosion erosion in elbows, welds, heat-affected zones, return bends on furance tubes, small connections, and thermowells.

Radiography can also be employed to indicate wall thickness, erosion, and weld integrity. Some disadvantages are: the physical limitations where it can be applied; difficulty in detecting cracks lying in the plane perpendicular to the rays; and the safety precautions involved in radioactive material handling.

pH Instrument

The pH instrument can be a useful tool for determining whether process conditions are such that increased corrosion can occur. The engineer must have a good understanding of the construction materials and their limitations during pH swings. The main advantage of the pH instrument is that it will show immediate changes in process conditions. Disadvantages are: the extent of corrosion occurring in the process equipment cannot be determined by the instrument; and pH instrumentation requires a high level of maintenance.

Infrared Thermographic

The infrared thermographic camera is used to identify hot spots on process equipment. The camera works on the theory that the hotter the object, the higher the frequency of radiation. The infrared (IR) detector in the camera detects the IR rays emitted by the object and converts them to an electronic video signal. The signal is then sent to a display unit showing the thermal image and temperature range. The temperature range of the equipment is -30 to +1000°C (-20° to +1830°F). The camera can be used to determine hot spots on reactors, furances and furance tubes, stacks, structural members, and rotating equipment.

Corrosion Coupons

Corrosion coupons are the most widely used tools to monitor corrosion in process equipment. Figure 23 and 24 show several typical methods of mounting corrosion coupons. Coupons can be made in any size or shape so as to be retrievable from process equipment without shutting the process unit down. This method of monitoring corrosion has the advantage that the corrosion observed has actually occurred on the sample and allows for a visual examination, physical measurements, and the chemical analysis of corrosion products. Corrosion coupons can be mounted in different configurations to study different types of corrosion mechanisms, such as crevice corrosion, galvanic attack, and stress corrosion.

Normally, the corrosion coupons are carefully weighed and measured prior to assembly on the corrosion test rack. Once the coupons

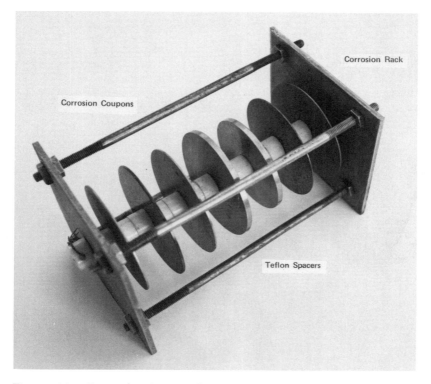

Figure 23. Corrosion test rack.

are assembled, they should be stored in a desiccator until installation.
During assembly the coupons are arranged so that they are electrically
insulated from the corrosion rack in order to avoid galvanic attack.
Upon completion of the test the corrosion rack is disassembled and the
corrosion coupons are cleaned, weighed, and measured. The formula
for calculating the uniform corrosion rate is:

$$\text{Mils per year} = \frac{534W}{DAT}$$

where
  W = weight loss, mg
  D = density of specimen, g/cm^3
  A = area of specimen, $in.^2$
  T = exposure time, h

The main advantages to corrosion coupons are: low cost; the ability
to test several different materials at once; and closer resemblance to

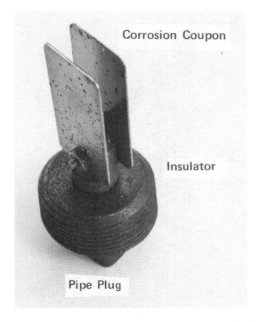

Figure 24. Corrosion coupons mounted on a pipe plug.

actual conditions of equipment. The main disadvantages include: test
location of coupons is limited; time involved in preparing and evaluating
coupons to determine corrosion rates after a sufficient exposure time.
In addition, this method of monitoring does not allow the engineer to
evaluate the effects of varying process operating conditions.

OFF-STREAM MONITORING EQUIPMENT

Acoustic Emission Testing

Acoustic emission testing (AET) is used most often to requalify pressure vessels and tanks during a hydrostatic test. Acoustic emission
is defined as the pressure of stress waves generated in metals by the
energy released as the metal deforms or fractures. AET sensors are
placed at various locations on the surface of the equipment being
tested to pick up signals emitted from flaws and discontinuities. These
signals are then amplified and through use of computerized equipment
are analyzed to determine the severity of the defects and their locations. The main advantages of AET are: flaws can be detected early
and therefore prevent catastrophic failures; and AET has a potential
use for on-stream monitoring, particularly in heavy vessel sections as

the equipment is heated or cooled. The main disadvantages are: high cost; and highly skilled test personnel are required to interpret the findings.

Eddy Current Inspection

Eddy current inspection is based on the principle of electromagnetic induction and is used to identify or differentiate between a wide variety of physical, structural, and metallurgical conditions in conductive ferromagnetic and nonferromagnetic metal components. Eddy current inspection can be used to detect cracks, voids, and inclusions, to sort dissimilar metals and detect differences in their chemical composition, and to measure the depth of nonconductive coatings on conductive metals. Some disadvantages of eddy current inspection are: liftoff factor, fill factor, edge and skin effects, and permeability variations in magnetic materials.

The Krautkramer-Branson Model 700 Probolog (Fig. 25) is a portable instrument for the nondestructive inspection of nonferrous exchanger tubes. The instrument operates on the principle of eddy current. The magnetic field generated by a coil to which an alternating current is applied will cause eddy currents to flow in metal objects near the coil. The eddy current density will be greatest at the surface of the metal adjacent to the coil and will decrease rapidly in the metal as the thickness increases. Any discontinuity in the tube wall (such as a hole, pit, or crack) will interrupt or distort the normal eddy current flow pattern. As the probe is pulled through a tube, results can be monitored on a storage oscilloscope and a strip chart records the tube condition. Figure 26 illustrates a sample strip chart recording corresponding to tube wall discontinuities. The Probolog can be used in the field to determine the condition of a heat exchanger tube bundle. Disadvantages are: the tubes must be clean; it is a slow process since only one tube can be examined at a time (2 to 3 min per tube); and only nonferrous tubes can be evaluated.

The Eddy Current I.D. tube testing system (Fig. 27) is manufactured by the Magnaflux Corporation, and is used for testing ferrous and nonferrous materials. The instrument operates on a principle similar to the Probolog. For testing ferromagnetic tubing, the permeability must be minimized. A constant magnetic field within the tube wall must be induced.

The Magnaflux eddy current system (Fig. 27) consists of six essential parts. The ED-800 eddy current instrument, part A, supplies the electrical energy to the I.D. probe, part B. An air gun, part C, is used to move the probe to the end of the tube. The probe puller, part D, is used to pull the probe back through the tubes at a uniform rate. As the probe is uniformly pulled through the tube, it monitors changes in the electrical field. Any change is amplified by the instru-

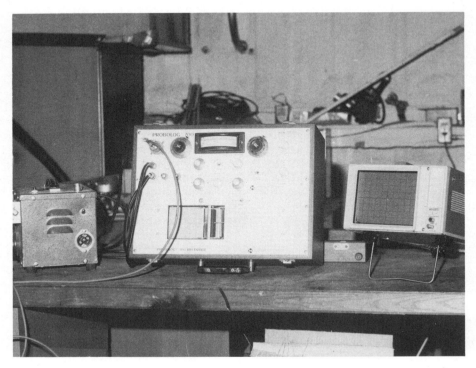

Figure 25. Krautkramer-Branson Model 700 Probolog. (Courtesy of Krautkramer-Branson, Stratford, Conn.)

ment and simultaneously displayed on the ED-800 cathode ray tube and the dual channel strip chart recorder, part E. A separate power supply, part F, energizes an electromagnet in the probe when inspecting ferromagnetic tubing. Figure 28 shows the probe being inserted into a tube opening at the tube sheet. The advantage of using eddy current instruments for evaluating tube condition is that in many instances the corrosion is localized in a tube bundle. Using this type of instrument allows the engineer to determine the extent of damage and consider whether plugging, partial retubing, or complete retubing is necessary. When the strip chart recorder is used, a permanent record is available; the rate or corrosion can be monitored and it becomes feasible to predict when failures will occur. This advance warning allows for tubes to be purchased and on hand prior to failure, thus avoiding or reducing the downtime for repairs/replacement.

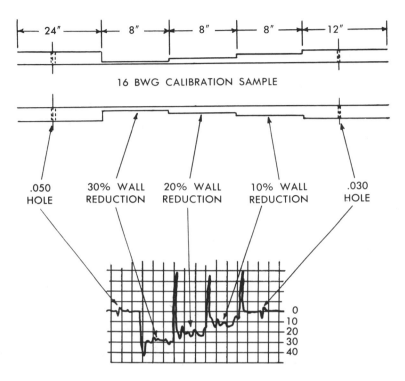

Figure 26. Calibration sample versus strip chart. (Courtesy of Krautkramer-Branson, Stratford, Conn.)

Figure 27. Magnaflux eddy current I.D. tube testing system. (Courtesy of Magnaflux Corporation, Chicago.)

Figure 28. Typical application of Magnaflux eddy current test equipment. (Courtesy of Magnaflux Corporation, Chicago.)

Magnetic Particle Inspection

Magnetic particle inspection is a nondestructive means for detecting
surface and slightly subsurface discontinuities in ferromagnetic materials. This method of nondestructive testing consists of four basic
operations: (1) establishing a suitable magnetic field in the test object, (2) applying magnetic particles to the surface of the test object,
(3) examining the test object surface for accumulations of the particles
(indications), and (4) evaluating the serviceability of the test object.
Magnetic particle inspection depends on the magnetic properties of the
test objects and is suitable only for metallic materials that can be in-
tensely magnetized. Nonferromagnetic materials that cannot be strongly magnetized cannot be inspected by this method. Inspection surface
areas must be clean and free of random dirt, rust, and scale. Any

Table 1. Off-Stream Monitoring Equipment

Type	Application	Advantages	Limitations
Borescope	Visual aid to inspect tubes, pumps, compressors, etc.	Actual view of equipment surface	Access to equipment
Chemical Spot testing	Identify alloy constituents of unknown materials	Economical, quick	Not precisely quantitative
TV camera	Visual aid to inspect tubes, pipes, vessels, etc.	Actual view of equipment condition	Access to equipment
Holography	Three-dimensional imaging	No physical contact with equipment	Type of flaw not easily defined
Texas Nuclear Analyzer	Positive material indentification	Accuracy and reliability of findings	Size and shape of test specimen, accessibility, temperature

part made of ferromagnetic material can be inspected by the magnetic particle method; there are no restrictions as to the shape and size of the part. Magnetic particle inspection can be accomplished using either the wet or dry method. The wet method of magnetic particle inspection is used with stationary equipment and is best suited for the detection of fine surface discontinuities such as fatigue cracks. The dry method or dry particle inspection are most sensitive for use on very rough surfaces and for detecting defects beneath the surface. The dry method of magnetic particle is normally used with portable equipment.

Liquid Penetrant Inspection

Liquid penetrant inspection is basically a very simple process. First a liquid penetrant is applied to the surface of the part. It is permitted to remain on the surface for a specified time, during which it penetrates into any defects open at the surface. The excess penetrant that remains on the surface is removed. Then an absorbent, light-colored, powdered material called a developer is applied to the surface. The developer acts as a blotter and draws out a portion of the penetrant which had previously seeped into the surface openings. As the penetrant is drawn out, it diffuses into the coating of the developer, forming indications that are much wider than the surface openings with which they are associated. The inspector then views the part and looks for these colored indications, noting the surface condition, against the background of the developing powder. The penetrant method is a quick and low cost method of nondestructive testing to determine surface defects. The most common type of penetrant inspection is accomplished using a colored visible dye. Fluorescent penetrants have been developed for use with a black light and provide greater sensitivity. The main disadvantages of penetrant inspection are: it can only be used to identify surface defects; and the surface must be clean and fairly smooth.

Miscellaneous Equipment

Table 1 lists several other tools which are commercially available to aid in monitoring off-stream corrosion.

SUMMARY

This chapter has only scratched the surface of the many nondestructive tools available to the engineer. These instruments play an important role in monitoring on-stream and off-stream corrosion. Corrosion

costs the chemical and refining industries billion of dollars annually.
The challenge to the engineer is to use the tools and techniques available to control and monitor corrosion to avoid unscheduled shutdowns,
and to educate others as to the true value of corrosion controls. The
days of bringing equipment down only for inspection are rapidly coming
to an end; much greater emphasis is now placed on keeping units
running. The engineer must understand the processes and rely on
corrosion monitoring to ensure safe and continual operation of process
equipment.

ACKNOWLEDGMENTS

The author acknowledges the Rohrback Corporation, Petrolite Corporation, Magnaflux Corporation, and Krautkramer-Branson for supplying
photographs and literature on the nondestructive testing equipment
described in this chapter.

An expression of sincere appreciation is due to A. J. Pron and
B. H. Elliott for their able assistance in preparing illustrations and
comments regarding the text.

GLOSSARY

Angle Testing An ultrasonic testing method in which the angle of in-
cidence is greater than zero.

Angle Transducer A transducer used in angled testing in which the
sound beam is set to some predetermined angle to achieve a special
effect (e.g., setting up shear or surface waves in the tested
piece).

Back Reflection The ultrasonic echo from the back surface of the
part.

Beam A directed flow of energy into space or matter.

Black Light Light energy just below the visible range of violet light,
often predominantly of about 3650 Å. This wave length reacts
strongly on certain dyes to make them fluoresce in a range visible
to the eye.

Cobalt 60 A radioisotope of the element cobalt.

Color-Contrast Dye A dye that can be used in a penetrant to impart
sufficient color intensity to give good color contrast in indications
against the background of the surface being tested, when viewed
under white light.

Color-Contrast Penetrant A penetrant incorporating a dye—usually
nonfluorescent—sufficiently intense to give good visibility to flaw
indications under white light.

Conductivity This is the inverse of resistance and refers to the ability of a conductor to carry current.

Contact Testing Testing with transducer assembly in direct contact with material through a thin layer of couplant.

Corrosion The deterioration of a metal by chemical or electrochemical reaction with its environment.

Couplant A substance used between the search unit and that surface to permit transmission of ultrasonic energy.

Defect A discontinuity, the size, shape, orientation, or location of which makes it detrimental to the useful service of the part in which it occurs.

Developer (Penetrant) A finely divided material applied over the surface of a part to help bring out penetrant indications.

Discontinuity Any interruption in the normal physical structure or configuration of a part, such as cracks, laps, seams, inclusions, or porosity. A discontinuity may or may not affect the usefulness of a part.

Echo *See* Reflection.

Electrochemical Corrosion Corrosion that occurs when current flows between cathodic and anodic area on metallic surfaces.

Electromagnet When ferromagnetic material is surrounded by a coil carrying current it becomes magnetized and is called an electromagnet.

Embrittlement Reduction in the normal ductility of a metal due to a physical or chemical change.

Erosion Destruction of metals or other materials by the abrasive action of moving fluids, usually accelerated by the presence of solid particles or matter in suspension. When corrosion occurs simultaneously, the term *erosion corrosion* is often used.

Ferromagnetic Materials The materials which are most strongly affected by magnetism are called ferromagnetic materials because iron and steel exhibit the strongest magnetic characteristics of all substances.

Flaw An imperfection in an item or material which may or may not be harmful.

Fluorescent Dye A dye that becomes fluorescent, giving off light, when it is exposed to short-wavelength radiation such as ultraviolet or near-ultraviolet light.

Fluorescent Penetrant A penetrant incorporating a fluorescent dye to improve the visibility of indications at the flaw.

Gamma Rays High-energy, short-wavelength electromagnetic radiation emitted by a nucleus. Energies of gamma rays are usually between 0.010 and 10 MeV. X-rays also occur in this energy range, but are of nonnuclear origin. Gamma radiation usually accompanies alpha and beta emissions and always accompanies fission. Gamma rays are very penetrating and are best attenuated by dense materials such as lead and depleted uranium.

Hydrogen Embrittlement A condition of low ductility in metals resulting from the absorption of hydrogen.

Indication In nondestructive inspection, a response, or evidence of a response, that requires interpretation to determine its significance.

Interface A common boundary between two surfaces.

Iridium 192 A radioactive isotope of the element iridium which has a half-life of 75 days. It is used extensively as a source of gamma radiation.

Node A point in a standing wave where some chracteristic of the wave field has essentially zero amplitude.

Nondestructive Testing (NDT) Testing to detect internal and concealed defects in materials using techniques that do not damage or destroy the items being tested.

Particle Motion Movement of particles in an article brought about by the action of a transducer.

Penetrant This is the fluid—usually a liquid but it can be a gas—that is caused to enter the discontinuity in order to produce an indication at the flaw.

Permeability (Magnetic) The ease with which a magnetic field or flux can be set up in a magnetic circuit. It is not a constant value for a given material, but is a ratio. At any given value of magnetizing force, permeability is B/H, the ratio of flux density B to magnetizing force H.

Reflection The indication of reflected energy.

Sensitivity The smallest quantitative increment detectable in using a measuring instrument.

Shear Wave A wave in which the particles of the medium vibrate in a direction perpendicular to the directio of propagation.

Source The origin of radiation; an x-ray tube or a radioisotope.

Source Material In atomic energy law, any material, except special nuclear material, which contains 0.05% or more of uranium, thorium, or any combination of the two.

Transducer Any device that is capable of converting energy from one form to another.

Ultrasonic Testing A nondestructive method of testing materials by transmitting high frequency sound waves through them.

X-Ray Penetrating electromagnetic radiation emitted when the inner orbital electrongs of an atom are excited and release energy. Thus the radiation is nonnuclear in origin and is generated in practice by bombarding a metallic target with high-speed electrons.

REFERENCES

1. P. H. Hutton, Acoustic emission in metals as a NDT tool, 27th Nat. Conf. Soc. Nondestructive Test., Cleveland, Ohio. October 18, 1967.

2. L. M. Callow, J. A. Richardson, and J. L. Dawson, Corrosion
    monitoring using polarization resistance measurements, *Br.
    Corros. J. 11*(3), 1976.
3. P. A. Burda, Linear polarization method for corrosion rate
    measurements in limestone slurry scrubber, *Mater. Perform.*,
    June 1975.
4. L. W. Holloway and D. M. Foreman, Eddy current inspection of
    tubular equipment containing nonferrous material, Baton Rouge
    Sect. NACE, Baton Rouge, La., November 27, 1972.
5. D. R. Fincher and A. C. Nestle, New developments in monitoring
    corrosion control, NACE Int. Corros. Forum, Anaheim, Calif.,
    March 19-23, 1973.
6. J. H. Lanaman and L. Azar, A new instrument for measurement
    of corrosion by electrical resistance techniques, NACE Int.
    Corros. Forum, Anaheim, Calif., March 19-23, 1973.
7. K. J. Miller, Jr., Infrared scanning for nondestructive testing,
    NACE Int. Corros. Forum, Anaheim, Calif., March 19-23, 1973.
8. J. C. Bovankovich, On-line corrosion monitoring, NACE Int.
    Corros. Forum, Anaheim, Calif., March 19-23, 1973.
9. *Petrolite Instruments General Catalog*, Stafford, Tex., 1977.
10. *Corrosion Measurement and Control Systems*, Rohrback Corpora-
    tion, Santa Fe Springs, Calif., 1975.
11. *Nondestructive Testing Handbook* (Robert C. McMaster, ed.),
    Ronald Press, New York, 1959.
12. S. Elonka, Nondestructive testing, *Power*, March 1958.
13. J. B. Harrell, Corrosion monitoring in the CPI, *Chem. Eng.
    Prog.*, March 1978.
14. *Instrument Catalog*, Krautkramer-Branson, Inc., Stratford,
    Conn., 1976.
15. E. C. French, Flush-mounted probe measures pipe corrosion,
    Oil Gas J., November 1975.
16. G. Johnson and T. W. McFarland. Better data with eddy current,
    Hydrocarbon Process., January 1978.
17. T. W. McFarland, Nondestructive testing of ferromagnetic and
    nonferromagnetic heat exchanger tubing from the ID for wall
    thinning and cracks, Joint Pet. Mech. Eng. Press. Vessels
    Piping Conf., Mexico City, September 1976.

10
Corrosion Testing Techniques

CHARLES G. ARNOLD / Dow Chemical Company, Texas Division, Freeport, Texas

INTRODUCTION

The most reliable corrosion testing technique is that of exposing actual parts in their intended service environment and then evaluating the corrosion that took place. In most cases, this technique is not used because it is too expensive and time consuming; and the results of the evaluation are often subjective, making it difficult to compare metals. Reliable corrosion data can be obtained using a number of techniques if the corrosion tests are designed and conducted properly. The simplest techniques involve the determination of a change in dimensions or weight and observation of the corroded surface. More complex techniques involve measurement of hydrogen diffusion or electrical resistance or determining electrochemical characteristics. The use of these techniques, together with their advantages and disadvantages, are discussed in the remainder of this chapter.

DIMENSION CHANGE

Ultrasonic Thickness Measurement

One of the most straightforward corrosion testing techniques is ultrasonic thickness measurement of parts, either under service conditions or after they have been removed from service. The thickness of the part is measured at the start of the test period and after regular service intervals. From a practical standpoint, the accuracy of ultrasonic thickness measurement that could involve different instruments and technicians is ±0.010 in. (more accurate measurements can be made using a single instrument and a highly qualified technician). The difference between thickness measurements can be divided by the time interval to obtain the corrosion rate. The change in thickness should be plotted versus time as shown in Fig. 1 so that changes in corrosion rate can be more easily traced to changes in the environment. This technique is not accurate enough for most laboratory testing, but it is the most accurate technique available that can be used to measure the thickness of a part while it is in service. It is extremely useful when one wants to determine the corrosion rate of a pressure vessel or pipeline that is being used and to determine the real effects of process changes or inhibitor additions. An instrument with a cathode ray tube should be used if there is a possibility of nonuniform corrosion because digital instruments can give misleading readings [1].

Eddy Current

The amount of corrosion that has taken place in a nonferromagnetic tube can be measured nondestructively using an eddy current instrument and a probe that fits inside the tube. This procedure makes it possible to measure the corrosion that has taken place in tubes while they are still in a heat exchanger so that their remaining life can be predicted. These instruments must be calibrated on a tube of known thickness of the same metal as the tube to be tested. After calibration, changes in thickness can be measured with an accuracy of ±2%. Nonuniform corrosion can be detected very easily.

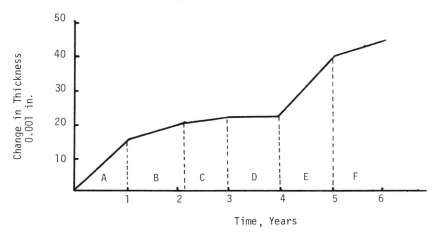

Figure 1. Change in thickness versus time. A, plant startup–corrosion rate equals 15 mpy; B, continuous operation–corrosion rate in the second year equals 5 mpy, overall corrosion rate equals 10 mpy; C, process changed–corrosion rate in the third year equals 2 mpy, overall corrosion rate equals 7 mpy; D, plant shutdown; E, vessel put in pilot plant–corrosion rate in the fourth year equals 18 mpy, overall corrosion rate equals 8 mpy; F, continued pilot plant service–corrosion rate in the fifth year equals 5 mpy, overall corrosion rate equals 6 mpy. If the vessel had an initial corrosion allowance of 0.125 in., it now has 0.080 in. remaining, or an expected life of 13 years based on the overall corrosion rate or 16 years based on the last year's corrosion rate.

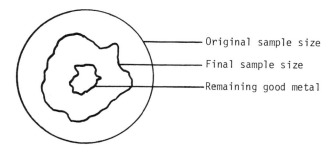

Figure 2. Microscopic corrosion measurement is often used in dealloying and high temperature corrosion testing to determine how much good metal is left.

Microscopic Examination

It is often necessary to examine a polished cross section to determine
how much unaffected metal is left when testing for either dealloying or
high temperature corrosion. In both of these types of corrosion, considerable damage can take place due either to the removal of some of
the elements from the solid alloy or to inward diffusion of a corrodant
such as oxygen or sulfur. The amount of damage that has taken place
will not be proportional to either a change in external dimensions or
weight. It is necessary to prepare a cross section of corroded metal
for metallographic examination and determine microscopically how much
metal is left, as shown in Fig. 2. The determination of a penetration
rate is simplified if the test specimen is either round or has flat sides,
but complex geometries such as airfoils can be used.

WEIGHT CHANGE

Weight loss corrosion testing is used more than any other technique.
When results of other techniques are in question, they are usually
verified with weight loss testing. More materials or changes in conditions can be evaluated for a given amount of money with weight loss
testing than with any other technique. In addition, weight loss testing
can be used in the laboratory or field. In weight loss corrosion testing, a prepared specimen is exposed to the test environment for a
period of time and then removed to determine how much metal has been
lost. The exposure time, weight loss, surface area exposed, and the
density of the metal are used to calculate the corrosion rate of the
metal. Over the years, a number of expressions have been developed
to express corrosion rate, and these are covered very well by Fontana
and Greene [2]. Penetration rate, expressed in either mils per year
(mpy) or millimeters per year (mm/y), is one of the best methods of
expressing corrosion rate because it can be used by designers to
determine the corrosion allowance of a part.

Calculations

Corrosion rates may be calculated using the equation

$$\text{mpy} = \frac{\text{WL} \times 22,273}{\text{D} \times \text{A} \times \text{T}} \tag{1}$$

where
   WL = weight loss, g
   D = density, g/cm^3
   A = area, $in.^2$
   T = time, days

The weight loss should be measured to the nearest milligram (0.0001 g). Densities for most alloys can be found in either the *Metals Handbook* [3] or *Properties of Some Metals and Alloys* [4]. Dimensions should be measured accurately enough so that the area can be calculated to ±1%. The area of all exposed surfaces, including thin edges, should be included in the A term. Specimen area that is masked by gaskets or holders should be omitted from the A term. Time should be measured to an accuracy of ±1% of the total exposure time.

Specimen Preparation

Most authors recommend that one of the final steps in specimen preparation be abrasion with 120-grit abrasive paper or cloth. This step helps the investigator obtain reproducible results, and it should be used when effects of changing the environment (inhibitor concentration, amount of oxidizer, etc.) are being studied on a metal or group of metals. If grinding is used, new abrasive paper should be used for different types of alloys, and wet grinding should be used for alloys that work harden easily (austenitic stainless steels, nickel-base alloys, titanium, etc.). It is necessary to grind or machine all edges that have been cut or sheared since preferential attack can start at the rough surfaces, which contain numerous fissures.

When corrosion testing is being conducted to determine which metals are most suitable for a given service, it is best to test them with a surface condition that is representative of how they will be used in practice. In these cases, "as-rolled" surfaces with sheared edges should be used on the corrosion coupons. When materials are susceptible to stress corrosion cracking, cold-worked or stressed specimens should be considered. Welded coupons should be used whenever there is a possibility that the weld could be anodic to the base metal (dissimilar weld metals) or when there is a possibility of heat-affected zone corrosion (stainless steels and nickel base alloys with a carbon content greater than 0.03%).

Specimen Holders

The test specimens should be physically separated from each other and supported on inert racks. Glass racks are often used in laboratory tests. Metal racks are usually used in the field and are sometimes used in the laboratory. The test specimens must be insulated from a metal rack as shown in Fig. 3. Some examples of how coupon racks can be installed in the field are: welded in place in a pipeline, nozzle, or vessel as shown in Fig. 4, connected to a pancake and held in place with a blind flange (Fig. 5), and installed through a valve using a packing gland (Fig. 6). The insulators should be stable

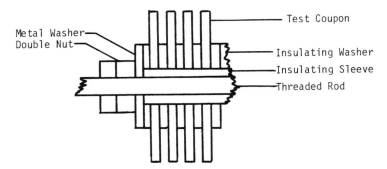

Figure 3. Corrosion coupons insulated from each other and the coupon rack.

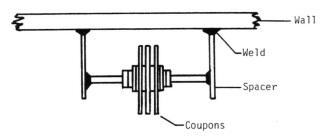

Figure 4. Specimen rack welded to the wall.

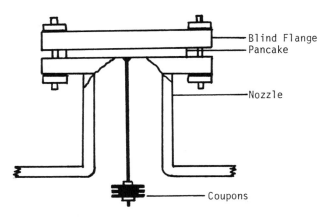

Figure 5. Coupons installed in a nozzle using a blind flange as a backup.

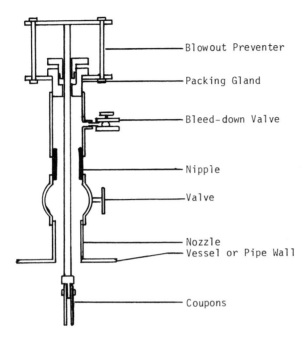

Figure 6. Coupons installed through a valve. A bleed-down valve is often used with high temperatures, high pressures, and so on, to ensure that the primary valve is closed and tight before removing the packing gland. A blowout preventor is used when the internal pressure is high.

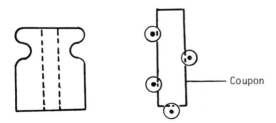

Ceramic Insulator 1 - 1 1/2 in. high

Figure 7. Atmospheric corrosion coupons held with ceramic electrical insulators.

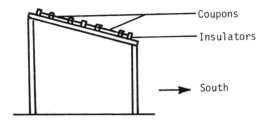

Figure 8. Atmospheric corrosion test rack.

in the test environment, and Teflon is used quite often at low to
moderate temperatures and ceramics at high temperatures.

Specimen holders for atmospheric corrosion testing are often made
using ceramic electrical insulators as shown in Fig. 7. Atmospheric
corrosion coupons are usually mounted either at an angle of 20 to 30°
from horizontal (Fig. 8) or vertically.

Test Conditions

Corrosion testing to determine which metal is best for a given process
should be conducted under actual field conditions whenever possible.
Coupons should be exposed under all of the conditions for which a
metal is needed. Example of where coupons should be located in a
tower of a chemical plant or refinery are: the feedstream, tower overhead, tower bottoms, and reboiler outlet line. When it is not possible
to test the metal in all locations, the areas of highest expected corrosion (highest velocity, highest temperature, etc.) should be chosen
as exposure sites.

When field testing is not possible, laboratory testing is sometimes
helpful in screening metals. When laboratory testing is required, it
should be kept in mind that very small changes in the environment can
produce large changes in the corrosion rates of some metals, especially
those which can be either active or passive and that laboratory test
results may not represent field conditions. Field conditions should be
duplicated as nearly as possible in the laboratory and the corrodant
volume-to-exposed metal area ratio should be as high as practical. In
addition, when trace amounts of materials that may act as corrodants,
oxidizers, inhibitors, or chelating agents are present, it may be necessary to change the corrodant sample frequently. Condensate is often
more corrosive than the bulk of the sample, so it becomes necessary
to fabricate test devices similar to that shown in Fig. 9 for corrosion

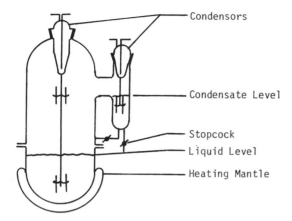

- Condensors
- Condensate Level
- Stopcock
- Liquid Level
- Heating Mantle

Figure 9. Apparatus for measuring corrosion rates in a liquid, its hot vapor, and condensate.

testing. Materials that are being evaluated for use as a heat transfer surface are often tested using appartus like that shown in Fig. 10. When using a heat transfer corrosion test appartus, the temperature of the heater should be kept constant so that the test conditions can be duplicated. This is usually accomplished by running a blank to determine the heater voltage necessary to maintain either the correct

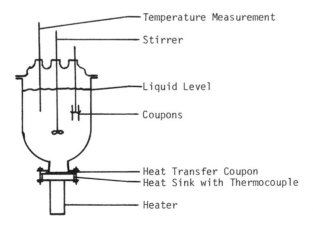

- Temperature Measurement
- Stirrer
- Liquid Level
- Coupons
- Heat Transfer Coupon
- Heat Sink with Thermocouple
- Heater

Figure 10. Simple heat transfer test apparatus.

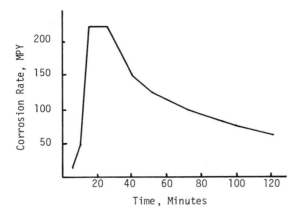

Figure 11. Corrosion rate versus time. The average corrosion rate during the 2-h batch reaction was in excess of 100 mpy. A 2-week exposure test gave a corrosion rate of less than 10 mpy. This shows that long-term weight loss data can be misleading.

metal or liquid temperature. Laboratory tests to determine the materials of construction for a batch process are very often misleading since the products that are present and thus the corrosivity of the mixture changes as the batch process reactions take place. Figure 11 is a corrosion rate versus time plot that is typical of a batch process.

New metals, inhibitors, and treatment chemicals are usually screened with carefully controlled laboratory tests so that the test results can be compared with the results of other tests. A large amount of corrosion data is available from manufacturers of new products or metals, but it still is usually necessary for users to conduct their own laboratory screening tests to determine if field tests should be conducted. New metals should be evaluated for their resistance to pitting, crevice corrosion, intergranular corrosion, and stress corrosion cracking before extensive field testing is started. Inhibitors and other treatment chemicals should be tested at several concentrations under different conditions of pH, temperature, and aeration in the laboratory to blanket the conditions under which they can either decrease or increase the corrosion of metals.

The duration of the tests is determined by the type and purpose of the test. When the test specimens are welded or bolted into a process vessel or piping, the test time is the time between two or more shutdowns of the equipment. When the coupons are inserted through a valve using a packing gland in field tests or when laboratory tests are conducted, the time of the test can and should be controlled. The

following equation [5] is often used to determine the duration of
a test:

$$\text{Duration of test (hours)} = \frac{2000}{\text{corrosion rate (mpy)}} \qquad (2)$$

A moderate corrosion rate of 10 mpy would require an exposure time
of 200 h, or a little over 8 days. A low corrosion rate of 2 mpy would
require an exposure of over 41 days. Commonly used laboratory test
times are 2, 7, and 14 days. Field tests are usually 14, 30, or 90 days
and sometimes last over 1 or more years. Atmospheric corrosion tests
usually last at least 1 year and very often extend to 3 to 5 years.

Specimen Cleaning

The specimens should be examined thoroughly before they are cleaned.
A binocular microscope should be used when possible. The location
and type of deposits and localized corrosion should be recorded and
photographs should be taken when they would be of value. Samples
of deposits and corrosion products should be removed for chemical
analysis when necessary.

The corrosion products should be removed completely from the
coupons using an appropriate cleaning solution [6] and they should be
weighed to determine the weight loss. The corrosion rate is calculated
using equation (1).

Evaluation of Results

The cleaned and weighed coupon should be examined thoroughly and
a binocular microscope should be used if any localized corrosion is
either visible or possible. Edges should be examined for end grain
attack, gasket areas for crevice corrosion, and the identifying stamp-
ing for stress corrosion cracking. The size and depth of pits and
crevice corrosion should be recorded.

A cross section of dealloyed specimens should be examined microscopically to determine the depth of the attack.

Data Recording

The type and amount of data that are recorded for a corrosion test
will depend on the purpose of the test. The checklist below is included as a guideline for what data should be recorded.

1. *Corrosive media*: overall concentration and variation in concentration during the test.

2. *Volume of test solution*: laboratory tests.
3. *Temperature*: average, variation, and was it a heat transfer test?
4. *Aeration*: technique or conditions for laboratory tests—process exposed to the atmosphere in field testing
5. *Agitation*: technique for laboratory tests and, velocity for field tests
6. *Apparatus and type of test rack*
7. *Test time*
8. *Test metals*: chemical composition, trade name, product type (plate, sheet, rod, casting, etc.), metallurgical condition (cold rolled, hot rolled, quenched and tempered, solution heat treated, stabilized, etc.), and the size and shape of the coupons
9. *Exposure location*
10. *Cleaning technique*
11. *Weight loss*
12. *Type and nature of localized corrosion*: stress corrosion cracking, intergranular corrosion, pitting (maximum and average depth), crevice corrosion, and so on
13. *Corrosion rate*: may be misleading if localized corrosion is present

Specialized Tests

Most industrial groups have developed specialized corrosion tests to fit their needs. Most of these test methods are listed in the *Manual of Industrial Corrosion Standards and Control* [7].

A number of stress corrosion cracking tests have been developed [8-10] and these are detailed in the literature. The U bend is the most widely used stress corrosion cracking test and a typical sample is shown in Fig. 12. Stress corrosion cracking specimens are usually acid cleaned, degreased, and weighed before exposure, but the base and weld metal surfaces are usually not ground since the object is to test the metal in a condition very near to how it would be put in service. Sheared edges and punched holes are usually used. After exposure, the samples are examined carefully, cleaned, weighed, and examined again. The sheared and stamped edges and the ends of welds should be examined very carefully for cracks. The specimens are then flattened and the bend is examined again to determine if any additional cracks were opened up. Dye penetrant inspection is frequently used to help locate small cracks. Weight loss corrosion rate is sometimes reported, but it can be misleading on welded specimens since the weld has a different composition than the base metal and galvanic corrosion occurs.

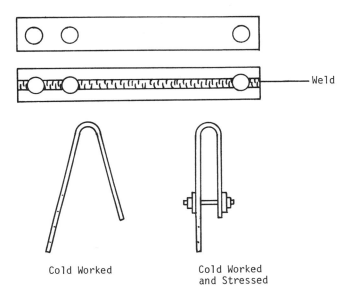

Cold Worked

Cold Worked
and Stressed

Figure 12. Stress corrosion cracking specimen. The hole that is not used in this figure is for mounting the specimen on a coupon rack.

The area under washers, insulators, and gaskets should be examined to determine if crevice corrosion is a problem. A more rigorous method of measuring crevice corrosion is to use a slotted washer or nut [11] as shown in Fig. 13, which allows the results to be analyzed statistically.

Galvanic corrosion tests may be conducted by bolting, wiring, or welding specimens together. If the specimens are bolted or wired together, it is possible to measure a weight loss corrosion rate, but the coupons should be examined carefully to ensure that crevice corrosion is not clouding the results. The test results will be subjective when welded specimens are used.

The effects of velocity on corrosion can be studied in a number of ways. The simplest is by rotating a metal disk in the test liquid as shown in Fig. 14. The next step up is a paddle wheel test, in which a number of coupons are bolted to a disk (Fig. 15) which is rotated. Other techniques involve putting coupons in nozzles or venturis or vibrating them [12].

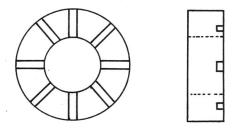

Figure 13. Slotted washer for crevice corrosion. The grooves are 0.020 to 0.030 in. deep and 0.040 to 0.050 in. wide. The groove size is not critical. The area of crevices should be one-tenth or less of the total coupon area.

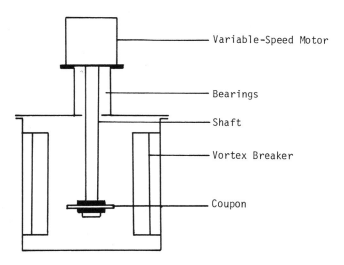

Variable-Speed Motor

Bearings

Shaft

Vortex Breaker

Coupon

Figure 14. Rotating disk corrosion test apparatus.

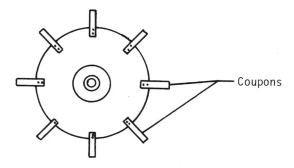

Coupons

Figure 15. Paddle wheel test.

HYDROGEN DIFFUSION

The atomic hydrogen that is produced at the cathode of some corrosion reactions can diffuse through steel and most other metals if it does not combine to form hydrogen molecules. When sulfides are present, as in sour service and some refinery streams, the atomic hydrogen produced by the corrosion reaction readily diffuses through steel. If hydrogen diffusion is detected, corrosion is taking place. Hydrogen diffusion can be measured using either a hydrogen probe (pressure measurement) or a hydrogen monitoring system (electrochemical).

Pressure Measurement

The cross section of a hydrogen probe is shown in Fig. 16. The body of the probe is made from pipe that has a section machined down to a

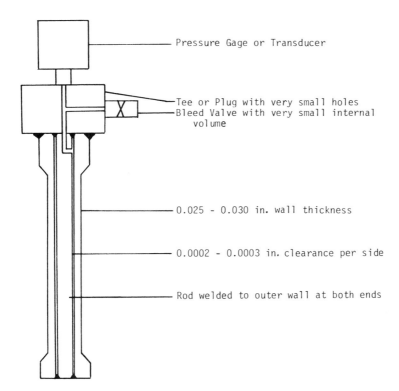

Figure 16. Hydrogen probe.

very thin wall thickness. A rod is inserted into the pipe so that the
free volume is very small. The hydrogen diffuses into the free volume, and the increase in pressure is easily detected using either a
pressure gage or a transducer. This type of probe has been used
for some time in refineries to help in process modification and inhibitor addition to prevent hydrogen blistering and embrittlement.

Electrochemical detection

The hydrogen that diffuses through a pipe wall can be detected electrochemically using a "patch probe," shown in Fig. 17. Palladium
foil is used as the working electrode because hydrogen diffuses
through it very rapidly, and it has very good corrosion resistance
in the sulfuric acid electrolyte. A potentiostat is used to maintain the
palladium foil at a potential where the hydrogen atoms are oxidized,
and the amount of current necessary to oxidize the hydrogen is a
measure of the amount of hydrogen that diffuses into the cell. These
systems have proven reliable at temperatures up to 105°C (220°F).
The amount of hydrogen that diffuses through the wall will be propro-
tional to the corrosion rate in a given system [13], but identical diffusion rates may result from different corrosion rates if the system
changes. Temperature will increase the rate of hydrogen diffusion and
wall thickness will decrease it.

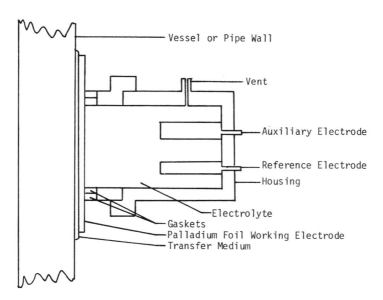

Figure 17. Electrochemical hydrogen probe.

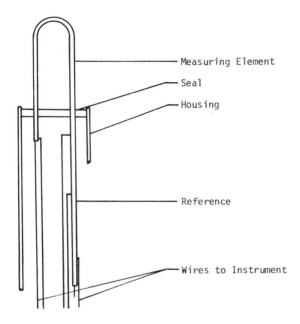

Figure 18. Electrical resistance corrosion probe.

ELECTRICAL RESISTANCE

The electrical resistance technique of corrosion measurement was developed [14] as a method of measuring low corrosion rates without removing the products of corrosion. Very small changes in resistance occur when corrosion takes place and changes in temperature can result in larger changes in resistance which will give erroneous corrosion readings. To overcome the problems encountered from changes in temperature, a noncorroding reference section of the test element was included in the probe as shown in Fig. 18. This reference element is able to handle small or gradual changes in temperature, but since it is encased in the tip of the probe, its temperature would still fall behind that of the test element. Numerous modifications have been made to probes over the years, but it is still not possible to measure very small changes in corrosion rate with a single reading unless you are absolutely sure that no changes in temperature have occurred.

Probes (Fig. 19) are available, with test elements made from all of the common alloys used for the fabrication of process equipment [15], that can operate at temperatures up to 400°C (750°F) at a pressure of up to 4000 psi. Probes are also available that can be inserted into a process stream through a 0.75- or 1-in. full port valve.

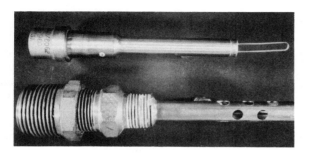

Probe Insert with Wire
Test Element

Probe Housing

Figure 19. Electrical resistance corrosion probe.

   In measuring corrosion using an electrical resistance probe, a reading is first taken on the test and check element. The probe is then
inserted into the test environment and allowed to come to the test
temperature (usually 30 to 60 min). Another reading is taken on the
test and check element and corrosion is allowed to take place for a few
hours. Another new set of readings is taken, and if they have
changed, the corrosion rate is calculated using the following equation:

$$\text{Corrosion rate (mpy)} = \frac{\text{change in reading X probe multiplier X 365}}{\text{change in time X 1000}} \quad (3)$$

"Change in reading" is the current reading minus the initial reading
   in the test environment or the previous reading. If the change in
   reading is negative, the results are not corrosion. They are either
   temperature of a conductive film or the test element.
"Change in time" is in days.
"Probe multiplier" is a number marked on each probe that reflects the
   life of the probe. Higher numbers mean that the probe will last
   longer and be less sensitive.
"1000" is the maximum instrument reading.
"365" is the number of days per year.

It is very useful to calculate both the overall corrosion rate (or corrosion rate over the total exposure time) and the corrosion rate between
readings to determine if the corrosion rate is changing with time.
Sometimes it is useful to plot the instrument readings versus time to
follow changes as they are occurring, as shown in Fig. 20.
   Manual and automatic instruments are available to measure the
change in electrical resistance of the probe, and two instruments are
available that will calculate the corrosion rate of the test element.
When automatic instruments are used, the lead wires from the probe
to the instrument are part of the circuit and their change in

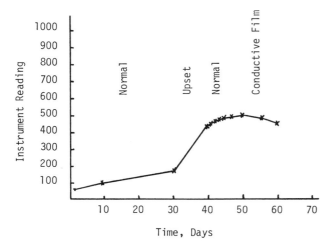

Figure 20. Electrical resistance corrosion probe readings.

temperature can affect the resistance readings [16], and thus the calculated corrosion rates. This can trip the alarm on an automatic instrument and cause it to display incorrect corrosion rates.

When the limitations of the system are taken into account, the electrical resistance technique can be used to measure very small corrosion rates while they are occurring. It is also very useful in detecting changes in corrosion rate and spotting upsets in process plants.

ELECTROCHEMICAL TECHNIQUES

Electrochemical techniques are very useful in determining what is actually happening to a metal at any given time. The three techniques most often used involve: polarization curves, linear polarization curves, and zero-resistance ampmeters.

Polarization Curves

Polarization curves may be determined galvanostatically, potentiostatically, or potentiodynamically. During a galvanostatic measurement, a predetermined amount of current is passed from an auxillary or counterelectrode to a working or test electrode, and the resulting shift in potential of the test electrode away from a reference electrode is measured using a high-input-impedance voltmeter, as shown in Fig.

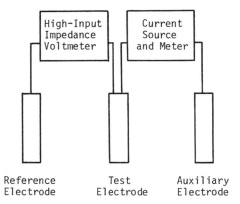

Figure 21. Galvanostatic polarization circuit. The input impedance of the volt meter should be 100 MΩ or greater. The current meter should have ranges from 10 μA to 1 A and should have an accuracy of at least 0.5%. The current source can be batteries or a power supply with less than 100 μV ripple.

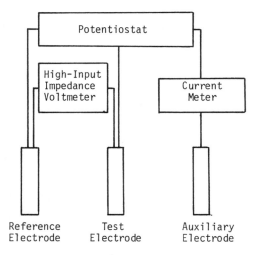

Figure 22. Potentiostatic polarization circuit. The potentiostat should be able to maintain the test electrode within 1 mV of its set value over a ±1-V range and have an output current of at least 1 A.

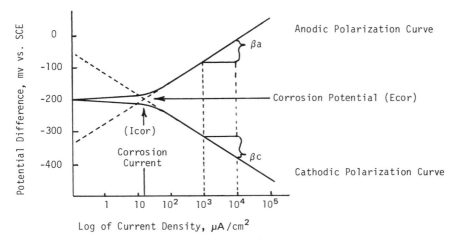

Figure 23. Anodic and cathodic polarization curves for a metal that is corroding actively. The Tafel slopes on βa and βc are the slopes of the anodic and cathodic polarization curves.

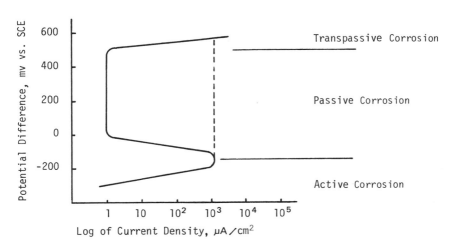

Figure 24. Anodic polarization curve for an active-passive metal. The solid lines are for potentiostatic polarization. Galvanostatic polarization will follow the dashed line and the passive area will not be seen.

21. Potentiostatic measurements are made using a bipolar operational power supply (potentiostat) to supply the current necessary, in either direction, to move the test electrode away from its open circuit or rest potential as shown in Fig. 22. Adding a sweep unit or wave generator to the potentiostatic allows the test electrode to sweep through a range of potentials.

When active corrosion is taking place and the log of the applied current density is plotted versus the potential difference of the test electrode away from a reference electrode, the corrosion current of a metal may be estimated as shown in Fig. 23. The corrosion or exchange current can be used to calculate the corrosion rate using Faraday's law. Figure 24 shows the shape of the anodic polarization curve of an active-passive metal measured with a potentiostat. The dashed lines in Fig. 24 show what would be measured using a galvanostat. If the cathodic polarization curve intersects the anodic curve only in the passive region, the corrosion rate of the test metal will be very low (Fig. 25). Deep pitting can occur when corrosion occurs in the trans-passive zone. Crevice corrosion (often mistaken for shallow pitting) occurs when the metal is corroding both passively and actively at different locations at the same time.

In the laboratory, good reference electrodes can be used, and it is almost always possible to measure the polarization curves accurately. High temperatures or pressures sometimes make it impossible to place a reference electrode near the test electrode under field conditions. When this occurs, either a corroding or remote electrode may be used. When a corroding electrode is used, it is not possible to determine the potential of the test electrode, but it is possible to determine if the

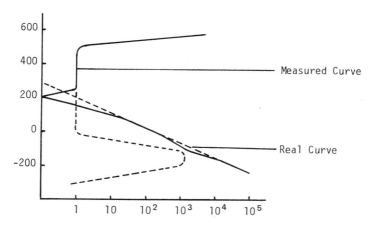

Figure 25. Anodic and cathodic polarization curves for an active-passive metal that is corroding passively.

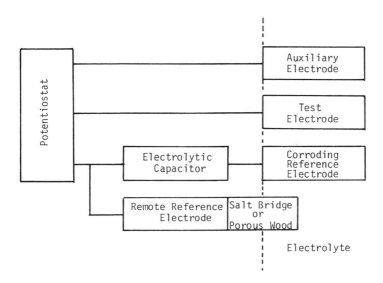

Figure 26. Remote reference electrode.

corrosion is active, passive, or transpassive, and if small changes in
the process could change the type of corrosion that is taking place.
A remote reference electrode can be used by placing a corroding
reference electrode near the test electrode and connecting it to the
potentiostat using a large electrolytic capacitor. The remote reference electrode is then connected directly to the potentiostat, as shown
in Fig. 26. In almost all cases, the tafel slopes can be measured for
use with the linear polarization technique, as described in the next
section.

Linear Polarization

The linear polarization technique makes it possible to measure the
corrosion rate of a metal at any instant. This technique will only work
when electrodes are exposed to an electrolyte that has a continuous
path between them. The technique is based on the Stern and Geary
[17] equation:

$$\text{Icor} = \frac{\Delta I}{\Delta E} \ \frac{\beta a \beta c}{2.3(\beta a + \beta c)} \tag{4}$$

where
   Icor, βa, and βc are as shown in Fig. 23
   ΔI = applied current

ΔE = change in potential from the freely corroding potential (since this is the change potential, a corroding reference can be used).

Equation (4) holds only when there is a linear relationship between ΔI and ΔE and usually only over a range of ± 10 to 50 mV for ΔE. When βa and βc are known and ΔE is preset, it becomes possible to manufacture a meter that will read out directly in corrosion rate since Icor is directly related to the amount of metal loss through Faraday's law. Equation (4) can be reduced to

$$C.R. = \Delta I \; K \tag{5}$$

where

    C.R. = corrosion rate, mpy

    K   = meter constant

Both manual and automatic linear polarization instruments are available that can be used under laboratory and field conditions. An automatic instrument installed in a cooling water system will usually pay for itself in 6 to 12 months by reducing the treatment chemical usage [18]. The corrosion rate of any metal can be measured. As with the electrical resistance technique, probes are available for laboratory and field use. Linear polarization probes normally use a corroding reference electrode. Standard reference electrodes and platinum counterelectrodes are often used in laboratory tests.

Most of the instruments that are available go through the following steps during their operation:

1. Measure potential from reference to test.
2. Add or subtract ΔE from the potential measured in step 1.
3. Change the potential of the test electrode to the potential determined in step 2 by applying current between the test and auxiliary electrodes.
4. Hold the test electrode at the potential determined in step 2 for a predetermined time.
5. Display the corrosion rate. Some instruments have an alarm circuit that can be activated during this step.
6. Go to the next probe and repeat steps 1 through 5. Instruments are available that will handle up to 10 probes and can polarize them all in one direction and then all in the opposite direction.

This sequence of events is shown in Fig. 27 for an instrument with two active probes and a meter proven on a third channel. When the polarization curves are recorded on a strip chart recorder, it is possible to detect when nonuniform (pitting, crevice, etc.) corrosion is taking place, as shown in Fig. 28. The corrosion rate measured in the cathodic direction will be slightly higher than that measured in the anodic direction when passive corrosion is taking place and the

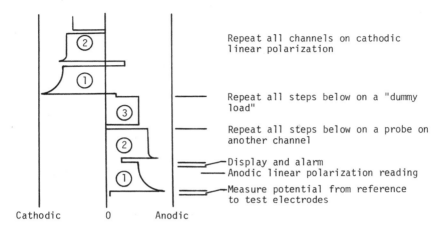

Repeat all channels on cathodic
linear polarization

Repeat all steps below on a "dummy
load"

Repeat all steps below on a probe on
another channel

Display and alarm
Anodic linear polarization reading

Measure potential from reference
to test electrodes

Cathodic   0   Anodic

Figure 27. Linear polarization readings on two probes and a meter prover or dummy load. The probes are on stations 1 and 2, and the meter prover is on station 3.

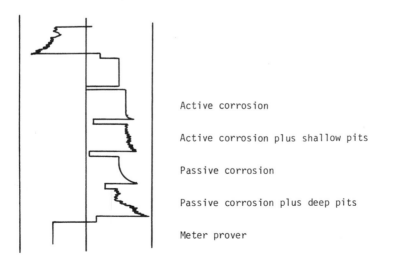

Active corrosion

Active corrosion plus shallow pits

Passive corrosion

Passive corrosion plus deep pits

Meter prover

Figure 28. Linear polarization nonuniform corrosion detection.

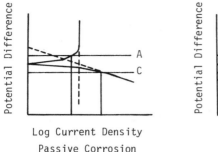
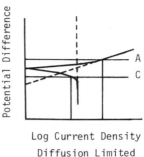

Passive Corrosion   Diffusion Limited

Figure 29. Linear polarization measurements in the passive (anodic < cathodic rate) and cathodic (anodic > cathodic rate) regions.

opposite will be true if diffusion is limiting the rate of reaction, as shown in Fig. 29.

Instrument manufacturers size their electrodes for a "known" valence change (usually +2 for iron) and $\beta a \beta c / (\beta a + \beta c)$ can be measured at the end of a test and the readings of corrosion rate can be factored or new electrodes can be made using an area that will make the instrument direct reading.

Zero-Resistance Ampmeter

A zero-resistance ampmeter is a potentiostat that has been programed for zero potential difference between the reference and test electrode, and in addition, the current lead to the counter electrode is connected to the reference electrode. This makes it possible to measure the amount and direction of the current that flows between two electrodes that are electrically short-circuited. When galvanic corrosion measurements are being made on two materials of similar chemical composition, the test should be allowed to run for several days because the direction of current flow often reverses [19]. The galvanic current cannot be used to calculate directly the corrosion rate of the anode since it is only a measure of how much faster the anode is corroding than the cathode.

REFERENCES

1. W. C. Minton, Wall thickness measurements—quick and easy, NACE Corrosion/70, Paper No. 76, March 1970, Philadelphia.
2. M. G. Fontana and N. D. Green, *Corrosion Engineering*,

McGraw-Hill, New York, 1967, Chap. 2.

3. *Metals Handbook*, Vol. 1, 8th ed., American Society for Metals,
    Metals Park, Ohio, 1961.

4. *Properties of Some Metals and Alloys*, International Nickel Com-
    pany, New York, 1968.

5. A. Wachter and R. S. Treseder, Corrosion testing evaluation of
    metals for process equipment, *Chem. Eng. Prog. 43*, 315-326,
    1947.

6. *Laboratory Corrosion Testing of Metals for the Process Industry*,
    NACE Standard TM-01-69, National Association of Corrosion
    Engineers, Houston, 1969.

7. F. H. Cocks, *Manual of Industrial Corrosion Standards and Con-
    trol*, STP 534, American Society for Testing and Materials,
    Philadelphia, 1973.

8. G. M. Ugiansky and J. H. Payer, *Stress Corrosion Cracking—
    The Slow Strain-Rate Technique*, STP 665, American Society for
    Testing and Materials, Philadelphia, 1979.

9. H. L. Craig, Jr., *Stress Corrosion Cracking of Metals—A State
    of the Art*, STP 518, American Society for Testing and Materials,
    Philadelphia, 1972.

10. *Stress Corrosion Testing*, STP 425, American Society for Testing
    and Materials, Philadelphia, 1967.

11. D. B. Anderson, *Statistical Aspects of Crevice Corrosion in Sea-
    water, Galvanic and Pitting Corrosion—Field and Laboratory
    Studies*, STP 576, American Society for Testing and Materials,
    Philadelphia, 1976, pp. 231-242.

12. K. G. Compton, Seawater corrosion tests, in *Handbook on Corro-
    son Testing and Evaluation*, Wiley, New York, 1971, pp. 507-
    529.

13. R. L. Martin and E. C. French, Corrosion monitoring in sour
    systems using electrochemical hydrogen patch probes, *J. Pet.
    Technol.*, 1566-1570, November 1978.

14. A. Dravnieks and H. A. Cataldi, Industrial applications of a
    method for measuring small amounts of corrosion without removal
    of corrosion products, *Corrosion 10*, 224, July 1954.

15. *Corrosometer Fixed Probe Selection Guide*, Rohrback Instruments,
    Santa Fe Springs, Calif.

16. D. L. Scholten, Process plant uses of corrosion rate measure-
    ments, Paper No. 40, NACE *Corrosion/76*, March 1976, Houston, Texas.

17. M. Stern and A. L. Geary, Electrochemical polarization: I. A
    theoretical analysis of the shape of polarization curves, *J.
    Electrochem. Soc. 104*: 56, 1957.

18. C. G. Arnold, Instrumentation in cooling water systems, Cooling
    Tower Inst. Annu. Meet., February 11, 1975, Houston, Texas.

19. C. G. Arnold, Galvanic corrosion measurement of weldments,
    Paper No. 71, NACE Corrosion/80, Chicago, March 3-7, 1980.

11
Selecting Materials of Construction

LEWIS W. GLEEKMAN / Materials and Chemical Engineering Services,
Southfield, Michigan

PRINCIPLES

In engineering practice the selection of materials of construction depends on the item being designed and the service requirements imposed on the item. Simple as this statement is, there are many compli-
cations involved in the choice of the preferred material of construction
for a particular item. It almost goes without saying that one would
not use aluminum foil as a structural roof deck on a building. Conversely, one would not use poured concrete as the load-carrying member on a jet aircraft. These two extremes point to the fact that in
selecting materials one has to balance many requirements; these requirements, somewhat in order of decreasing importance, include the
following:

Cost
Availability

Ability to be fabricated
Ability to be joined
Mechanical properties as a function of temperature
Strength-to-weight ratio
Physical properties (including thermal and sound properties) as a
     function of temperature
Electrical properties as a function of temperature
Corrosion properties as a function of environment

At the present time the most common starting material is steel. The
reason for this lies in the fact that steel is relatively inexpensive; it
is readily available; it is fabricated by a wide variety of techniques,
including casting and wrought methods; it is easily joined by a wide
variety of inexpensive techniques to itself and to other materials; and
its mechanical properties are well established and can be changed as a
function of composition and temperature. Indeed, by alloying, steel
can meet the extremes of relatively high, as well as relatively low,
temperatures. Under most circumstances its thermal and electrical
properties are well known, and its ability to carry sound is quantitatively recorded. Although the strength-to-weight ratio of steel is not
as favorable as those of certain aluminum alloys and titanium alloys,
when combined with cost, steel often is quite favorable. Finally, the
corrosion properties of steel in most environments have been determined
and techniques of protecting steel against corrosion in these environments are also widely developed.
     However, this has not always been the case. It must be recognized
that steel has been widely used as an item of commerce only since about
the middle of the nineteenth century. Prior to that, before the development of the Bessemer process and modern high-speed rolling mills,
cast iron was the common material of construction. It was well known
that cast iron had poor tensile strength, but the engineers of the pre-
1850s were able to design around the poor properties of cast iron in
tension (but great in compression), and thus cast iron was widely used
in certain aspects of the chemical process industry where; indeed, it
was used right up to the 1970s. Even before cast iron, wood was a
common material of construction, and in modified form, it is still widely
used for certain items in industry. Consider, for example, the construction of an atmospheric storage tank handling a product that is
corrosive only under aerated conditions; for example, a sodium chloride brine solution. The number of materials that can be used for
such a storage tank to handle this product is mind-boggling when it
is considered that virtually every material (with one or more modifications) could be used for this purpose. Wood-staved brine storage
tanks (particularly cypress and redwood) can give long service provided that the vertical staves of the wood storage tank are reinforced
with steel hoops. It is the steel hoops that are usually the weak link
in the construction of wood-staved tanks since part of the hoop (where

the hoop rests against the wood) is subject to crevice corrosion as
brine exudes through the wood and wets the hoops. Rubber, particularly in its reinforced form, has been used for brine storage tanks
either in the form of pond liners or in the form of pillow storage tanks
with all-welded seams. Rubber has also been used as a lining for steel
tanks.

Steel itself has been used for brine storage tanks; where, the only
problem is that of corrosion at the liquid level. Some companies in
the past made a practice of letting the steel corrode in an initial high
level of brine in the tank and then dropping the level an inch or two
each year to get around corrosion in the higher portions of the tank.
Ultimately, when the corrosion has proceeded down the tank with
lowering liquid levels to the point where the capacity of the tank was
insufficient, new steel plates would be installed to replace the corroded
plates. This is not as unusual a practice as it may seem; one of the
major chemical companies was doing this as recently as the 1960s. The
more common approach at the present time, due to economic considerations such as loss of production and the relatively high cost of maintenance labor, is to put a lining on the interior of the tank that will
offer protection to the steel substrate. This lining, in many cases,
can be a nominal 5-mil vinyl coat, or in certain instances, can be a
relatively heavy coal tar epoxy or sprayed glass flakes dispersed in
polyester resin.

In fact, with the advances in the fiberglass industry, it is quite
common presently to find brine storage tanks or other such tanks made
from fiber-reinforced plastic (FRP); this material not only resists the
corrosive effects of brine on the interior but also protects the exterior
against corrosion from atmospheric contaminents or overflowing conditions of the tank. The fiberglass industry has advanced to the point
where vessels can be field fabricated to provide large-diameter, large-
volume vessels, somewhat analogous to the field erection of steel storage tanks. If temperature and trace element considerations demanded,
the tank can be constructed of titanium either as a discrete metal or
as a lining on steel (either loose lined or explosive clad or metallurgically roll-bonded).

One could go down the list of present materials of construction item
by item and point out that in one form or other, and for one or more
specific conditions, these materials could be used for brine storage
tanks. This could include glass, concrete, acid brick and resin mortar, and even paper with resin impregnants; proper construction for
the known strength considerations allow such materials to be used for
this service.

With such a variety of materials of construction available, there is
obviously a need to differentiate among the materials to arrive at the
one which is most economical both in terms of first cost and in terms
of annual cost (including maintenance cost) over a period of time.

Experience has taught that many materials have widely used or unique properties that extensively govern their utilization for particular facets of design; thus the design engineer need not reinvent the wheel for each particular problem, but frequently need only check handbooks or textbooks or, at worse, do a comprehensive literature search in the technical journals in the appropriate fields. Whatever is forthcoming from the recorded experiences of others needs to be tempered with developments that may have taken place based on recent experiences. As simple a thing as fluctuations in the precious metals market can establish whether titanium-0.2% palladium alloy, with its great resistance to crevice corrosion in the presence of hot brine solutions, is a feasible material of construction or whether one should use instead titanium-nickel (1% nickel) alloy, where the price of the material would not be influenced by the demand for noble metals such as gold, platinum, or palladium.

Another example would be that certain cast alloys with great corrosion resistance in particular environments have been modified to produce an equivalent wrought alloy with the same corrosion resistance but with somewhat different metallurgy; this points to the fact that designers need not restrict themselves to relatively heavy cast metal sections but instead can think of fabrications involving somewhat thinner wrought sections.

This whole matter of cast versus wrought construction is an ever-fluctuating one and often depends on the number of items being produced in a given shape. For one-of-a-kind units, cast construction is often expensive. On the other hand, 50 or more similar items can often be economically justified as sand casting, and 500 or more of a given item can be economically justified as permanent mold castings. In the manufacture of a new valve, the prototype valve could well be made from bar stock of a given material with whatever machining was necessary to accomplish the configuration and desired shape. After the prototype valve had proven itself, it would be feasible to consider fabricating one or more of the components as castings to minimize machining time and scrap from the machining operation. This decision would also require determining that the cast alloy had equivalent corrosion resistance for the anticipated environments in which the wrought alloy had originally been tested.

Historically, one can point to the use of titanium pump impellers in chlorohydrin-HCl solutions in the manufacture of either ethylene or propylene oxide. Here problems had been encountered with the use of a nickel-chrome-molybdenum cast alloy (Hastelloy C) because of sensitization that occurred during the welding operation. The first titanium impeller was made by hot working wrought stock into blades and welding the blades to a previously machined back plate which itself had been made from wrought stock. After several years of successful operation in the hot corrosive environment it was decided that three

other pumps should be equipped with titanium impellers. At that time
one of the titanium companies then proposed to die cast the titanium
in steel molds, even though only four impellers were going to be
ordered. The economics of this means of manufacture was chosen because of future anticipated business.

STRENGTH-TO-WEIGHT RATIO

Strength-to-weight ratio had been of historical significance to designers of lighter-than-air vehicles such as blimps and zeppelins, where
the lifting capacity of the vehicle depended on the buoyancy imparted
by helium (or occasionally, hydrogen). Therefore, the weight of the
structural members giving rigidity to the vehicle became of critical
consideration relative to the live-load carrying capacity of the vehicle.
Duraluminum was originally a German development based on magnesium
which found wide use in the construction of zeppelins. As the switch
took place from the lighter-than-air vehicle to powered aircraft, weight
was also a consideration in the air frame, and the strength-to-weight
ratio factor was important in the early stages of aircraft design.
Strength-to-weight ratio became even more important when considerations of live load of passengers, freight, or in the case of military
aircraft, bomb loads, again came into play. Aluminum alloys were
developed which had higher strength than the conventional alloys and
yet had the relatively low density of aluminum itself. Great improvements came about through the use of titanium alloys, where, in spite
of an approximate doubling of the weight of aluminum, the strength
advantage became more than fourfold greater than that of the aluminum
alloys, thereby reducing the strength-to-weight ratio of a given part
when using titanium alloy.
     Yet further advances have come into being in recent years, as car-
bon fibers impregnated with epoxy resin have given a structure that
can be readily formed and has extremely high strength that can be
readily formed and has extremely high strength for its relatively low
weight. Even certain synthetic fibers of the aramid type (trademarked
Kevlar by Dupont), when used as a reinforcement with either epoxy or
polyester resins, result in a material with an extremely high strength-
to-weight ratio. Such materials are not limited to aircraft, but their
use is now being directed toward the entire transportation field, in
view of the present consideration toward reducing the weight of automobiles and other vehicles in order to achieve a conservation of fuel.
The construction of the Corvette body out of molded fiberglass has
not only reduced the weight of the vehicle but has achieved a step
toward a noncorroding material of construction relative to the corrosive effects of salt, moisture, and other contaminants in the atmosphere.

CORROSION CONSIDERATIONS

Since this is a handbook directed toward corrosion control and prevention, it would not be appropriate to emphasize the other considerations
involved in selection of materials, such as the mechanical, physical,
and electrical properties. Instead, it should be pointed out there are
three relatively simple actions that can be taken to stop corrosion:
(1) change material, (2) change environment, and (3) protect material.
Although these may oversimplify the process of corrosion control, they
do indicate to the practicing engineer steps to be used in stopping
corrosion.

Relative to changing materials to minimize or stop corrosion is the
practice of using a more-corrosion-resistance material, usually achieved
by alloying, or occasionally achieved by upgrading the material to a
more resistant material, based on reported experiments and experience.
In addition to changing by alloying, one can also change to nonmetallic
materials such as plastics (with and without reinforcement), elastomers,
ceramics, stoneware, glass, carbon, graphite, and woods, among
others. The matter of changing materials is not merely a function of
selecting material that has approved corrosion resistance; other factors, such as thermal and electrical properties, ease of fabrication,
and all the other aspects previously indicated, must be considered.

It should be recongized that every material has at least one Achilles'
heel, if not two or more. These weaknesses in a material must always
be considered in their selection for a given part. In fact, the design
engineer would do well when considering any material to think immediately of the drawbacks associated with that material and assure that
these factors are not applicable to the given situation. Listed in Table
1 are some of the more common weaknesses of a variety of engineering
materials.

When one speaks of changing the environment to reduce corrosion,
the easiest and most obvious method is to lower the temperature. Since
corrosion processes are chemical reactions, every 10°C (18°F) decrease
in temperature reduces the reaction rate by half. Thus if one lowers
the temperature, the rate of corrosivity will be retarded. In addition,
the atmosphere can be changed by the presence of certain gases; some
metals are corroded in the absence of air and others in the presence
of air. Other environmental changes involve agitation, aeration, and
velocity, all of which have a decided influence on many materials. Also,
a change fairly easily brought about in some processes is that of ad-
justing the pH, which is a measure of the acidity or basicity of the
solution. For most materials, increasing the pH to the alkaline side
reduces the corrosivity. The presence of trace additives is a change
in environments that can be either good or bad for certain metal systems. It is necessary to know the behavior of the metal toward various
additives to establish whether a positive or negative effect will be
brought about. For example, the presence of copper ions improves

Table 1. Engineering Materials and Weaknesses

Material	
Mild steel	Caustic embrittlement of stressed areas
	High-temperature oxidization
	Alternate wetting and drying—"rusting"
Austenitic stainlesses	Chloride-containing solutions, especially in crevices and on welds or stressed areas
	Pitting
	High-temperature fusion from certain sodium and vanadium compounds
Nickel and Monel	Contact of heated surfaces with sulfur and its compounds; ammonia, mercury
Copper	Essentially the same as nickel
Aluminum	Galvnanic effects with iron, copper, and other nobler metals; mercury
Titanium (C.P.)	Crevice corrosion in hot brine solutions; nonoxidizing environments; dry chlorine
Hastelloy C	Reducing high-temperature environments
Hastelloy B	Oxidizing high-temperature environments
Thermoplastics	Temperatures above 120°C (250°F) (exceptions being most fluorinated polymers); some organic solvents
Thermosetting polymers	Oxidizing effects (including concentrated H_2SO_4) on epoxies; strong hot alkalies on many polyesters; certain solvents; loss of properties on long-term temperature exposure
Rubber	Temperature; solvents; certain oxidizing chemicals depending on specific elastomers
Wood	Temperature; strong acids and alkalies and oxidizing chemicals; certain impregnating resins can be used to overcome these effects
Glass	Fragility unless protected by "armor" cladding of FRP; strong hot caustic and HF (HCl under certain conditions)

the corrosion resistance of stainless steels in certain environments, whereas the presence of copper irons will cause localized pitting and attack of aluminum alloys in other solutions. There is yet the wider option of adding inhibitors to a solution which suppress the corrosivity of the main species in the solution.

Protecting a material to reduce corrosion often means isolating the metallic surface from the corrosive environment. This can be done with either organic or metallic coatings on the surface. Organic coatings can be thick or thin and can be paint film, solid lining, or plastics in the form of tape or sheet or powder fused to the surface. Metallic coatings are often applied as an electroplated material, although some metals can be deposited by chemical means in an electrodeless deposition similar to silvering glass to make a mirror. A metal may be coated by diffusion at moderate temperatures, such as the application of zinc in galvanizing or aluminum in aluminizing, or metallizing the surface by spray application of partially melted materials. Selection of any one of these protective methods depends on such considerations as service factors, environmental conditions, economics, ease of application of material, and complexity of parts. In addition, one can protect the metal surface by applying an electrical potential under conditions that make the surface cathodic or anodic depending on the environment and the metal being protected. The more widely used of these two methods is that of cathodic protection, often applied to underground pipelines, tank bottoms, or water boxes on exchangers. The application of cathodic protection is sometimes done with a sacrificial anode of magnesium or zinc or aluminum or by the application of applied potential from a rectifier or battery, where one uses a more permanent anode of graphite or Duriron or platinized titanium.

Types of Corrosion

Although the various types of corrosion are discussed in detail elsewhere in this handbook, they bear repetition here because it is only as one recognizes the type of corrosion and the materials that are susceptible to specific corrosion attack (under the specific environmental conditions and/or construction and operating conditions) that one can avoid costly errors in materials selection.

Macroscopic types
Uniform
Galvanic
Crevice
Pitting
Erosion (velocity)
Microscopic types
Selective leaching (parting or "dezincification")

Intergranular
Stress corrosion cracking

NONMETALLIC MATERIALS

Metallic contamination of products has become a very important consideration and in some instances is responsible for limiting the use of
metals for process equipment. The use of nonmetallic material, in
particular plastics and elastomers, has proved very successful in
avoiding metallic contamination. For most aqueous environments, nonmetallic materials will often provide greater chemical resistance than
metallic construction, at lower cost.

Plastics

Traditionally, the use of plastics has been limited to relatively low
temperature services and to low-pressure applications, except where
they were used as a lining bonded to or otherwise supported by a
strong substrate. This is true of the nonreinforced plastics categorized as thermoplastic materials, specifically those materials that
are softened by heat. However, the development of reinforcing materials, combined with the class of plastics known as thermoset materials
(materials that are not softened by heat), and the further development
of many thermoplastic materials that have relatively high thermal-dis-
tortion temperatures, has meant that present-day technology uses
plastics for moderately high temperatures and for high-pressure applications.
    Notable among the thermoplastic materials are polyethylene, poly-
propylene, polyvinyl chloride, the styrene-synthetic rubber blends,
the acrylics, and the fluorocarbons. Notable among the thermosetting
reinforced materials are the polyesters, epoxy, and the furane resins
as custom-made reinforced materials, and the phenolic and epoxy
resins molded, filament wound, and/or extruded with reinforcement.
All these materials are available as piping, sheet stock, and miscellaneous molded and fabricated items. These materials, particularly polyvinyl chloride, polypropylene, and reinforced polyesters, are now
being used extensively for ventilating ductwork in handling corrosive
fumes: In many instances they have proved to be economically superior
to metals such as stainless steel, lead, and galvanized steel. Not only
are these materials used for the ductwork but they can be used to
fabricate the scrubbers, pumps, blowers, fan wheels, and virtually all
the components of a system. They are generally not subject to pitting,
stress corrosion cracking, and other forms of corrosion common to
metal. However, there are design limitations and they usually cannot
be substituted for metals part for part. A point that should not be

overlooked is that nonmetallics do not require painting for protection against external corrosion. Plastic materials are replacing metals in many applications because of their better resistance to chemical exposures and improved service life and economy.

The most chemically resistant plastic commercially available today is tetrafluoroethylene (Teflon or Halon). This is a thermoplastic material which is practically unaffected by all acids, alkalies, and organics at temperatures up to about 260°C (500°F). It has proved to be an outstanding material for gaskets, packing diaphragms, O rings, seals, and other relatively small molded items. Its chemical inertness makes normal bonding and cementing operations difficult and impractical. It can be cemented to metal and other materials by using special sheets that have rough backing surfaces which provide mechanical adherence through any one of a number of resin cements. Techniques have also been worked out that make it possible to heat-seal sheets of material together. Loose linings, including nozzle linings, may now be installed in tanks, ductwork, and other straight-sided and nonintricate equipment.

Kynar, a chlorofluoroethylene, also possesses excellent chemical resistance to almost all acids and alkalies at temperatures up to about 180°C (350°F). It, too, is difficult to bond to itself and other materials, but the use of tape-bonded laminated construction has widened the use of the material. It can be extruded readily and so is available in the form of solid pipe and also as a lining material for steel pipe.

Polyethylene is the lowest-cost plastic that has excellent resistance to a wide variety of chemicals. Its greatest use has been as piping and tubing in corrosive services, but large quantities are also used as thin sheet or film liners in drums or other packages. Polyethylene tape with pressure-sensitive adhesives on one side is receiving increasing attention and is used as a wrapping material to protect conduit and pipe from corrosion. Unfortunately, its mechanical properties are relatively poor, particularly at temperatures above 50°C (120°F), and it must therefore be supported for most applications. It can be readily joined to itself by heat sealing and fusing, and a wide variety of equipment has been satisfactorily made using heat-sealing techniques. Weathering resistance of the unfilled grades is poor, but the carbon-filled grades have good resistance to sunlight and are satisfactory for outside use.

The unplasticized polyvinyl chloride materials have excellent resistance to oxidizing acids other than nitric and sulfuric and to nonoxidizing acids in all concentrations and are satisfactory for use at temperatures up to about 65°C (150°F). They are also resistant to both weak and strong alkaline materials and to solutions of most chemical salts. The resistance to aromatic and aliphatic hydrocarbons is generally good, but resistance to chlorinated hydrocarbons is poor. They are not satisfactory for use with ketone or ester solvents. They are self-extinguishing with regard to flame resistance and have been known to give off large volumes of smoke when in a fire. They are resistant to sunlight and outdoor weathering. Two general types are available, regular and high impact. The latter has appreciably better impact resistance but somewhat lower strength and lower overall chemical resistance. Both are readily fabricated and can be joined by fusion and solvent-welding techniques.

The styrene-synthetic rubber blend materials, which are a mixture of styrene-acrylonitrile polymer and butadiene-acrylonitrile have good resistance in nonoxidizing weak acids but are not satisfactory for handling oxidizing acids. As is the case with most common thermoplastic materials, the upper useful temperature limit is about 65°C (150°F). Resistance to strong alkaline solutions is fair and to weak alkaline chemical-salt solutions is generally good. They are not satisfactory for use with aromatic or chlorinated hydrocarbons and they possess only fair resistance to aliphatic hydrocarbons. They are not satisfactory for use with ketone and ester solvents. These materials normally will burn, but fire-retardant grades are commercially available. Resistance to outdoor weathering is generally good. They can be readily fabricated and can be joined with solvent-welding techniques.

The use of resin systems that are applied as liquids, reinforced, and then converted to solids by catalytic action has increased tremendously in the last 20 years. These include polyester resins with the peroxide catalysts, epoxy resins with the basic amine catalysts, and furane resins with acid catalysts. The nature of furane resins is such that they are available only as a black material, which sometimes limits their application. Polyester and epoxy materials are available as translucent materials that can be pigmented to any desired color. Epoxy resins have excellent resistance to nonoxidizing and weak acids and to alkaline materials but poor resistance to strong oxidizing acids. The upper temperature limit is generally about 90°C (200°F). The epoxy resins also have resistance to aromatic and aliphatic hydrocarbons but have only fair resistance to chlorinated hydrocarbons, ketones, and ester solvents. Polyester resins have good resistance to nonoxidizing and oxidizing acids, both weak and concentrated, with moderate resistance to many alkaline solutions and excellent resistance to chemical salt solutions. The upper temperature limit is again 90°C (200°F), although there is a loss of physical properties at this high a temperature. Polyester resins have good resistance to aromatic and aliphatic hydrocarbons but only fair resistance to the strong solvents. The material will burn, but fire-retardant grades, based on the addition of antimony trioxide and the use of halogenated compounds in the manufacture of the polyester resins, have been developed. Resistance to sunlight and outdoor weathering is good. Overall resistance of the resin is often improved for severe chemical services by the use of a chemically resistant synthetic-fiber cloth such as Orlon acrylic fiber or Dacron polyester fiber. The glass-fiber laminates are not satisfactory for use in hydrofluoric acid, and it has been found that they are subject to attack and penetration in other acids and alkalies, including hydrochloric acid under certain conditions. Furane resins have broad chemical resistance to acids and alkalies as well as to many solvents. Otherwise, the material is similar to the epoxy and polyester laminates.

Urethane resins, based on the isocyanate molecule, are finding increasing use in maintenance engineering work, primarily as an insulation material in the form of a foam and as an abrasion-resistance material in the form of a compounded elastomer. Although urethane foams have been made to meet certain requirements of the various regulatory agencies with regard to flame retardance, extreme caution is recommended in the use of these materials when the application parameters differ from the flammability parameters. The insulation properties are outstanding, as is the ease of fabrication, since urethane foam can be applied as a mixture of two solutions, each sprayed simultaneously onto a substrate. Other resins systems, including the epoxy, can also be spray-applied to form a foam, although their properties are somewhat different from those of the urethane foam. Top coating of the urethane foam with a flame-retardant coating is virtually requisite, along with the additional protection given to the foam from ultraviolet degradation.

Rubber and Elastomers

Natural rubber has been used for many years as a material for molded and lined equipment for chemical service. It can be compounded for maximum resistance for a number of service conditions and has proved to be a very useful material for many conditions that are highly corrosive to metals. Natural rubber compounds will resist a wide variety of chemical solutions, including all concentrations of hydrochloric acid, phosphoric acid, sulfuric acid up to about 50% concentration, saturated salt solutions such as ferric chloride, brine, bleaching solutions, and most plating solutions. They are readily attacked by strong oxidizing acids such as nitric and chromic and by aliphatic, aromatic, and chlorinated solvents. Maximum temperature at which rubber compounds can be used varies with the chemical and the strength of solutions. The temperature limitation for continuous exposure for most soft rubber compounds is about 60°C (140°F) and that for hard rubber is about 80°C (180°F). However, heat-resisting compounds are available which may be used at somewhat higher temperatures. Soft rubber, especially, compounded for maximum temperature resistance, may be used for continuous exposures under some chemical conditions up to 90°C (200°F), and hard rubber may be compounded for service temperatures as high as 110°C (230°F).

A special number of synthetic rubbers and elastomeric materials have been developed with special characteristics that extend the overall usefulness of the elastomers for corrosion-resistant equipment. Notable among these are Buna S(GR-S), Buna N(GR-A), butyl (GR-1), neoprene (GR-M), Hypalon (chlorsulfonated polyethylene), and Thiokol compounds. In addition, polymers of ethylene and propylene have been developed with elastomeric properties. Like natural rubber, each of these may be compounded in several ways to maximize resistance to specific chemical exposures. Natural rubber and other elastomers are frequently used in combination with brick linings for temperature conditions that are above that allowed for elastomer material alone; they have proved to be excellent membrane linings for such construction.

Brick Linings

Brick-lining protection can be used for many conditions that are severely corrosive even to high alloy materials. It should be considered for tanks, vats, stacks, vessels, and similar equipment items. Brick shapes commonly used for such construction are made of carbon, red shale, or acid-proof refractory materials. Carbon bricks are useful for handling alkaline conditions as well as acid, while the shale and the acid-proof refractory materials are used primarily for acid solutions. Carbon can also be used where sudden temperature changes are involved that would cause spalling of the other two materials. Red shale brick generally are not used at temperatures above 150°C (300°F) because of poor spalling resistance. Acid-proof refractories are sometimes used at temperatures up to 870°C (1600°F).

There are a number of cement materials that are regularly used for brick-lined construction. The most commonly used are sulfur, silicate-base, and resin-base. The resin cements include the phenolic epoxy- and furane-resin-base materials, which are used at temperatures up to 180°C (350°F). The carbon-filled phenolic resin cements have excellent resistance to all nonoxidizing acids, salts, and most other solvents. The carbon-filled furane resins have excellent resistance to all nonoxidizing acids and alkalies, salts, and organic solvents. The silica-filled resin compositions are available in all types of resins and are almost equally resistant except to hydrofluoric acid and alkalies. Sulfur-based cements are limited to a maximum temperature of about 90°C (200°F). In general, they have excellent chemical resistance to nonoxidizing acids and salts but are not suitable for use in the presence of alkalies or organic solvents. The sodium silicate-based cements have good resistance to all inorganic acids except hydrofluoric at temperatures up to about 400°C (750°F). The potassium silicate-based cements are useful at somewhat higher temperatures, the upper limit depending on specific conditions and requirements.

Concrete

Concrete is a material of construction not usually used under severe corrosive conditions other than as a substrate. For example, there are tanks, vessels, and so on, whose shape and size make concrete an economical material of construction, provided that there is a barrier that separates the corrosive environment from the concrete. Such a barrier sometimes is an elastomer or plastic sheet cemented in place, and often it is a protective coating applied by spray or trowel. For weathering atmospheres, concrete is protected against abnormal deterioration by the use of either a clear penetrating coating or a protective pigmented coating. The most common of the clear penetrating coatings are the silicone resins, where the water repellency of the silicone and the penetrating characteristics of the vehicle prolong the life of the concrete. Since concrete is inherently alkaline in nature (until it has weathered and reacted with the natural acids of the environment such as carbonic acid), it is necessary to use an alkaline-resistant protective coating on new concrete. Such materials include vinyls, chlorinated rubber, and epoxies. Particularly *not* to be used on fresh concrete (unless acid-etched) are the oil-base paints; the presence of free alkali in the concrete will cause the oil base to saponify and possibly be removed by rain or other weathering factors. The two major reasons for protecting concrete are (1) appearance and (2) improved longevity based on the fact that most concrete structures are reinforced, usually with steel in one form or another. If the concrete is not dense or is not protected, there is the possibility that moisture or other chemicals will penetrate the surface; under conditions of severe freezing and thawing the concrete will then spall, ultimately either exposing the reinforcing steel or allowing the moisture and its contaminants to attack the reinforcing material. When steel is attacked in a crevice condition such as exists when reinforcing rod or mesh is embedded in concrete, rust is formed, which results in an expansion of force, further lifting the concrete and further exposing the reinforcing material.

In building construction of reinforced concrete, it is necessary to know whether the moisture has a driving force causing it to go from outside the building into the inside or whether the conditions inside the building are such in terms of moisture transmission and moisture-vapor driving force, together with the temperature driving force, to cause the moisture to go from the inside of the building outward. In the latter case, it is particularly requisite to have a breathing coating such as a silicone, on the outside of the building, since nonbreathing mastic coating or other pigmented protective coating will be lifted as the moisture vapor from the inside attempts to reach the lower moisture potential on the outside.

The use of low alloy high-strength steel on certain highway construction has caused some minor concern among engineers and architects with regard to the reinforced concrete used as the piers upon which the steel beams rest. The staining, aging, and weathering of the steel have, in the absence of protective coating (which is no longer required because of the good weathering corrosion resistance of the steel), caused staining of the concrete piers. One either makes the concrete less absorptive or uses a "pigmented" concrete that blends with the rust from the steel.

Protective Coatings

Protective coatings are probably the most widely used and, at the same time, the most controversial material employed for minimizing corrosion of steel and certain other materials. Because of its importance, the subject of protective coating and painting is discussed in a separate chapter (Chapter 7). It is important here to emphasize that it is unwise, and generally uneconomical, to try to use steel equipment with a chemically resistant coating for containing chemicals that are quite corrosive to the steel. This results from the fact that it is almost impossible to avoid some pinholes or holidays in the coating. Rapid attack of the steel will occur at such points, and continued maintenance attention will be required. This is the reason for the more stringent requirements on coatings for continuous-immersion service such as tank linings. Such requirements include thickness (sometimes minimum and sometimes maximum), number of coats, freedom from pinholes, and degree of cure.

The chemically resistant coatings, such as the baked phenolics, baked epoxies, and the air-dried epoxy, vinyl, and neoprene coatings, are ideal for minimizing contamination of chemicals handled in steel equipment. They should not be used where 100% protection from corrosion is required. An excellent material for immersion service has been developed that consists of flakes of glass dispersed in a polyester resin. This is applied by spray to a properly prepared surface and the wet coating is rolled with a paint roller to orient the glass flakes in a plane parallel to the substrate and to provide maximum resistance to chemical attack.

Glass Linings

Glass-lined equipment is available from a chemically resistant standpoint for handling all acids except hydrofluoric and concentrated phosphoric acid (at ambient and elevated temperatures) and many alkaline conditions at ambient and slightly higher temperatures. The glass lining is resistant to all concentrations of hydrochloric acid at temperatures up to 150°C (300°F), to dilute concentrations of sulfuric acid at their boiling points, to concentrated solutions of H_2SO_4 up to about 230°C (450°F), and to all concentrations of nitric acid up to their boiling

point. An acid-resistant glass with improved alkali resistance is
commercially available for use under alkaline conditions up to pH 12
at temperatures of 90°C (200°F). Equipment items such as tanks,
pressure vessels, and reactors, pipelines, valves, and accessory
equipment are available. Improved resistance to impact has been
developed for the glass linings. Methods of field repair of glass linings have been developed which include the use of cover plates and
plugs of tantalum in combination with resin cements and Teflon.

Wood

All woods are affected adversely by acids, particularly the strong
oxidizing acids, but they are regularly used in dilute hydrochloric
acid solutions at ambient temperature. Improved corrosion resistance
can be imparted to wood by impregnating the wood under pressure
conditions with certain resin solutions that include asphalt, phenolic,
and furane. This greatly extends the area of application of woods in
corrosion services. Strong alkaline solutions, particularly caustic,
generally cause disintegration and cannot be used with impregnated
wood. Weak solutions can be used with wood equipment with reasonably good service life.

Inhibitors

The corrosion of iron and other metals in aqueous solutions can frequently be minimized or inhibited by the addition of soluble chromates,
phosphates, molybdates, silicates, and amines or other chemicals,
singly or in combination. Such materials are called inhibitors and are
generally attractive for use in recirculating systems or closed systems.
They are also used in neutral or very slightly acid solutions. Sodium
silicate has also been effectively used as a inhibitor for aluminum in
alkaline solutions. The concentration of an inhibitor for maximum control depends on the solution, composition, temperature, velocity,
metal system, and the presence of dissimilar metals in contact in the
solution. Care should be taken in the selection and application of
inhibitors, since in some instances they can cause increased localized
attack.

   Although chromate treatment is widely used, it does require attention to keep concentration at the required minimum for specific environmental conditions. In addition, there is the always danger of pollution
from loss of chromate to the surrounding environment.

   One of the most common uses of inhibitors is in brine systems.
When calcium or sodium chloride brine is used in steel equipment, it
is generally recommended that sodium dichromate be used. Where
chromates cannot be used, disodium phosphate is recommended for
sodium chloride brines. Where aluminum equipment is used in service,

it is recommended that 1% as much sodium dichromate be used as there is chloride present.

For recirculating-water systems made of steel, it has been found that 0.2 ml of sodium silicate (40° Baume) per liter is effective in inhibiting corrosion. Sodium dichromate at 0.01% concentration is also effective and can be used where toxicity effects are not important.

For preventing corrosion of steel in ferrous-base materials, particularly in protecting machine parts and storage and equipment, the use of volatile (VLI) or vapor-phase (VPI) corrosion inhibitors has been found to be effective. These materials are amine nitrite salts. They can also be used to protect steel process equipment when idle or in standby condition. These materials are also available as crystals or as impregnated paper. The inhibitors are slightly volatile at atmospheric temperature; the protection obtained results from the diffusion and condensation of the vapors on the surface of the items being protected.

Cathodic Protection

Fundamentally, cathodic protection is the use of an impressed current to prevent or minimize the corrosion of metal by making the metal a cathode in the system. The current is provided either by the use of rectifiers or by sacrifical galvanic anodes. Graphite, titanium, and high silicon iron are used as anodes in conjunction with rectifiers. The most commonly used sacrifical materials are magnesium, zinc, and aluminum. Cathodic protection is recognized as a proved method of control of corrosion of steel and other metals under many environmental conditions. It has been used successfully for minimizing corrosion of equipment such as buried pipeline, water storage tanks, condensers, heat exchangers, and dock piling.

Index